Federated Learning

Alexander Jung

Federated Learning

From Theory to Practice

Alexander Jung
Department of Computer Science
Aalto University
Helsinki, Finland

ISBN 978-981-95-1008-5 ISBN 978-981-95-1009-2 (eBook)
https://doi.org/10.1007/978-981-95-1009-2

This Springer imprint is published by the registered company Springer Nature Singapore Pte Ltd.
The registered company address is: 152 Beach Road, #21-01/04 Gateway East, Singapore 189721, Singapore

Preface

This book offers a hands-on introduction to building and understanding federated learning (FL) systems. FL enables multiple devices—such as smartphones, sensors, or local computers—to collaboratively train machine learning (ML) models, while keeping their data private and local. It is a powerful solution when data cannot or should not be centralized due to privacy, regulatory, or technical reasons.

The book is designed for students, engineers, and researchers who want to learn how to design scalable, privacy-preserving FL systems. Our main focus is on personalization: enabling each device to train its own model while still benefiting from collaboration with relevant devices. This is achieved by leveraging similarities between the learning tasks associated with devices. We represent these similarities as weighted edges of a federated learning network (FL network).

The key idea is to represent real-world FL systems as networks of devices, where nodes correspond to devices and edges represent communication links and data similarities between them. The training of personalized models for these devices can be naturally framed as a distributed optimization problem. This optimization problem is referred to as generalized total variation minimization (GTVMin) and ensures that devices with similar learning tasks learn similar model parameters.

Our approach is both mathematically principled and practically motivated. While we introduce some advanced ideas from optimization theory and graph-based learning, we aim to keep the book accessible. Readers are guided through the core ideas step-by-step, with intuitive explanations. Throughout, we maintain a focus of building FL systems that are trustworthy—robust against attacks, privacy-friendly, and secure.

Audience We assume a basic background in undergraduate-level mathematics, including calculus and linear algebra. Familiarity with concepts such as convergence, derivatives, and norms will be helpful but not strictly necessary. No prior experience with ML or optimization is required, as we build up most concepts from first principles.

The book is intended for advanced undergraduates, graduate students, and practitioners who are looking for a practical, principled, and privacy-friendly approach to decentralized ML.

Structure The book begins by introducing the key motivations and challenges of FL. We then move on to introduce the notion of an FL network and explain how they capture the structure of distributed ML applications. The core chapters develop the GTVMin formulation and explore how to solve it using various distributed optimization techniques. Later chapters focus on practical concerns such as robustness and privacy protection of GTVMin-based systems. A comprehensive glossary is also included to better support the reader.

Acknowledgments The development of this book has greatly benefited from feedback and insights gathered during the course *CS-E4740 Federated Learning* at Aalto University, taught between 2023 and 2025. I am grateful to Bo Zheng, Olga Kuznetsova, Diana Pfau, and Shamsiiat Abdurakhmanova for their thoughtful comments on early drafts. Special thanks go to Mikko Seesto for his careful proofreading of the manuscript and to Konstantina Olioumtsevits for her meticulous revision of the glossary. Some of the figures in the glossary have been prepared with the help of Salvatore Rastelli and Juliette Gronier.

This work was supported by:

- The Research Council of Finland (grants 331197, 363624, 349966)
- The European Union (grant 952410)
- The Jane and Aatos Erkko Foundation (grant A835)
- Business Finland, as part of the project *Forward-Looking AI Governance in Banking and Insurance (FLAIG)*

Competing Interests The authors have no competing interests to declare that are relevant to the content of this manuscript.

Helsinki, Finland Alexander Jung

Contents

Lists of Symbols

Sets and Functions

$a \in \mathcal{A}$	The object a is an element of the set $\mathcal{A}$.
$a := b$	Depending on the context, we use the symbol $:=$ either to mean a definition or to mean an assignment (e.g., within a pseudocode for an algorithm.
$\mathcal{A} \subseteq \mathcal{B}$	$\mathcal{A}$ is a subset of $\mathcal{B}$.
$\mathcal{A} \subset \mathcal{B}$	$\mathcal{A}$ is a strict subset of $\mathcal{B}$.
$\mathbb{N}$	The natural numbers $1, 2, \ldots$.
$\mathbb{R}$	The real numbers x [1].
$\mathbb{R}_{+}$	The non-negative real numbers $x \geq 0$.
$\mathbb{R}_{++}$	The positive real numbers $x > 0$.
$\|x\|$	The absolute value of a real number $x \in \mathbb{R}$.
$\{0, 1\}$	The set consisting of the two real numbers 0 and 1.
$[0, 1]$	The closed interval of real numbers x with $0 \leq x \leq 1$.
$\underset{\mathbf{w}}{\operatorname{argmin}} f(\mathbf{w})$	The set of minimizers for a real-valued function $f(\mathbf{w})$.
$\mathbb{S}^{(n)}$	The set of unit-norm vectors in $\mathbb{R}^{n+1}$.
$\log a$	The logarithm of the positive number $a \in \mathbb{R}_{++}$.
$h(\cdot) : \mathcal{A} \rightarrow \mathcal{B} : a \mapsto h(a)$	A function (map) that accepts any element $a \in \mathcal{A}$ from a set $\mathcal{A}$ as input and delivers a well-defined element $h(a) \in \mathcal{B}$ of a set $\mathcal{B}$. The set $\mathcal{A}$ is the domain of the function h and the set $\mathcal{B}$ is the co-domain of h. ML aims at finding (or learning) a function h (i.e., a hypothesis) that reads in the features $\mathbf{x}$ of a data point and delivers a prediction $h(\mathbf{x})$ for its label y.
$\nabla f(\mathbf{w})$	The gradient of a differentiable real-valued function $f : \mathbb{R}^d \rightarrow \mathbb{R}$ is the vector $\nabla f(\mathbf{w}) = \left(\frac{\partial f}{\partial w_1}, \ldots, \frac{\partial f}{\partial w_d}\right)^T \in \mathbb{R}^d$ [2, Ch. 9].

Matrices and Vectors

$\mathbf{x} = (x_1, \ldots, x_d)^T$	A vector of length d, with its j-th entry being x_j.
$\mathbb{R}^d$	The set of vectors $\mathbf{x} = (x_1, \ldots, x_d)^T$ consisting of d real-valued entries $x_1, \ldots, x_d \in \mathbb{R}$.
$\mathbf{I}_d, \mathbf{I}$	A square identity matrix of size $d \times d$. If the size is clear from context, we drop the subscript.
$\|\mathbf{x}\|_2$	The Euclidean (or ℓ_2) norm of the vector $\mathbf{x} = (x_1, \ldots, x_d)^T \in \mathbb{R}^d$ defined as $\|\mathbf{x}\|_2 := \sqrt{\sum_{j=1}^{d} x_j^2}$.
$\|\mathbf{x}\|$	Some norm of the vector $\mathbf{x} \in \mathbb{R}^d$ [3]. Unless specified otherwise, we mean the Euclidean norm $\|\mathbf{x}\|_2$.
$\mathbf{x}^T$	The transpose of a matrix that has the vector $\mathbf{x} \in \mathbb{R}^d$ as its single column.
$\mathbf{X}^T$	The transpose of a matrix $\mathbf{X} \in \mathbb{R}^{m \times d}$. A square real-valued matrix $\mathbf{X} \in \mathbb{R}^{m \times m}$ is called symmetric if $\mathbf{X} = \mathbf{X}^T$.
$\mathbf{0} = (0, \ldots, 0)^T$	The vector in $\mathbb{R}^d$ with each entry equal to zero.
$\mathbf{1} = (1, \ldots, 1)^T$	The vector in $\mathbb{R}^d$ with each entry equal to one.
$\lambda_j(\mathbf{Q})$	The j-th eigenvalue (sorted in either ascending or descending order) of a positive semi-definite (psd) matrix $\mathbf{Q}$. We also use the shorthand λ_j if the corresponding matrix is clear from context.
$(\mathbf{v}^T, \mathbf{w}^T)^T$	The vector of length $d + d'$ obtained by concatenating the entries of vector $\mathbf{v} \in \mathbb{R}^d$ with the entries of $\mathbf{w} \in \mathbb{R}^{d'}$.
$\mathrm{span}\{\mathbf{B}\}$	The span of a matrix $\mathbf{B} \in \mathbb{R}^{a \times b}$, which is the subspace of all linear combinations of the columns of $\mathbf{B}$, $\mathrm{span}\{\mathbf{B}\} = \{\mathbf{Ba} : \mathbf{a} \in \mathbb{R}^b\} \subseteq \mathbb{R}^a$.
$\mathbf{A} \otimes \mathbf{B}$	The Kronecker product of $\mathbf{A}$ and $\mathbf{B}$ [4].

Probability Theory

$\mathbb{E}_p\{f(\mathbf{z})\}$	The expectation of a function $f(\mathbf{z})$ of a random variable (RV) $\mathbf{z}$ whose probability distribution is $p(\mathbf{z})$. If the probability distribution is clear from context, we just write $\mathbb{E}\{f(\mathbf{z})\}$.
$p(\mathbf{x}; \mathbf{w})$	A parametrized probability distribution of an RV $\mathbf{x}$. The probability distribution depends on a parameter vector $\mathbf{w}$. For example, $p(\mathbf{x}; \mathbf{w})$ could be a multivariate normal distribution with the parameter vector $\mathbf{w}$ given by the entries of the mean vector $\mathbb{E}\{\mathbf{x}\}$ and the covariance matrix $\mathbb{E}\left\{(\mathbf{x} - \mathbb{E}\{\mathbf{x}\})(\mathbf{x} - \mathbb{E}\{\mathbf{x}\})^T\right\}$.

$\mathcal{N}(\mu, \sigma^2)$	The probability distribution of a Gaussian random variable (Gaussian RV) $x \in \mathbb{R}$ with mean (or expectation) $\mu = \mathbb{E}\{x\}$ and variance $\sigma^2 = \mathbb{E}\{(x-\mu)^2\}$.
$\mathcal{N}(\boldsymbol{\mu}, \mathbf{C})$	The multivariate normal distribution of a vector-valued Gaussian RV $\mathbf{x} \in \mathbb{R}^d$ with mean (or expectation) $\boldsymbol{\mu} = \mathbb{E}\{\mathbf{x}\}$ and covariance matrix $\mathbf{C} = \mathbb{E}\{(\mathbf{x}-\boldsymbol{\mu})(\mathbf{x}-\boldsymbol{\mu})^T\}$.

Machine Learning

r	An index $r = 1, 2, \ldots$ that enumerates data points.
m	The number of data points in (i.e., the size of) a dataset.
$\mathcal{D}$	A dataset $\mathcal{D} = \{\mathbf{z}^{(1)}, \ldots, \mathbf{z}^{(m)}\}$ is a list of individual data points $\mathbf{z}^{(r)}$, for $r = 1, \ldots, m$.
d	The number of features that characterize a data point.
x_j	The j-th feature of a data point. The first feature is denoted x_1, the second feature x_2, and so on.
$\mathbf{x}$	The feature vector $\mathbf{x} = (x_1, \ldots, x_d)^T$ of a data point whose entries are the individual features of a data point.
$\mathcal{X}$	The feature space $\mathcal{X}$ is the set of all possible values that the features $\mathbf{x}$ of a data point can take on.
$\mathcal{B}$	A mini-batch (or subset) of randomly chosen data points.
B	The size of (i.e., the number of data points in) a mini-batch.
y	The label (or quantity of interest) of a data point.
$y^{(r)}$	The label of the r-th data point.
$\mathbf{x}^{(r)}$	The feature vector of the r-th data point within a dataset.
$(\mathbf{x}^{(r)}, y^{(r)})$	The features and label of the r-th data point.
$\mathcal{Y}$	The label space $\mathcal{Y}$ of an ML method consists of all potential label values that a data point can carry. The nominal label space might be larger than the set of different label values arising in a given dataset (e.g., a training set). ML problems (or methods) using a numeric label space, such as $\mathcal{Y} = \mathbb{R}$ or $\mathcal{Y} = \mathbb{R}^3$, are referred to as regression problems (or methods). ML problems (or methods) that use a discrete label space, such as $\mathcal{Y} = \{0, 1\}$ or $\mathcal{Y} = \{cat, dog, mouse\}$, are referred to as classification problems (or methods).
η	Learning rate (or step size) used by gradient-based methods.
$h(\cdot)$	A hypothesis map that reads in features $\mathbf{x}$ of a data point and delivers a prediction $\hat{y} = h(\mathbf{x})$ for its label y.
$\mathcal{H}$	A hypothesis space or model used by an ML method. The hypothesis space consists of different hypothesis maps $h : \mathcal{X} \rightarrow \mathcal{Y}$, between which the ML method must choose.
$d_{\text{eff}}(\mathcal{H})$	The effective dimension of a hypothesis space $\mathcal{H}$.

$L((\mathbf{x}, y), h)$	The loss incurred by predicting the label y of a data point using the prediction $\hat{y} = h(\mathbf{x})$. The prediction $\hat{y}$ is obtained by evaluating the hypothesis $h \in \mathcal{H}$ for the feature vector $\mathbf{x}$ of the data point.
$\mathcal{R}\{h\}$	A regularizer that assigns a hypothesis h a measure for the anticipated increase in average loss when h is applied to data points outside the training set.
E_v	The validation error of a hypothesis h, which is its average loss incurred over a validation set.
$\widehat{L}(h\|\mathcal{D})$	The empirical risk or average loss incurred by the hypothesis h on a dataset $\mathcal{D}$.
E_t	The training error of a hypothesis h, which is its average loss incurred over a training set.
t	A discrete-time index $t = 0, 1, \ldots$ used to enumerate sequential events (or time instants).
α	A regularization parameter that controls the amount of regularization.
$\mathbf{w}$	A parameter vector $\mathbf{w} = (w_1, \ldots, w_d)^T$ of a model, e.g., the weights of a linear model or in an artificial neural network (ANN).
$h^{(\mathbf{w})}(\cdot)$	A hypothesis map that involves tunable model parameters $w_1, \ldots, w_d$ stacked into the vector $\mathbf{w} = (w_1, \ldots, w_d)^T$.
$\mathbf{\Phi}(\cdot)$	A feature map $\mathbf{\Phi} : \mathcal{X} \to \mathcal{X}' : \mathbf{x} \mapsto \mathbf{x}' := \phi(\mathbf{x}) \in \mathcal{X}'$.

Federated Learning

$\mathcal{G} = (\mathcal{V}, \mathcal{E})$	An undirected graph whose nodes $i \in \mathcal{V}$ represent devices within an FL network. The undirected edges $\{i, i'\} \in \mathcal{E}$, each having a positive weight $A_{i,i'}$, represent either some form of connectivity between devices or statistical similarities between their local datasets.
$i \in \mathcal{V}$	A node that represents some device within an FL network. The device can access a local dataset and train a local model.
$\mathcal{C}$	Given a graph we denote by $\mathcal{C} \subseteq \mathcal{V}$ a subset (or cluster) of nodes which are connected by many edges with large weights.
$\|\partial\mathcal{C}\|$	The boundary of a cluster, which is the sum $\sum_{i \in \mathcal{C}, i' \notin \mathcal{C}} A_{i,i'}$.
$\mathcal{G}^{(\mathcal{C})}$	The induced subgraph of $\mathcal{G}$ using the nodes in $\mathcal{C} \subseteq \mathcal{V}$.
$\mathbf{L}^{(\mathcal{G})}$	The Laplacian matrix of a graph $\mathcal{G}$.
$\mathbf{L}^{(\mathcal{C})}$	The Laplacian matrix of the induced graph $\mathcal{G}^{(\mathcal{C})}$.
$\mathcal{N}^{(i)}$	The neighborhood of a node i in a graph $\mathcal{G}$.
$d^{(i)}$	The weighted degree $d^{(i)} := \sum_{i' \in \mathcal{N}^{(i)}} A_{i,i'}$ of a node i in a graph $\mathcal{G}$.
$d_{\max}^{(\mathcal{G})}$	The maximum weighted node degree of a graph $\mathcal{G}$.
$\mathcal{D}^{(i)}$	The local dataset $\mathcal{D}^{(i)}$ carried by node $i \in \mathcal{V}$ of an FL network.

m_i	The number of data points (i.e., sample size) contained in the local dataset $\mathcal{D}^{(i)}$ at node $i \in \mathcal{V}$.
$\mathbf{x}^{(i,r)}$	The features of the r-th data point in the local dataset $\mathcal{D}^{(i)}$.
$y^{(i,r)}$	The label of the r-th data point in the local dataset $\mathcal{D}^{(i)}$.
$\mathbf{w}^{(i)}$	The local model parameters of device i within an FL network.
$L_i(\mathbf{w})$	The local loss function used by device i to measure the usefulness of some choice $\mathbf{w}$ for the local model parameters.
$\mathcal{R}^{(i)}\{h\}$	A regularizer used for model training by device i within an FL network. This regularizer typically depends on the model parameters of other devices $i' \in \mathcal{V} \setminus \{i\}$.
$d^{(i,i')}$	A quantitative measure for the variation (or discrepancy) between trained local models at nodes i, i'.
$\text{stack}\{\mathbf{w}^{(i)}\}_{i=1}^{n}$	The vector $\left(\left(\mathbf{w}^{(1)}\right)^T, \ldots, \left(\mathbf{w}^{(n)}\right)^T\right)^T \in \mathbb{R}^{dn}$ that is obtained by vertically stacking the local model parameters $\mathbf{w}^{(i)} \in \mathbb{R}^d$.

Chapter 1
Introduction to Federated Learning

Abstract This chapter introduces the foundations and motivations of federated learning (FL), a distributed machine learning approach that enables model training directly on devices where data is generated. It emphasizes the relevance of FL in modern settings where decentralized data-such as from smartphones, wearables, and weather stations-exhibits structural relationships and privacy constraints. FL addresses challenges associated with data centralization by leveraging local computation and network-based optimization. Key drivers of FL include privacy, robustness to failures and attacks, parallelization, and personalization. Structurally, the chapter begins by motivating FL through real-world examples and then outlines core mathematical tools such as Euclidean spaces, gradient-based methods, and fixed-point iterations. Finally, it offers a roadmap of the book, organized into three parts: a review of machine learning principles, a formal treatment of FL models and methods, and a discussion of ethical and practical requirements for trustworthy AI. The chapter closes with exercises to reinforce foundational concepts in linear algebra and optimization.

We are surrounded by devices, such as smartphones and wearables, that generate decentralized collections of local datasets [5–9]. These local datasets often exhibit an intrinsic network structure, arising from functional dependencies or statistical similarities (see Sect. 7.3).

For example, contact networks underpin pandemic modeling, network medicine maps disease relationships via comorbidities [10], and social sciences leverage social graphs to relate the data of connected individuals [11]. Similarly, weather stations of the Finnish Meteorological Institute (FMI) produce local datasets with statistical properties influenced by geographic proximity.

Federated learning (FL) is an umbrella term for distributed optimization methods that train machine learning (ML) models directly at the locations of data generation [12–16]. Unlike traditional ML workflows that centralize data before training, FL leverages in situ computations. Figure 1.1 contrasts these approaches.

From an engineering perspective, this book is about building federated learning systems by formulating them as network-based optimization problems. The core idea is to represent a real-world FL setup via a federated learning network (FL

A. Jung, *Federated Learning*, https://doi.org/10.1007/978-981-95-1009-2_1

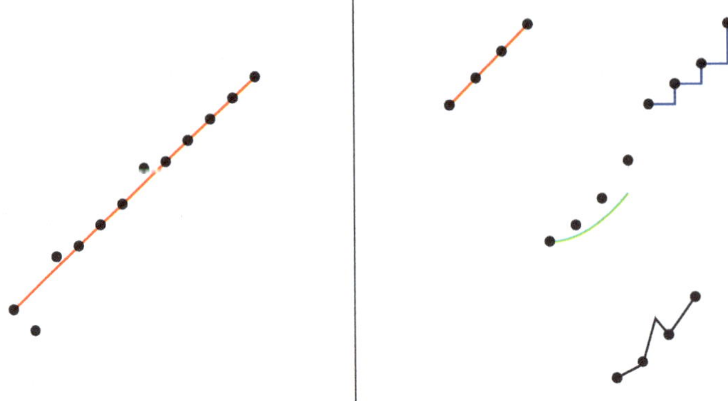

Fig. 1.1 Left, a basic ML method uses a single dataset to train a single model. Right, decentralized collection of devices with the ability to access data and train models locally

network), where nodes correspond to devices with local datasets and models and edges reflect communication capabilities or statistical similarity. We then pose FL as an optimization problem over this FL network, which we call generalized total variation minimization (GTVMin). GTVMin balances local model performance with smoothness of model parameters across connected nodes.

Different choices for how to measure model variation across the FL network lead to different flavors of FL methods. The overarching goal is to derive these methods in a principled way by applying distributed optimization methods. All FL algorithms we study can be seen as fixed-point iterations for solving an instance of GTVMin.

Beyond methodology, FL is also driven by several practical motivations:

- **Privacy.** By exchanging only updates to model parameters, FL avoids raw data transmission and thus mitigates privacy risks (see Chap. 9).
- **Robustness.** FL systems can tolerate stragglers and are more resilient to cyber-attacks, such as data poisoning (see Chap. 10).
- **Parallelism.** We can interpret the interconnected devices of an FL network as a parallel computer. One example of such a parallel computer is a mobile network constituted by smartphones that can communicate via radio links. This parallel computer allows to speeding up computations required for the training of ML models (see Chap. 4).
- **Democratization.** FL enables collective learning using low-cost, widely available devices—rather than relying on centralized high-end hardware [17, 18].
- **Communication Efficiency.** In remote or bandwidth-limited scenarios, training locally can be cheaper than transmitting raw datasets [19].
- **Personalization.** FL naturally supports training personalized models that adapt to device-specific data distributions (see Chap. 6).

1.1 Core Techniques in Federated Learning

To build and analyze FL algorithms, this book draws on core mathematical concepts:

Euclidean space Our main mathematical structure for the study and design of FL systems is the Euclidean space $\mathbb{R}^d$. We expect familiarity with the algebraic and geometric structure of $\mathbb{R}^d$ [20, 21]. For example, we often use the spectral decomposition of positive semi-definite (psd) matrices that naturally arise in the formulation of FL applications. We will also use the geometric structure of $\mathbb{R}^d$, which is defined by the inner-product $\mathbf{w}^T\mathbf{w}' := \sum_{j=1}^{d} w_j w'_j$ between two vectors $\mathbf{w}, \mathbf{w}' \in \mathbb{R}^d$ and the induced norm $\|\mathbf{w}\|_2 := \sqrt{\mathbf{w}^T\mathbf{w}} = \sqrt{\sum_{j=1}^{d} w_j^2}$.

Calculus A main toolbox for the design of FL algorithms are variants of gradient descent (GD). The common idea of gradient-based methods is to approximate a function $f(\mathbf{w})$ locally by a linear function. This local linear approximation is determined by the gradient $\nabla f(\mathbf{w})$. We, therefore, expect some familiarity with multivariable calculus [2].

Fixed-Point Iterations Each algorithm that we discuss in this book can be interpreted as a fixed-point iteration of some operator $\mathcal{P} : \mathbb{R}^d \rightarrow \mathbb{R}^d$. These operators depend on the local datasets and personal models used within an FL system. A prime example of such an operator is the gradient step of gradient-based methods (see Chap. 4). The computational properties of these FL algorithms are determined by the contraction properties of the underlying operator [22].

1.2 Book Structure and Roadmap

This book is organized into three parts:

- **Part I: ML Refresher.** Chapters 2 and 4 review basic ML concepts and optimization methods. These chapters serve both to refresh prerequisite knowledge and to highlight techniques like regularization and gradient descent that underpin FL.
- **Part II: FL Theory and Methods.** Chapter 3 introduces the FL network and formulates the core optimization principle, i.e., GTVMin. Chapters 4 and 5 show how to apply optimization methods to derive scalable and personalized FL algorithms. Chapter 6 explores main FL variants as special cases of GTVMin, and Chap. 7 discusses methods for constructing meaningful edge structures in FL networks.
- **Part III: trustworthy artificial intelligence (trustworthy AI).** Chapters 8–10 explore key requirements for trustworthy AI systems, including privacy protection and robustness against data poisoning. These chapters link FL methodology to emerging ethical and regulatory demands in AI deployment.

1.3 Exercises

1.1 Complexity of Matrix Inversion. Choose your favorite computer architecture (represented by a mathematical model) and think about how much computation is required—in the worst case—by the most efficient algorithm that can invert any given invertible matrix $\mathbf{Q} \in \mathbb{R}^{d \times d}$. Try also to reflect on how practical your chosen computer architecture is, i.e., is it possible to buy such a computer in your nearest electronics shop?

1.2 Vector Spaces and Euclidean Norm. Consider data points, each characterized by a feature vector $\mathbf{x} \in \mathbb{R}^d$ with entries $x_1, x_2, \ldots, x_d$.

- Show that the set of all feature vectors forms a vector space under standard addition and scalar multiplication.
- Calculate the Euclidean norm of the vector $\mathbf{x} = (1, -2, 3)^T$.
- If $\mathbf{x}^{(1)} = (1, 2, 3)^T$ and $\mathbf{x}^{(2)} = (-1, 0, 1)^T$, compute $3\mathbf{x}^{(1)} - 2\mathbf{x}^{(2)}$.

1.3 Matrix Operations in Linear Models. Linear regression methods learn model parameters $\widehat{\mathbf{w}} \in \mathbb{R}^d$ via solving the optimization problem:

$$\widehat{\mathbf{w}} = \arg\min_{\mathbf{w} \in \mathbb{R}^d} \|\mathbf{y} - \mathbf{X}\mathbf{w}\|_2^2,$$

with some matrix $\mathbf{X} \in \mathbb{R}^{m \times d}$, and some vector $\mathbf{y} \in \mathbb{R}^m$.

- Derive a closed-form expression for $\widehat{\mathbf{w}}$ that is valid for *arbitrary* matrix $\mathbf{X}$, and vector $\mathbf{y}$.
- Discuss the conditions under which $\mathbf{X}^T\mathbf{X}$ is invertible.
- Compute $\widehat{\mathbf{w}}$ for the following dataset:

$$\mathbf{X} = \begin{pmatrix} 1 & 2 \\ 3 & 4 \\ 5 & 6 \end{pmatrix}, \quad \mathbf{y} = \begin{pmatrix} 7 \\ 8 \\ 9 \end{pmatrix}.$$

- Compute $\widehat{\mathbf{w}}$ for the following dataset: The rth row of $\mathbf{X}$, for $r = 1, \ldots, 28$, is given by the temperature recordings (with a 10-minute interval) during day r/Mar/2023 at FMI weather station *Kustavi Isokari*. The rth row of $\mathbf{y}$ is the maximum daytime temperature during day $r + 1$/Mar/2023 at the same weather station.

1.4 Eigenvalues and Positive Semi-definiteness. The convergence properties of widely used ML methods rely on the properties of psd matrices. Let $\mathbf{Q} = \mathbf{X}^T\mathbf{X}$, where $\mathbf{X} \in \mathbb{R}^{m \times d}$.

1. Prove that $\mathbf{Q}$ is psd.
2. Compute the eigenvalues of $\mathbf{Q}$ for $\mathbf{X} = \begin{pmatrix} 1 & 2 \\ 3 & 4 \end{pmatrix}$.
3. Compute the eigenvalues of $\mathbf{Q}$ for the matrix $\mathbf{X}$ used in Exercise 1.3 that is constituted by FMI temperature recordings.

Chapter 2
Machine Learning Foundations for FL

Abstract This chapter reviews foundational machine learning techniques that support the development of federated learning systems. It refreshes core concepts essential for understanding later chapters, while introducing the empirical risk minimization (ERM) principle as a unifying framework. The chapter begins by outlining the components of machine learning systems-data points, models, and loss functions-and explains how ERM combines them to train models. It then explores the computational aspects of ERM, focusing on gradient-based optimization methods. Statistical properties of ERM solutions are analyzed using a simple probabilistic model. Model validation is introduced as a practical tool for diagnosing overfitting, and different forms of regularization are presented, including data augmentation, loss penalization, and model pruning. Finally, the chapter explains how regularization can be used to couple learning across devices in a federated system, laying the groundwork for personalized and collaborative learning methods. Structurally, the chapter proceeds from basic machine learning principles to more advanced ideas relevant for distributed learning and concludes with exercises designed to reinforce theoretical understanding and practical skills in linear regression and regularization.

This chapter covers basic ML techniques instrumental for FL. Content-wise, this chapter is more extensive compared to the following chapters. However, this chapter should be considerably easier to follow than the following chapters as it mainly refreshes pre-requisite knowledge.

In Sect. 2.3, we begin with the basic components of any ML method: data, a model, and a loss function. We also describe how these components are combined through empirical risk minimization (ERM) which is a main design principle for ML.

Section 2.2 then explores the computational aspects of ERM, focusing on gradient-based methods for parametric models. Section 2.3 discusses the statistical properties of ML methods and their analysis via probabilistic models. Section 2.4 introduces the idea of model validation and discusses simple rules for the diagnosis of ML methods.

A. Jung, *Federated Learning*, https://doi.org/10.1007/978-981-95-1009-2_2

Section 2.5 explains three fundamental forms of regularization: data augmentation, model pruning, and loss penalization. We will then show in Sect. 2.6 how to use regularization to couple the training of local models at different devices, resulting in FL.

2.1 Components of ML Systems: A Design Framework

ML revolves around learning a hypothesis map h out of a hypothesis space $\mathcal{H}$ that allows to accurately predict the label of a data point solely from its features. One of the most crucial steps in applying ML methods to a given application domain is the definition or choice of what precisely a data point is. Coming up with a good choice or definition of data points is not trivial as it influences the overall performance of an ML method in many different ways.

We will use weather prediction as a recurring example of an FL application. Here, data points represent the daily weather conditions around FMI weather stations. We denote a specific data point by $\mathbf{z}$. It is characterized by the following features:

- name of the FMI weather station, e.g., “TurkuRajakari”
- latitude lat and longitude lon of the weather station, e.g., $\mathsf{lat} := 60.37788$, $\mathsf{lon} := 22.0964$,
- timestamp of the measurement in the format YYYY-MM-DD HH:MM:SS, e.g., 2023-12-31 18:00:00

It is convenient to stack the features into a feature vector $\mathbf{x}$. The label $y \in \mathbb{R}$ of such a data point is the maximum daytime temperature in degree Celsius, e.g., -20. We indicate the features $\mathbf{x}$ and label y of a data point via the notation $\mathbf{z} = (\mathbf{x}, y)$.

Strictly speaking, a data point $\mathbf{z}$ is not the same as the pair of features $\mathbf{x}$ and label y. Indeed, a data point can have additional properties that are used neither as features nor as label. A more precise notation would then be $\mathbf{x}(\mathbf{z})$ and $y(\mathbf{z})$, indicating that the features $\mathbf{x}$ and label y are functions of the data point $\mathbf{z}$.

We predict the label of a data point with features $\mathbf{x}$ by the function value $h(\mathbf{x})$ of a hypothesis (map) $h(\cdot)$. The prediction will typically be not perfect, i.e., $h(\mathbf{x}) \neq y$. ML methods use a loss function $L\left((\mathbf{x}, y), h\right)$ to measure the error incurred by using the prediction $h(\mathbf{x})$ as a guess for the true label y. The choice of loss function crucially influences the statistical and computational properties of the resulting ML method [23, Ch. 2].

It seems natural to learn a hypothesis by minimizing the average loss—or empirical risk—on a given set of data points

$$\mathcal{D} := \left\{\left(\mathbf{x}^{(1)}, y^{(1)}\right), \ldots, \left(\mathbf{x}^{m}, y^{(m)}\right)\right\}.$$

This is known as ERM:

$$\hat{h} \in \underset{h \in \mathcal{H}}{\operatorname{argmin}} (1/m) \sum_{r=1}^{m} L\left(\left(\mathbf{x}^{(r)}, y^{(r)}\right), h\right). \tag{2.1}$$

As the notation in (2.1) indicates, there can be several different solutions to the optimization problem (2.1). We denote by $\hat{h}$ one of these solutions, i.e., $\hat{h}$ is an element of the solution set for (2.1).

Several important ML methods use a parametric model $\mathcal{H}$: Each hypothesis $h \in \mathcal{H}$ is defined by parameters $\mathbf{w} \in \mathbb{R}^d$, often indicated by the notation $h^{(\mathbf{w})}$. A prime example of a parametric model is the linear model [23, Sec. 3.1]:

$$\mathcal{H}^{(d)} := \left\{h^{(\mathbf{w})} : \mathbb{R}^d \mapsto \mathbb{R} : h^{(\mathbf{w})}(\mathbf{x}) = \mathbf{w}^T \mathbf{x}\right\}.$$

This book presents FL algorithms (see Chap. 5) that are flexible in the sense of allowing to use different types of ML models. However, for ease of exposition, we mainly focus on the special case of linear models. The restriction to linear models allows for a more comprehensive analysis of FL applications. On the flip side, the scope of our analysis is limited to FL applications involving local models that can be well approximated by linear models.

Several important ML methods are obtained from the combination of nonlinear feature learning and a linear model. For example,

- a deep net with the hidden layers representing a trainable feature map and the output layer implements a linear model [24], [23, Sec. 3.11.],
- a decision tree with a fixed topology that corresponds to a specific decision boundary and trainable predictions for each decision region [25], [23, Sec. 3.10],
- kernel methods [26], [23, Sec. 3.9].

Linear regression learns the parameters of a linear model by minimizing the average squared error loss:

$$\widehat{\mathbf{w}}^{(\mathrm{LR})} \in \underset{\mathbf{w} \in \mathbb{R}^d}{\operatorname{argmin}} (1/m) \sum_{r=1}^{m} \underbrace{\left(y^{(r)} - \mathbf{w}^T \mathbf{x}^{(r)}\right)^2}_{=L\left(\left(\mathbf{x}^{(r)}, y^{(r)}\right), h^{(\mathbf{w})}\right)}. \tag{2.2}$$

Note that (2.2) minimizes a smooth and convex function:

$$f(\mathbf{w}) := (1/m)\left[\mathbf{w}^T \mathbf{X}^T \mathbf{X} \mathbf{w} - 2\mathbf{y}^T \mathbf{X} \mathbf{w} + \mathbf{y}^T \mathbf{y}\right]. \tag{2.3}$$

Here, we use the feature matrix

$$\mathbf{X} := \left(\mathbf{x}^{(1)}, \ldots, \mathbf{x}^{(m)}\right)^T$$

and the label vector

$$\mathbf{y} := \left(y^{(1)}, \ldots, y^{(m)}\right)^T \tag{2.4}$$

of the training set $\mathcal{D}$.

Inserting (2.3) into (2.2) allows to formulate linear regression as

$$\widehat{\mathbf{w}}^{(\mathrm{LR})} \in \operatorname*{argmin}_{\mathbf{w} \in \mathbb{R}^d} \mathbf{w}^T \mathbf{Q} \mathbf{w} + \mathbf{w}^T \mathbf{q} \tag{2.5}$$
$$\text{with } \mathbf{Q} := (1/m)\mathbf{X}^T\mathbf{X}, \mathbf{q} := -(2/m)\mathbf{X}^T\mathbf{y}.$$

The matrix $\mathbf{Q} \in \mathbb{R}^{d \times d}$ is positive semi-definite (psd) with eigenvalue decomposition (EVD),

$$\mathbf{Q} = \sum_{j=1}^{d} \lambda_j \mathbf{u}^{(j)} \left(\mathbf{u}^{(j)}\right)^T. \tag{2.6}$$

The EVD (2.6) consists of orthonormal eigenvectors $\mathbf{u}^{(1)}, \ldots, \mathbf{u}^{(d)}$ and corresponding list of non-negative eigenvalues:

$$0 \leq \lambda_1 \leq \ldots \leq \lambda_d, \text{ with } \mathbf{Q}\mathbf{u}^{(j)} = \lambda_j \mathbf{u}^{(j)}.$$

The list of eigenvalues is unique for a given psd matrix $\mathbf{Q}$. In contrast, the eigenvectors $\mathbf{u}^{(j)}$ are not unique in general.

To train an ML model $\mathcal{H}$ means to solve ERM (2.1) (or (2.2) for linear regression); the dataset $\mathcal{D}$ is therefore referred to as a training set. Figure 2.1 shows an example of the objective function (2.2) and the resulting optimal model parameters.

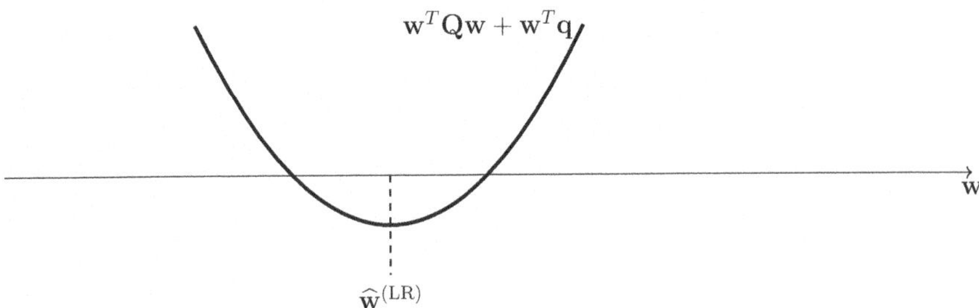

Fig. 2.1 ERM (2.1) for linear regression minimizes a convex quadratic function $\mathbf{w}^T\mathbf{Q}\mathbf{w} + \mathbf{w}^T\mathbf{q}$

The trained model results in the learned hypothesis $\hat{h}$. Two key questions for the analysis of a given ML method are:

- **Computational aspects.** How much compute do we need to solve (2.1)?
- **Statistical aspects.** How useful is the solution $\hat{h}$ to (2.1) in general, i.e., how accurate is the prediction $\hat{h}(\mathbf{x})$ for the label y of an **arbitrary** data point with features $\mathbf{x}$?

2.2 Computational Aspects of ERM

A principled approach to design ML methods is to apply some optimization method to solve (2.1) [27]. Most of these optimization methods operate in an iterative fashion: Starting from an initial choice $h^{(0)}$, they construct a sequence

$$h^{(0)}, h^{(1)}, h^{(2)}, \ldots,$$

which are hopefully increasingly accurate approximations to a solution $\hat{h}$ of (2.1). The computational complexity of such an ML method can be measured by the number of iterations required to guarantee some prescribed level of approximation.

For a parametric model and a smooth loss function, we can solve (2.2) by gradient-based methods: Starting from an initial parameters $\mathbf{w}^{(0)}$, we iterate the gradient step:

$$\begin{aligned} \mathbf{w}^{(k)} &:= \mathbf{w}^{(k-1)} - \eta \nabla f\big(\mathbf{w}^{(k-1)}\big) \\ &= \mathbf{w}^{(k-1)} + (2\eta/m) \sum_{r=1}^{m} \mathbf{x}^{(r)} \big(y^{(r)} - \big(\mathbf{w}^{(k-1)}\big)^{T} \mathbf{x}^{(r)}\big). \end{aligned} \tag{2.7}$$

How much computation do we need for one iteration of (2.7)? How many iterations do we need? We will try to answer the latter question in Chap. 4. The first question can be answered more easily for a typical computational infrastructure (e.g., "Python running on a commercial Laptop"). The evaluation of (2.7) then typically requires around $m\,d$ arithmetic operations (addition, multiplication).

It is instructive to consider the special case of a linear model that does not use any feature, i.e., $h(\mathbf{x}) = w$. For this extreme case, the ERM (2.2) has a simple closed-form solution:

$$\widehat{w} = (1/m) \sum_{r=1}^{m} y^{(r)}. \tag{2.8}$$

Thus, for this special case of the linear model, solving (2.8) is to sum m numbers $y^{(1)}, \ldots, y^{(m)}$. The amount of computation, measured by the number of elementary arithmetic operations, required by (2.8) is proportional to m.

2.3 Statistical Aspects of ERM

We can train a linear model on a given training set as ERM (2.2). But how useful is the solution $\widehat{\mathbf{w}}$ of (2.2) for predicting the labels of data points outside the training set? Consider applying the learned hypothesis $h^{(\widehat{\mathbf{w}})}$ to an arbitrary data point not contained in the training set. What can we say about the resulting prediction error $y - h^{(\widehat{\mathbf{w}})}(\mathbf{x})$ in general? In other words, how well does $h^{(\widehat{\mathbf{w}})}$ generalize beyond the training set.

A widely used approach to study the generalization of ML methods uses a simple probabilistic models: The idea is to interpret data points as independent and identically distributed (i.i.d.) random variables (RVs) with common probability distribution $p(\mathbf{x}, y)$. Under this independent and identically distributed assumption (i.i.d. assumption), we can evaluate the overall performance of a hypothesis $h \in \mathcal{H}$ via the expected loss (or risk):

$$\mathbb{E}\{L\left((\mathbf{x}, y), h\right)\}. \tag{2.9}$$

One example of a probability distribution $p(\mathbf{x}, y)$ relates the label y with the features $\mathbf{x}$ of a data point as

$$y = \overline{\mathbf{w}}^T\mathbf{x} + \varepsilon \text{ with } \mathbf{x} \sim \mathcal{N}(\mathbf{0}, \mathbf{I}), \varepsilon \sim \mathcal{N}(0, \sigma^2), \mathbb{E}\{\varepsilon\mathbf{x}\} = \mathbf{0}. \tag{2.10}$$

A simple calculation reveals the expected squared error loss of a given linear hypothesis $h(\mathbf{x}) = \mathbf{x}^T\widehat{\mathbf{w}}$ as

$$\mathbb{E}\{(y - h(\mathbf{x}))^2\} = \|\overline{\mathbf{w}} - \widehat{\mathbf{w}}\|^2 + \sigma^2. \tag{2.11}$$

Strictly speaking, (2.11) only holds for constant model parameters $\widehat{\mathbf{w}}$. However, the learned model parameters $\widehat{\mathbf{w}}$ are often the output of an ML method that is applied to a dataset $\mathcal{D}$. If we interpret the data points in $\mathcal{D}$ as i.i.d. realizations from some underlying probability distribution, we can replace the expectation on the LHS of (2.11) with the conditional expectation $\mathbb{E}\{(y - h(\mathbf{x}))^2 \middle| \mathcal{D}\}$ [28].

The first component in (2.11) is the estimation error $\|\overline{\mathbf{w}} - \widehat{\mathbf{w}}\|^2$ of an ML method that reads in the training set and delivers an estimate $\widehat{\mathbf{w}}$ (e.g., via (2.2)) for the parameters of a linear hypothesis. The second component σ^2 in (2.11) can be interpreted as the intrinsic noise level of the label y. We cannot hope to find a hypothesis with an expected loss below σ^2.

We next study the estimation error $\overline{\mathbf{w}} - \widehat{\mathbf{w}}$ incurred by the specific estimate $\widehat{\mathbf{w}} = \widehat{\mathbf{w}}^{(\mathrm{LR})}$ (2.5) delivered by linear regression methods. To this end, we first use the probabilistic model (2.10) to decompose the label vector $\mathbf{y}$ in (2.4) as

$$\mathbf{y} = \mathbf{X}\overline{\mathbf{w}} + \mathbf{n}\,, \text{ with } \mathbf{n} := \left(\varepsilon^{(1)}, \ldots, \varepsilon^{(m)}\right)^T. \tag{2.12}$$

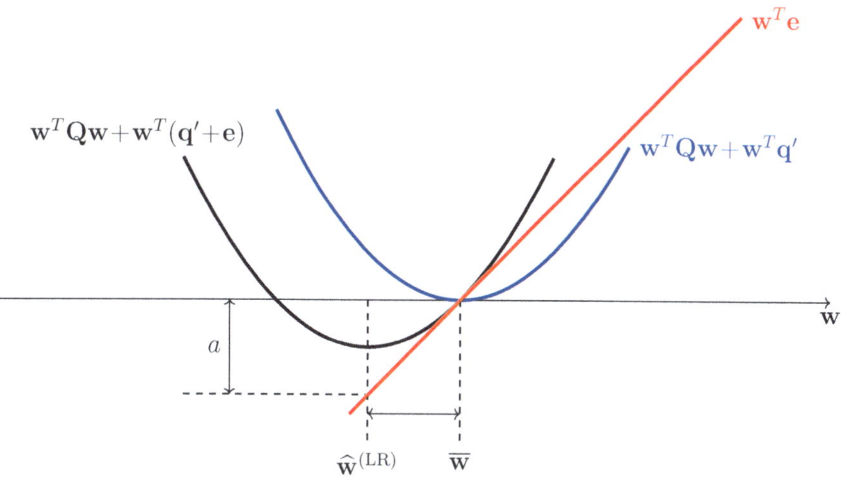

Fig. 2.2 The estimation error of linear regression is determined by the effect of the perturbation term $\mathbf{w}^T\mathbf{e}$ on the minimizer of the convex quadratic function $\mathbf{w}^T\mathbf{Q}\mathbf{w} + \mathbf{w}^T\mathbf{q}'$

Inserting (2.12) into (2.5) yields

$$\widehat{\mathbf{w}}^{(\mathrm{LR})} \in \operatorname*{argmin}_{\mathbf{w}\in\mathbb{R}^d} \mathbf{w}^T\mathbf{Q}\mathbf{w} + \mathbf{w}^T\mathbf{q}' + \mathbf{w}^T\mathbf{e} \tag{2.13}$$

$$\text{with } \mathbf{Q} := (1/m)\mathbf{X}^T\mathbf{X},\ \mathbf{q}' := -(2/m)\mathbf{X}^T\mathbf{X}\overline{\mathbf{w}}, \text{ and } \mathbf{e} := -(2/m)\mathbf{X}^T\mathbf{n}. \tag{2.14}$$

Figure 2.2 depicts the objective function of (2.13). It is a perturbation of the convex quadratic function $\mathbf{w}^T\mathbf{Q}\mathbf{w} + \mathbf{w}^T\mathbf{q}'$, which is minimized at $\mathbf{w} = \overline{\mathbf{w}}$. In general, the minimizer $\widehat{\mathbf{w}}^{(\mathrm{LR})}$ delivered by linear regression is different from $\overline{\mathbf{w}}$ due to the perturbation term $\mathbf{w}^T\mathbf{e}$ in (2.13).

The following result bounds the deviation between $\widehat{\mathbf{w}}^{(\mathrm{LR})}$ and $\overline{\mathbf{w}}$ under the assumption that the matrix $\mathbf{Q} = (1/m)\mathbf{X}^T\mathbf{X}$ is invertible.[1]

Proposition 2.1 *Consider a solution* $\widehat{\mathbf{w}}^{(\mathrm{LR})}$ *to the ERM instance* (2.13) *that is applied to the dataset* (2.12). *If the matrix* $\mathbf{Q} = (1/m)\mathbf{X}^T\mathbf{X}$ *is invertible, with minimum eigenvalue* $\lambda_1(\mathbf{Q}) > 0$,

$$\left\|\widehat{\mathbf{w}}^{(\mathrm{LR})} - \overline{\mathbf{w}}\right\|_2^2 \leq \frac{\|\mathbf{e}\|_2^2}{\lambda_1^2} \stackrel{(2.14)}{=} \frac{4}{m^2}\frac{\left\|\mathbf{X}^T\mathbf{n}\right\|_2^2}{\lambda_1^2}. \tag{2.15}$$

[1] Can you think of sufficient conditions on the feature matrix of the training set that ensure $\mathbf{Q} = (1/m)\mathbf{X}^T\mathbf{X}$ is invertible?

Proof Let us rewrite (2.13) as

$$\widehat{\mathbf{w}}^{(\mathrm{LR})} \in \operatorname*{argmin}_{\mathbf{w}\in\mathbb{R}^d} f(\mathbf{w}) \text{ with } f(\mathbf{w}) := \left(\mathbf{w}-\overline{\mathbf{w}}\right)^T \mathbf{Q}\left(\mathbf{w}-\overline{\mathbf{w}}\right) + \mathbf{e}^T\left(\mathbf{w}-\overline{\mathbf{w}}\right). \tag{2.16}$$

Clearly $f\left(\overline{\mathbf{w}}\right) = 0$ and, in turn, $f\left(\widehat{\mathbf{w}}\right) = \min_{\mathbf{w}\in\mathbb{R}^d} f(\mathbf{w}) \leq 0$. On the other hand,

$$\begin{aligned} f(\mathbf{w}) &\overset{(2.16)}{=} \left(\mathbf{w}-\overline{\mathbf{w}}\right)^T \mathbf{Q}\left(\mathbf{w}-\overline{\mathbf{w}}\right) + \mathbf{e}^T\left(\mathbf{w}-\overline{\mathbf{w}}\right) \\ &\overset{(a)}{\geq} \left(\mathbf{w}-\overline{\mathbf{w}}\right)^T \mathbf{Q}\left(\mathbf{w}-\overline{\mathbf{w}}\right) - \|\mathbf{e}\|_2 \|\mathbf{w}-\overline{\mathbf{w}}\|_2 \\ &\overset{(b)}{\geq} \lambda_1 \|\mathbf{w}-\overline{\mathbf{w}}\|_2^2 - \|\mathbf{e}\|_2 \|\mathbf{w}-\overline{\mathbf{w}}\|_2 . \end{aligned} \tag{2.17}$$

Step (a) used Cauchy–Schwarz inequality and (b) used the EVD (2.6) of $\mathbf{Q}$. Evaluating (2.17) for $\mathbf{w}=\widehat{\mathbf{w}}$ and combining with $f\left(\widehat{\mathbf{w}}\right) \leq 0$ yields (2.15).

The bound (2.15) suggests that the estimation error $\widehat{w}^{(\mathrm{LR})} - \overline{w}$ is small if $\lambda_1(\mathbf{Q})$ is large. This smallest eigenvalue of the matrix $\mathbf{Q} = (1/m)\mathbf{X}^T\mathbf{X}$ could be controlled by a suitable choice (or transformation) of features $\mathbf{x}$ of a data point. Trivially, we can increase $\lambda_1(\mathbf{Q})$ by a factor of 100 if we scale each feature by a factor of 10. However, this approach would also scale the error term $\left\|\mathbf{X}^T\mathbf{n}\right\|_2^2$ in (2.15) by a factor of 100. For some applications, we can find feature transformations that increase $\lambda_1(\mathbf{Q})$ but do not increase $\left\|\mathbf{X}^T\mathbf{n}\right\|_2^2$. We finally note that the error term $\left\|\mathbf{X}^T\mathbf{n}\right\|_2^2$ in (2.15) vanishes if the noise vector $\mathbf{n}$ is orthogonal to the columns of the feature matrix $\mathbf{X}$.

It is instructive to evaluate the bound (2.15) for the special case where each data point has the same feature value $x = 1$. Here, the probabilistic model (2.12) reduces to a "signal in noise" model [29]

$$y^{(r)} = x^{(r)}\overline{w} + \varepsilon^{(r)} \text{ with } x^{(r)} = 1, \tag{2.18}$$

with some true underlying parameter $\overline{w}$. The noise terms $\varepsilon^{(r)}$, for $r = 1, \ldots, m$, are realizations of i.i.d. RVs with probability distribution $\mathcal{N}(0, \sigma^2)$. The feature matrix then becomes $\mathbf{X} = \mathbf{1}$ and, in turn, $\mathbf{Q} = 1$, $\lambda_1(\mathbf{Q}) = 1$. Inserting these values into (2.15) results in the bound

$$\left(\widehat{w}^{(\mathrm{LR})} - \overline{w}\right)^2 \leq 4\,\|\mathbf{n}\|_2^2 / m^2.$$

For the labels and features in (2.18), the solution of (2.13) is given by

$$\widehat{w}^{(\mathrm{LR})} = (1/m)\sum_{r=1}^{m} y^{(r)} \overset{(2.18)}{=} \overline{w} + (1/m)\sum_{r=1}^{m} \varepsilon^{(r)}.$$

2.4 Validation and Diagnosis of ML

The above analysis of the generalization error started from postulating the probabilistic model (2.10) for the generation of data points. Strictly speaking, if the data points are not generated according to the probabilistic model, the bound (2.15) does not apply. Thus, we might want to use a more data-driven approach for assessing the usefulness of a learned hypothesis $\hat{h}$ obtained, e.g., from solving ERM (2.1).

Loosely speaking, validation tries to find out if a learned hypothesis $\hat{h}$ performs similarly well inside and outside the training set. A basic form of validation is to compute the average loss of a learned hypothesis $\hat{h}$ on some data points not included in the training set. We refer to these data points as the validation set.

Algorithm 1 summarizes a single iteration of a prototypical ML workflow that consists of model training and validation. The workflow starts with an initial choice of a dataset $\mathcal{D}$, model $\mathcal{H}$, and loss function $L(\cdot, \cdot)$. We then repeat Algorithm 1 several times. After each repetition, based on the resulting training error and validation error, we modify some of the design choices for the dataset, the model, and the loss function.

Algorithm 1 One iteration of ML training and Validation

Input: dataset $\mathcal{D}$, model $\mathcal{H}$, loss function $L(\cdot, \cdot)$

1: split $\mathcal{D}$ into a training set $\mathcal{D}^{(\text{train})}$ and a validation set $\mathcal{D}^{(\text{val})}$

2: learn a hypothesis via solving ERM

$$\widehat{h} \in \underset{h \in \mathcal{H}}{\operatorname{argmin}} \sum_{(\mathbf{x}, y) \in \mathcal{D}^{(\text{train})}} L((\mathbf{x}, y), h) \tag{2.19}$$

3: compute resulting training error

$$E_t := (1/|\mathcal{D}^{(\text{train})}|) \sum_{(\mathbf{x}, y) \in \mathcal{D}^{(\text{train})}} L\left((\mathbf{x}, y), \widehat{h}\right)$$

4: compute validation error

$$E_v := (1/|\mathcal{D}^{(\text{val})}|) \sum_{(\mathbf{x}, y) \in \mathcal{D}^{(\text{val})}} L\left((\mathbf{x}, y), \widehat{h}\right)$$

Output: learned hypothesis (or trained model) $\widehat{h}$, training error E_t and validation error E_v

---

We can diagnose an ERM-based ML method, such as Algorithm 1, by comparing its training error with its validation error. This diagnosis is further enabled if we know a baseline $E^{(\text{ref})}$. One important source for a baseline $E^{(\text{ref})}$ is probabilistic models for the data points.

Given a probabilistic model $p(\mathbf{x}, y)$, we can compute the minimum achievable risk (2.9). Indeed, the minimum achievable risk is precisely the expected loss of the Bayes estimator $\widehat{h}(\mathbf{x})$ of the label y, given the features $\mathbf{x}$ of a data point. The Bayes estimator $\widehat{h}(\mathbf{x})$ is fully determined by the probability distribution $p(\mathbf{x}, y)$ [30, Ch. 4].

A further potential source for a baseline $E^{(\text{ref})}$ is an existing, but for some reason unsuitable, ML method. This existing ML method might be computationally too expensive to be used for the ML application at hand. However, we might still use its statistical properties as a baseline.

We can also use the performance of human experts as a baseline. For example, if we develop an ML method to detect skin cancer from images, a possible baseline is the classification accuracy achieved by experienced dermatologists [31].

We can diagnose an ML method by comparing the training error E_t with the validation error E_v and the baseline $E^{(\text{ref})}$.

- $E_t \approx E_v \approx E^{(\text{ref})}$: The training error is on the same level as the validation error and the baseline. There seems to be little point in trying to improve the method further since the validation error is already close to the baseline. Moreover, the training error is not much smaller than the validation error which indicates that there is no overfitting.
- $E_v \gg E_t$: The validation error is significantly larger than the training error, which hints at overfitting. We can address overfitting either by reducing the effective dimension of the hypothesis space or by increasing the size of the training set. To reduce the effective dimension of the hypothesis space, we can use fewer features (in a linear model), a smaller maximum depth of decision trees or fewer layers in an artificial neural network (ANN). Instead of this coarse-grained discrete model pruning, we can also reduce the effective dimension of a hypothesis space continuously via regularization (see [23, Ch. 7]).
- $E_t \approx E_v \gg E^{(\text{ref})}$: The training error is on the same level as the validation error and both are significantly larger than the baseline. Thus, the learned hypothesis seems to not overfit the training set. However, the training error achieved by the learned hypothesis is significantly larger than the baseline. There can be several reasons for this to happen. First, it might be that the hypothesis space is too small, i.e., it does not include a hypothesis that provides a satisfactory approximation for the relation between the features and the label of a data point. One remedy to this situation is to use a larger hypothesis space, e.g., by including more features in a linear model, using higher polynomial degrees in polynomial regression, using deeper decision trees or ANNs with more hidden layers (deep net). Second, besides the model being too small, another reason for a large training error could be that the optimization algorithm used to solve ERM (2.19) is not working properly (see Chap. 4).
- $E_t \gg E_v$: The training error is significantly larger than the validation error. The idea of ERM (2.19) is to approximate the risk (2.9) of a hypothesis by its average loss on a training set $\mathcal{D} = \{(\mathbf{x}^{(r)}, y^{(r)})\}_{r=1}^{m}$. The mathematical underpinning for this approximation is the law of large numbers which characterizes the average of i.i.d. RVs. The accuracy of this approximation depends on the validity of two conditions: First, the data points used for computing the average loss "should behave" like realizations of i.i.d. RVs with a common probability distribution.

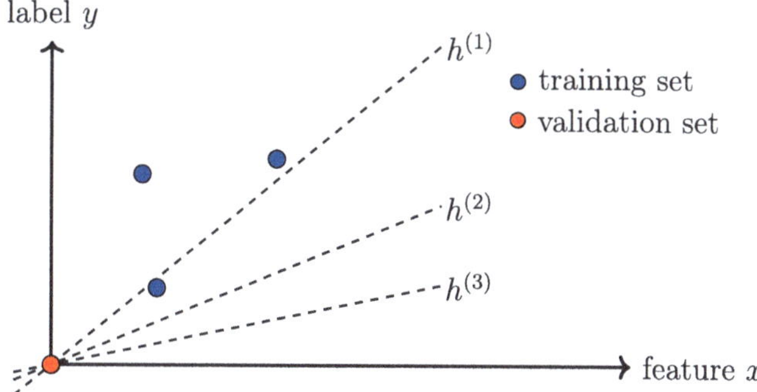

Fig. 2.3 An example of an unlucky split of a dataset into a training set and a validation set for the model $\mathcal{H} := \{h^{(1)}, h^{(2)}, h^{(3)}\}$

Second, the number of data points used for computing the average loss must be sufficiently large.
Whenever the training set or validation set differs significantly from realizations of i.i.d. RVs, the interpretation (and comparison) of the training error and the validation error of a learned hypothesis becomes more difficult. Figure 2.3 illustrates an extreme case of a validation set consisting of data points for which every hypothesis incurs a small average loss. Here, we might try to increase the size of the validation set by collecting more labeled data points or by using data augmentation. If the size of the training set and the validation set is large but we still obtain $E_t \gg E_v$, we should verify if the data points in these sets conform to the i.i.d. assumption. There are principled statistical tests for the validity of the i.i.d. assumption for a given dataset; see [32] and references therein.

2.5 Regularization

Consider an ERM-based method with hypothesis space $\mathcal{H}$ and training set $\mathcal{D}$. A key indicator for the performance of such an ML method is the ratio $d_{\text{eff}}(\mathcal{H})/|\mathcal{D}|$ between the model size $d_{\text{eff}}(\mathcal{H})$ and the number $|\mathcal{D}|$ of data points. The tendency of the ML method to overfit increases with the ratio $d_{\text{eff}}(\mathcal{H})/|\mathcal{D}|$.

Regularization techniques decrease the ratio $d_{\text{eff}}(\mathcal{H})/|\mathcal{D}|$ via three approaches:

- collect more data points, possibly via data augmentation (see Fig. 2.4),
- add a penalty term $\alpha\mathcal{R}\{h\}$ to average loss in ERM (2.1)

$$\hat{h} \in \underset{h \in \mathcal{H}}{\operatorname{argmin}} (1/m) \sum_{r=1}^{m} L\left(\left(\mathbf{x}^{(r)}, y^{(r)}\right), h\right) + \alpha\mathcal{R}\{h\}, \tag{2.20}$$

- shrink the hypothesis space, e.g., by adding constraints on the model parameters such as $\|\mathbf{w}\|_2 \leq 10$.

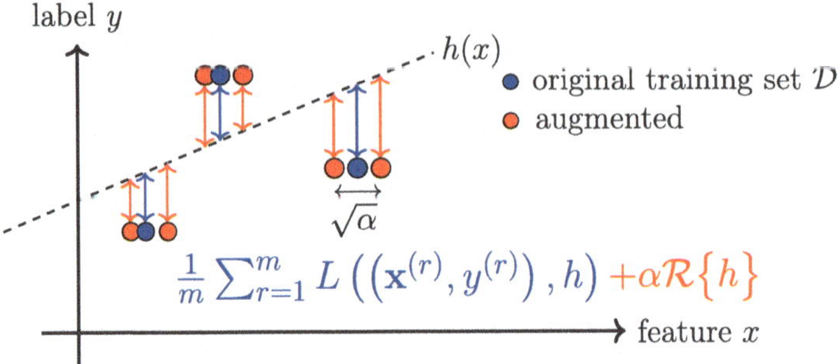

Fig. 2.4 Equivalence between data augmentation and loss penalization

As illustrated in Fig. 2.4, these three forms of regularization are closely related [23, Ch. 7]. For example, the regularized ERM (2.20) is equivalent to ERM (2.1) with a pruned hypothesis space $\mathcal{H}^{(\alpha)} \subseteq \mathcal{H}$. Using a larger α typically results in a smaller $\mathcal{H}^{(\alpha)}$.

One example of regularization by adding a penalty term is ridge regression. In particular, ridge regression uses the regularizer $\mathcal{R}\{h\} := \|\mathbf{w}\|_2^2$ for a linear hypothesis $h(\mathbf{x}) := \mathbf{w}^T\mathbf{x}$. Thus, ridge regression learns the parameters of a linear hypothesis via solving

$$\widehat{\mathbf{w}}^{(\alpha)} \in \underset{\mathbf{w}\in\mathbb{R}^d}{\operatorname{argmin}} \left[(1/m)\sum_{r=1}^{m}\left(y^{(r)} - \mathbf{w}^T\mathbf{x}^{(r)}\right)^2 + \alpha\,\|\mathbf{w}\|_2^2\right]. \tag{2.21}$$

The objective function in (2.21) can be interpreted as the objective function of linear regression applied to a modification of the training set $\mathcal{D}$: We replace each data point $(\mathbf{x}, y) \in \mathcal{D}$ by a sufficiently large number of i.i.d. realizations of

$$(\mathbf{x} + \mathbf{n}, y)\,, \text{ with } \mathbf{n} \sim \mathcal{N}(\mathbf{0}, \alpha\mathbf{I}). \tag{2.22}$$

Thus, ridge regression (2.21) is equivalent to linear regression applied to an augmentation $\mathcal{D}'$ of the original dataset $\mathcal{D}$. The augmentation $\mathcal{D}'$ is obtained by replacing each data point $(\mathbf{x}, y) \in \mathcal{D}$ with a sufficiently large number of noisy copies. Each copy of $(\mathbf{x}, y)$ is obtained by adding an i.i.d. realization $\mathbf{n}$ of a zero-mean Gaussian noise with covariance matrix $\alpha\mathbf{I}$ to the features $\mathbf{x}$ (see (2.22)). The label of each copy of $(\mathbf{x}, y)$ is equal to y, i.e., the label is not perturbed.

To study the computational aspects of ridge regression, let us rewrite (2.21) as

$$\widehat{\mathbf{w}}^{(\alpha)} \in \underset{\mathbf{w}\in\mathbb{R}^d}{\operatorname{argmin}}\, \mathbf{w}^T\mathbf{Q}\mathbf{w} + \mathbf{w}^T\mathbf{q},$$
$$\text{with } \mathbf{Q} := (1/m)\mathbf{X}^T\mathbf{X} + \alpha\mathbf{I},\ \mathbf{q} := (-2/m)\mathbf{X}^T\mathbf{y}. \tag{2.23}$$

Thus, like linear regression (2.5), also ridge regression minimizes a convex quadratic function. A main difference between linear regression (2.5) and ridge regression (for $\alpha > 0$) is that the matrix $\mathbf{Q}$ in (2.23) is guaranteed to be invertible for any training set $\mathcal{D}$. In contrast, the matrix $\mathbf{Q}$ in (2.5) for linear regression might be singular for some training sets.[2]

The statistical properties of the solutions to (2.23) crucially depend on the value of α. This choice can be guided by an error analysis using a probabilistic model for the data (see Proposition 2.1). Instead of using a probabilistic model, we can also compare the training error and validation error of the hypothesis $h(\mathbf{x}) = \left(\widehat{\mathbf{w}}^{(\alpha)}\right)^T \mathbf{x}$ learned by ridge regression with different values of α.

2.6 From ML to FL via Regularization

The main theme of this book is the analysis of FL systems that consists of a network of devices, indexed by $i = 1, \ldots, n$. Each device i trains a local (or personalized) model $\mathcal{H}^{(i)}$. One natural way to couple the model training of different devices is via regularization.

Assuming parametric models for ease of exposition, each device $i = 1, \ldots, n$ solves a separate instance of regularized empirical risk minimization (RERM) (2.20):[3]

$$\widehat{\mathbf{w}}^{(i)} \in \underset{\mathbf{w}^{(i)} \in \mathbb{R}^d}{\operatorname{argmin}} \underbrace{(1/m) \sum_{r=1}^{m} L\left(\left(\mathbf{x}^{(r)}, y^{(r)}\right), h\right)}_{=: L_i\left(\mathbf{w}^{(i)}\right)} + \alpha \mathcal{R}^{(i)}\left\{\mathbf{w}^{(i)}\right\}. \tag{2.24}$$

We can couple the instances of (2.24) at device i with other devices $i' \in \mathcal{V} \setminus \{i\}$ by using a regularizer $\mathcal{R}^{(i)}\left\{\mathbf{w}^{(i)}\right\}$ that depends on their model parameters $\mathbf{w}^{(i')}$. For example, we will study constructions for $\mathcal{R}^{(i)}\left\{\mathbf{w}^{(i)}\right\}$ that penalize deviations between the model parameters $\mathbf{w}^{(i)}$ and those at other devices, $\mathbf{w}^{(i')}$ for $i' \in \mathcal{N}^{(i)}$. Figure 2.5 illustrates a simple network of two devices, each training a personalized model via (2.24). Chapter 3 will discuss in more detail how to construct a useful regularizer $\mathcal{R}^{(i)}\left\{\mathbf{w}^{(i)}\right\}$.

[2] Consider the extreme case where all features of each data point in the training set $\mathcal{D}$ are zero.

[3] It will be convenient to avoid explicit reference to the local dataset of device i and instead work with the local loss function $L_i\left(\mathbf{w}^{(i)}\right)$. The FL algorithms studied in this book require only access to $L_i\left(\mathbf{w}^{(i)}\right)$.

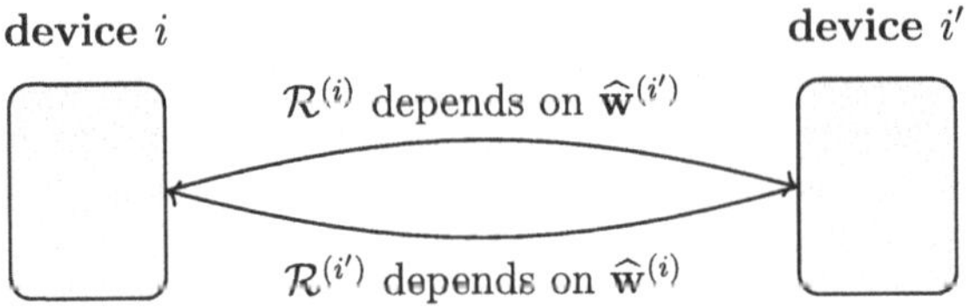

$$\widehat{\mathbf{w}}^{(i)} \in \underset{\mathbf{w}^{(i)}}{\operatorname{argmin}}\, L_i\left(\mathbf{w}^{(i)}\right) + \alpha\mathcal{R}^{(i)}(\mathbf{w}^{(i)}) \qquad \widehat{\mathbf{w}}^{(i')} \in \underset{\mathbf{w}^{(i')}}{\operatorname{argmin}}\, L_{i'}\left(\mathbf{w}^{(i')}\right) + \alpha\mathcal{R}^{(i')}(\mathbf{w}^{(i')})$$

Fig. 2.5 Two devices i and i' learn personalized model parameters. Each device executes a separate instance of regularized ERM with the regularizer depending on the model parameters of the other device

2.7 Exercises

2.1 Fundamental Limits for Linear Regression. Linear regression learns model parameters of a linear model to minimize the risk $\mathbb{E}\{(y - \mathbf{w}^T\mathbf{x})^2\}$ where $(\mathbf{x}, y)$ is a RV. In practice, we do not observe the RV $(\mathbf{x}, y)$ itself but a (realization of a) sequence of i.i.d. samples $(\mathbf{x}^{(t)}, y^{(t)})$, for $t = 1, 2, \ldots$. The minimax risk is a lower bound on the risk achievable by any learning method [33, Ch. 15]. Determine the minimax risk in terms of the probability distribution of $(\mathbf{x}, y)$.

2.2 Uniqueness of Eigenvectors. Consider the EVD $\mathbf{Q} = \sum_{j=1}^{d} \lambda_j \mathbf{u}^{(j)} (\mathbf{u}^{(j)})^T$ of a psd matrix $\mathbf{Q}$. The EVD consists of orthonormal eigenvectors $\mathbf{u}^{(j)}$ and non-negative eigenvalues λ_j, with $\mathbf{Q}\mathbf{u}^{(j)} = \lambda_j \mathbf{u}^{(j)}$, for $j = 1, \ldots, d$. Can you provide conditions on the eigenvalues $\lambda_1 \leq \ldots \leq \lambda_d$ such that the (unit-norm) eigenvectors are unique?

2.3 Penalty Term as Data Augmentation. Consider an ML method that trains a model with model parameters $\mathbf{w}$. The training uses ERM with squared error loss. Show that regularization of the model training via adding a penalty term $\alpha \|\mathbf{w}\|_2^2$ is equivalent to a specific form of data augmentation. What is the augmented training set?

2.4 Data Augmentation via Linear Interpolation. Consider an ML method that trains a model, with model parameters $\mathbf{w}$, from a training set $\mathcal{D}$. Each data point $\mathbf{z} \in \mathcal{D}$ is characterized by a feature vector $\mathbf{x} \in \mathbb{R}^d$ and label $y \in \mathbb{R}$, i.e., $\mathbf{z} = (\mathbf{x}, y)$. We augment the training set by adding, for each pair of two different data points $\mathbf{z}, \mathbf{z}' \in \mathcal{D}$, synthetic data points $\tilde{\mathbf{z}}^{(r)} := \mathbf{z} + (\mathbf{z}' - \mathbf{z})r/100$ and, for $r = 0, \ldots, 99$. Does this augmentation typically increase the training error?

2.5 Ridge Regression via Deterministic Data Augmentation. Ridge regression is obtained from linear regression by adding the penalty term $\alpha \|\mathbf{w}\|_2^2$ to the average squared error loss incurred by the hypothesis $h^{(\mathbf{w})}$ on the training set $\mathcal{D}$:

$$\min_{\mathbf{w}} (1/m) \sum_{r=1}^{m} \left(y^{(r)} - h\left(\mathbf{x}^{(r)}\right)\right)^2 + \alpha \|\mathbf{w}\|_2^2 . \tag{2.25}$$

Construct an augmented training set $\mathcal{D}'$ such that the objective function of (2.25) coincides with the objective function of plain linear regression using $\mathcal{D}'$ as training set. To construct $\mathcal{D}'$, add carefully chosen data points to the original training set $\mathcal{D} = \left\{ \left(y^{(1)}, \mathbf{x}^{(1)}\right), \ldots, \left(y^{(m)}, \mathbf{x}^{(m)}\right) \right\}$. Generalize the construction of $\mathcal{D}'$ to implement a generalized form of ridge regression:

$$\min_{\mathbf{w}} (1/m) \sum_{r=1}^{m} \left(y^{(r)} - h\left(\mathbf{x}^{(r)}\right)\right)^2 + \alpha \|\mathbf{w} - \widetilde{\mathbf{w}}\|_2^2 . \tag{2.26}$$

Here, we used some prescribed reference model parameters $\widetilde{\mathbf{w}}$. Note that (2.26) reduces to basic ridge regression (2.25) for the specific choice $\widetilde{\mathbf{w}} = \mathbf{0}$.

Chapter 3
A Design Principle for FL

Abstract This chapter introduces a core design principle for federated learning (FL) based on network optimization. FL systems are modeled as networks, where each node is a device with its own data and model. Edges between nodes represent communication or statistical similarity. To keep models consistent across the network, the chapter defines generalized total variation (GTV) as a measure of model differences. This leads to generalized total variation minimization (GTVMin), a learning framework that balances local accuracy with global consistency. The chapter explains how different choices in GTVMin affect learning behavior. It discusses how to optimize GTVMin using gradient or proximal methods and when these solutions are statistically reliable. Extensions to non-parametric models are handled by comparing predictions instead of parameters. The chapter also connects GTVMin to known ideas such as convex clustering and minimum-cost flow. Overall, it moves from defining FL networks to practical algorithms, making GTVMin a flexible tool for building scalable FL systems.

Chapter 2 reviewed ERM as a central design principle for traditional, centralized ML systems that rely on a single dataset to train a single model. This chapter extends these foundations to the distributed setting of FL, where learning takes place over a network of devices, each having their own datasets and models.

We begin in Sect. 3.1 by introducing the notion of an FL network—a mathematical abstraction for FL systems. Each node of an FL network represents a device that collects a local dataset and trains a local model, while the edges encode communication links and statistical similarities between local datasets.

Section 3.2 introduces the concept of GTV as a measure of discrepancy between local model parameters at connected nodes. This notion leads directly to Sect. 3.3, where we develop GTVMin as a principled regularization framework for training parametric local models in a federated setting. We then generalize this approach in Sect. 3.4 to accommodate non-parametric local models, broadening its applicability. Finally, Sect. 3.5 offers several interpretations of GTVMin that connect it to broader themes in applied mathematics and statistics, highlighting its conceptual and practical significance in FL design.

A. Jung, *Federated Learning*, https://doi.org/10.1007/978-981-95-1009-2_3

3.1 FL Networks

Consider an FL system that consists of devices, indexed by $i = 1, \ldots, n$, each with the ability to generate a local dataset $\mathcal{D}^{(i)}$ and to train a personalized model $\mathcal{H}^{(i)}$. These devices collaborate with each other via some communication network to learn a local hypothesis $h^{(i)} \in \mathcal{H}^{(i)}$. We measure the quality of $h^{(i)} \in \mathcal{H}^{(i)}$ via some loss function $L_i\left(h^{(i)}\right)$.

We now introduce the concept of an FL network as a mathematical model for FL applications. An FL network consists of an undirected weighted graph $\mathcal{G} = (\mathcal{V}, \mathcal{E})$ with nodes $\mathcal{V} := \{1, \ldots, n\}$ and undirected edges $\mathcal{E}$ between pairs of different nodes. The nodes $\mathcal{V}$ represent devices with varying amounts of computational resources.

An undirected edge $\{i, i'\} \in \mathcal{E}$ in an FL network represents a form of similarity between device i and device i'. The amount of similarity is represented by an edge weight $A_{i,i'}$. We can collect edge weights into an adjacency matrix $\mathbf{A} \in \mathbb{R}^{n \times n}$, with $A_{i,i'} = A_{i',i}$. Figure 3.1 depicts an example of an FL network.

Note that the undirected edges $\mathcal{E}$ of an FL network encode a symmetric notion of similarity between devices: If the device i is similar to the device i', i.e., $\{i, i'\} \in \mathcal{E}$, then also the device i' is similar to the device i. For some FL applications, an asymmetric notion of similarity, represented by directed edges, could be more accurate. However, the generalization of an FL network to directed graphs is beyond the scope of this book.

It can be convenient to replace a given FL network $\mathcal{G}$ with an equivalent fully connected FL network $\mathcal{G}'$ (see Fig. 3.2). The fully connected graph $\mathcal{G}'$ contains an

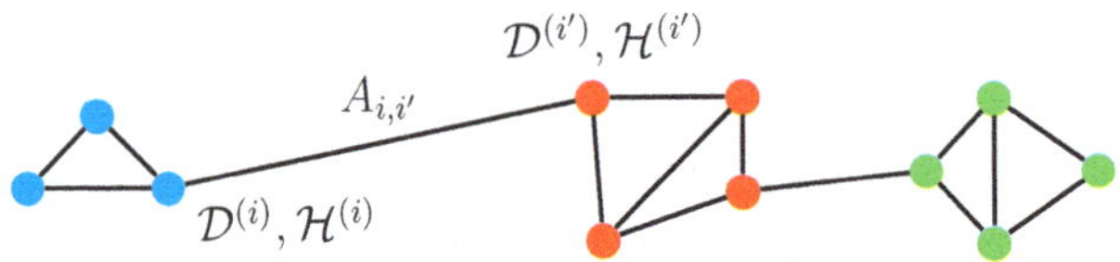

Fig. 3.1 Example of an FL network whose nodes $i \in \mathcal{V}$ represent devices. Each device i generates a local dataset $\mathcal{D}^{(i)}$ and trains a local model $\mathcal{H}^{(i)}$. Some devices i, i' are connected by an undirected edge $\{i, i'\}$ with a positive edge weight $A_{i,i'}$

Fig. 3.2 Left: An FL network $\mathcal{G}$ consisting of $n = 4$ nodes. Right, equivalent fully connected FL network $\mathcal{G}'$ with the same nodes and non-zero edge weights $A'_{i,i'} = A_{i,i'}$ for $\{i, i'\} \in \mathcal{E}$ and $A'_{i,i'} = 0$ for $\{i, i'\} \notin \mathcal{E}$

edge between every pair of two different nodes i, i':

$$\mathcal{E}' = \{\{i, i'\} : i, i' \in \mathcal{V}, i \neq i'\}.$$

The edge weights are chosen $A'_{i,i'} = A_{i,i'}$ for any edge $\{i, i'\} \in \mathcal{E}$ and $A'_{i,i'} = 0$ if the original FL network $\mathcal{G}$ does not contain an edge between nodes i, i'.

An FL network is more than the undirected weighted graph $\mathcal{G}$: It also includes the local dataset $\mathcal{D}^{(i)}$ and the local model $\mathcal{H}^{(i)}$ (or its model parameters $\mathbf{w}^{(i)}$) for each device $i \in \mathcal{V}$. The details of the generation and the format of a local dataset will not be important in what follows. A local dataset is just one possible means to construct a loss function in order to evaluate model parameters. However, to build intuition, we can think of a local dataset $\mathcal{D}^{(i)}$ as a labeled dataset:

$$\mathcal{D}^{(i)} := \left\{\left(\mathbf{x}^{(i,1)}, y^{(i,1)}\right), \ldots, \left(\mathbf{x}^{(i,m_i)}, y^{(i,m_i)}\right)\right\}. \tag{3.1}$$

Here, $\mathbf{x}^{(i,r)}$ and $y^{(i,r)}$ denote, respectively, the features and the label of the rth data point in the local dataset $\mathcal{D}^{(i)}$. Note that the size m_i of the local dataset can vary between different nodes $i \in \mathcal{V}$.

It is convenient to collect the feature vectors $\mathbf{x}^{(i,r)}$ and labels $y^{(i,r)}$ into a feature matrix $\mathbf{X}^{(i)}$ and label vector $\mathbf{y}^{(i)}$, respectively,

$$\mathbf{X}^{(i)} := \left(\mathbf{x}^{(i,1)}, \ldots, \mathbf{x}^{(i,m_i)}\right)^T, \text{ and } \mathbf{y} := \left(y^{(1)}, \ldots, y^{(m_i)}\right)^T. \tag{3.2}$$

The local dataset $\mathcal{D}^{(i)}$ can then be represented compactly by the feature matrix $\mathbf{X}^{(i)} \in \mathbb{R}^{m_i \times d}$ and the vector $\mathbf{y}^{(i)} \in \mathbb{R}^{m_i}$.

Besides the local dataset $\mathcal{D}^{(i)}$, each node $i \in \mathcal{G}$ also carries a local model $\mathcal{H}^{(i)}$. Our focus is on parametric local models with by model parameters $\mathbf{w}^{(i)} \in \mathbb{R}^d$, for $i = 1, \ldots, n$. The usefulness of a specific choice of the local model parameter $\mathbf{w}^{(i)}$ is then measured by a local loss function $L_i\left(\mathbf{w}^{(i)}\right)$, for $i = 1, \ldots, n$. Note that we can use different local loss functions $L_i(\cdot) \neq L_{i'}(\cdot)$ at different nodes $i, i' \in \mathcal{V}$.

We now have introduced all the components of an FL network. Strictly speaking, an FL network is a tuple $\left(\mathcal{G}, \{\mathcal{H}^{(i)}\}_{i\in\mathcal{V}}, \{L_i(\cdot)\}_{i\in\mathcal{V}}\right)$ consisting of an undirected weighted graph $\mathcal{G}$, a local model $\mathcal{H}^{(i)}$, and local loss function $L_i(\cdot)$ for each node $i \in \mathcal{V}$. In principle, all of these components are design choices that influence the computational and statistical properties of the FL algorithms presented in Chap. 5. To some extent, also the edges $\mathcal{E}$ in the FL network are a design choice.

The role (or meaning) of an edge $\{i, i'\}$ in an FL network is twofold: First, it represents a communication link that allows to exchange messages between devices i, i'. Second, an edge $\{i, i'\}$ indicates similar statistical properties of local datasets generated by devices i, i'. It then seems natural to learn similar hypothesis maps $h^{(i)}, h^{(i')}$. This is actually the main idea behind all the FL algorithms that we will discuss in the rest of this book. To make this idea precise, we next discuss how to obtain quantitative measures for how much local hypothesis maps $h^{(i)}$ vary across the edges $\{i, i'\} \in \mathcal{E}$ of an FL network.

3.2 Generalized Total Variation

Consider an FL network with nodes $i = 1, \ldots, n$, undirected edges $\mathcal{E}$ with edge weights $A_{i,i'} > 0$ for each $\{i, i'\} \in \mathcal{E}$. For each edge $\{i, i'\} \in \mathcal{E}$, we want to couple the training of the corresponding local models $\mathcal{H}^{(i)}, \mathcal{H}^{(i')}$. The strength of this coupling is determined by the edge weight $A_{i,i'}$. We implement the coupling by penalizing the variation (or discrepancy) between the model parameters $\mathbf{w}^{(i)}, \mathbf{w}^{(i')}$.

We can measure the variation between two trained local models $h^{(i)}, h^{(i')}$ across an edge $\{i, i'\} \in \mathcal{E}$ in different ways. For example, we can compare their predictions on a common test set $\mathcal{D}$ by computing

$$d^{(i,i')} := (1/|\mathcal{D}|) \sum_{\mathbf{x} \in \mathcal{D}} \left[h^{(i)}(\mathbf{x}) - h^{(i')}(\mathbf{x})\right]^2. \tag{3.3}$$

In principle, we can use a different test set in (3.3) for each edge $\{i, i'\}$ of $\mathcal{G}$. For example, the test set could be obtained by merging randomly selected data points from each local dataset $\mathcal{D}^{(i)}, \mathcal{D}^{(i')}$.

Our main focus will be FL applications that use parametric local models, i.e., each node learns local model parameters $\mathbf{w}^{(i)} \in \mathbb{R}^d$, for $i = 1, \ldots, n$. Here, we can measure the variation between $h^{\mathbf{w}^{(i)}}$ and $h\mathbf{w}^{(i')}$ directly in terms of the model parameters $\mathbf{w}^{(i)}, \mathbf{w}^{(i')}$ at the nodes of an edge $\{i, i'\}$. In particular, we use a penalty function $\phi : \mathbb{R}^d \rightarrow \mathbb{R}$ of the difference between the model parameters:

$$d^{(i,i')} := \phi\big(\mathbf{w}^{(i)} - \mathbf{w}^{(i')}\big). \tag{3.4}$$

The penalty function ϕ will be mainly a design choice. Our main requirement is that ϕ is monotonically increasing with respect to some norm in the Euclidean space $\mathbb{R}^d$ [16, 34]. This requirement ensures symmetry, i.e., $\phi\big(\mathbf{w}^{(i)} - \mathbf{w}^{(i')}\big) = \phi\big(\mathbf{w}^{(i')} - \mathbf{w}^{(i)}\big)$, allowing its use as a measure of variation across an undirected edge $\{i, i'\} \in \mathcal{E}$.

Summing up the edge-wise variations (weighted by the edge weights) yields the GTV of a collection of local model parameters:

$$\sum_{\{i,i'\} \in \mathcal{E}} A_{i,i'} \phi\big(\mathbf{w}^{(i)} - \mathbf{w}^{(i')}\big). \tag{3.5}$$

Our main focus will be on the special case of (3.5), obtained for $\phi(\cdot) := \|\cdot\|_2^2$,

$$\sum_{\{i,i'\} \in \mathcal{E}} A_{i,i'} \left\|\mathbf{w}^{(i)} - \mathbf{w}^{(i')}\right\|_2^2. \tag{3.6}$$

The choice of penalty $\phi(\cdot)$ has a crucial impact on the computational and statistical properties of the FL algorithms presented in Chap. 5. Our main choice during the rest of this book will be the penalty function $\phi(\cdot) := \|\cdot\|_2^2$. This choice often allows to formulate FL as the minimization of a smooth convex function, which can be done via simple gradient-based methods (see Chap. 7). On

Fig. 3.3 Left, example of an FL network $\mathcal{G}$ with three nodes $i = 1, 2, 3$. Right, Laplacian matrix $\mathbf{L}^{(\mathcal{G})} \in \mathbb{R}^{3\times 3}$ of $\mathcal{G}$

the other hand, choosing ϕ to be a norm results in FL algorithms that require more computation but less training data [34].

The connectivity of an FL network $\mathcal{G}$ can be characterized locally—around a node $i \in \mathcal{V}$—by its node degree:

$$d^{(i)} := \sum_{i' \in \mathcal{N}^{(i)}} A_{i,i'}. \tag{3.7}$$

Here, we used the neighborhood $\mathcal{N}^{(i)} := \{i' \in \mathcal{V} : \{i, i'\} \in \mathcal{E}\}$ of node $i \in \mathcal{V}$. A global characterization for the connectivity of $\mathcal{G}$ is the maximum node degree:

$$d_{\max}^{(\mathcal{G})} := \max_{i \in \mathcal{V}} d^{(i)} \overset{(3.7)}{=} \max_{i \in \mathcal{V}} \sum_{i' \in \mathcal{N}^{(i)}} A_{i,i'}. \tag{3.8}$$

Besides inspecting the node degrees, we can study the connectivity of $\mathcal{G}$ also via the eigenvalues and eigenvectors of its Laplacian matrix $\mathbf{L}^{(\mathcal{G})} \in \mathbb{R}^{n\times n}$.[1] The Laplacian matrix of an undirected weighted graph $\mathcal{G}$ is defined element-wise as

$$L_{i,i'}^{(\mathcal{G})} := \begin{cases} -A_{i,i'} & \text{for } i \neq i', \{i, i'\} \in \mathcal{E} \\ \sum_{i'' \neq i} A_{i,i''} & \text{for } i = i' \\ 0 & \text{else.} \end{cases} \tag{3.9}$$

Figure 3.3 illustrates the Laplacian matrix of a small graph.

The Laplacian matrix is symmetric and psd, which follows from the identity

$$\mathbf{w}^T (\mathbf{L}^{(\mathcal{G})} \otimes \mathbf{I}) \mathbf{w} = \sum_{\{i,i'\} \in \mathcal{E}} A_{i,i'} \left\| \mathbf{w}^{(i)} - \mathbf{w}^{(i')} \right\|_2^2$$

$$\text{for any } d \in \mathbb{N}, \mathbf{w} := \underbrace{\left(\left(\mathbf{w}^{(1)}\right)^T, \ldots, \left(\mathbf{w}^{(n)}\right)^T \right)^T}_{=:\text{stack}\left\{\mathbf{w}^{(i)}\right\}_{i=1}^n} \in \mathbb{R}^{d \cdot n}. \tag{3.10}$$

[1] The study of graphs via the eigenvalues and eigenvectors of associated matrices is the main subject of spectral graph theory [35, 36].

As a psd matrix, $\mathbf{L}^{(\mathcal{G})}$ possesses an EVD

$$\mathbf{L}^{(\mathcal{G})} = \sum_{i=1}^{n} \lambda_i \mathbf{u}^{(i)} \big(\mathbf{u}^{(i)}\big)^T, \tag{3.11}$$

with orthonormal eigenvectors $\mathbf{u}^{(1)}, \ldots, \mathbf{u}^{(n)}$ and corresponding list of eigenvalues

$$0 = \lambda_1\big(\mathbf{L}^{(\mathcal{G})}\big) \leq \lambda_2\big(\mathbf{L}^{(\mathcal{G})}\big) \leq \ldots \leq \lambda_n\big(\mathbf{L}^{(\mathcal{G})}\big). \tag{3.12}$$

We just write λ_i instead of $\lambda_i\big(\mathbf{L}^{(\mathcal{G})}\big)$ if the Laplacian matrix $\mathbf{L}^{(\mathcal{G})}$ is clear from context. The eigenvalue $\lambda_i\big(\mathbf{L}^{(\mathcal{G})}\big)$ corresponds to the eigenvector $\mathbf{u}^{(i)}$, i.e., $\mathbf{L}^{(\mathcal{G})}\mathbf{u}^{(i)} = \lambda_i\big(\mathbf{L}^{(\mathcal{G})}\big)\mathbf{u}^{(i)}$ for $i = 1, \ldots, n$.

It is important to note that the ordered list of eigenvalues (3.12) is uniquely determined for a given Laplacian matrix. In contrast, the eigenvectors $\mathbf{u}^{(i)}$ in (3.11) are not unique in general.[2]

The ordered eigenvalues $\lambda_i\big(\mathbf{L}^{(\mathcal{G})}\big)$ in (3.12) can be computed (or characterized) via the Courant–Fischer–Weyl min-max characterization (CFW) [3, Thm. 8.1.2.]. Two important special cases of this characterization are [35, 36]

$$\begin{aligned} \lambda_n\big(\mathbf{L}^{(\mathcal{G})}\big) &\stackrel{\text{CFW}}{=} \max_{\substack{\mathbf{v}\in\mathbb{R}^n \\ \|\mathbf{v}\|=1}} \mathbf{v}^T \mathbf{L}^{(\mathcal{G})} \mathbf{v} \\ &\stackrel{(3.10)}{=} \max_{\substack{\mathbf{v}\in\mathbb{R}^n \\ \|\mathbf{v}\|=1}} \sum_{\{i,i'\}\in\mathcal{E}} A_{i,i'}\big(v_i - v_{i'}\big)^2 \end{aligned} \tag{3.13}$$

and

$$\begin{aligned} \lambda_2\big(\mathbf{L}^{(\mathcal{G})}\big) &\stackrel{\text{CFW}}{=} \min_{\substack{\mathbf{v}\in\mathbb{R}^n \\ \mathbf{v}^T\mathbf{1}=0 \\ \|\mathbf{v}\|=1}} \mathbf{v}^T \mathbf{L}^{(\mathcal{G})} \mathbf{v} \\ &\stackrel{(3.10)}{=} \min_{\substack{\mathbf{v}\in\mathbb{R}^n \\ \mathbf{v}^T\mathbf{1}=0 \\ \|\mathbf{v}\|=1}} \sum_{\{i,i'\}\in\mathcal{E}} A_{i,i'}\big(v_i - v_{i'}\big)^2. \end{aligned} \tag{3.14}$$

By (3.10), we can compute the GTV of a collection of model parameters via the quadratic form $\mathbf{w}^T\big(\mathbf{L}^{(\mathcal{G})} \otimes \mathbf{I}_{d\times d}\big)\mathbf{w}$. This quadratic form involves the vector $\mathbf{w} \in \mathbb{R}^{nd}$ which is obtained by stacking the local model parameters $\mathbf{w}^{(i)}$ for

[2] Consider the scenario where the list (3.12) contains repeated entries, i.e., several ordered eigenvalues are identical.

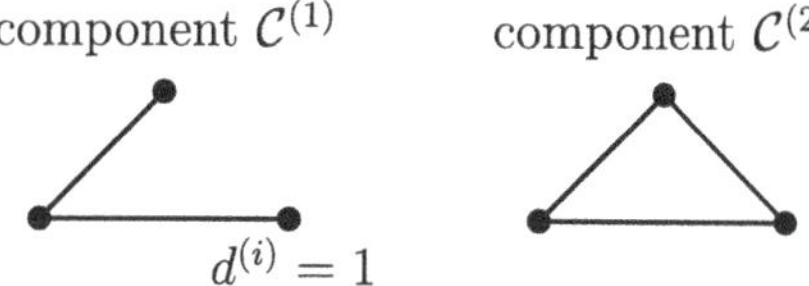

Fig. 3.4 An FL network $\mathcal{G}$ that consists of $n = 6$ nodes forming two connected components $\mathcal{C}^{(1)}, \mathcal{C}^{(2)}$

$i = 1, \ldots, n$. Another consequence of (3.10) is that any collection of identical local model parameters, stacked into the vector

$$\mathbf{w} = \text{stack}\{\mathbf{c}\} = \left(\mathbf{c}^T, \ldots, \mathbf{c}^T\right)^T, \text{ with some } \mathbf{c} \in \mathbb{R}^d, \tag{3.15}$$

is an eigenvector of $\mathbf{L}^{(\mathcal{G})} \otimes \mathbf{I}$ with corresponding eigenvalue $\lambda_1 = 0$ (see (3.12)). Thus, the Laplacian matrix of any FL network is singular (non-invertible).

The second eigenvalue λ_2 of $\mathbf{L}^{(\mathcal{G})}$ provides a great deal of information about the connectivity structure of $\mathcal{G}$.[3] Indeed, much of spectral graph theory is devoted to the analysis of λ_2, which is also referred to as algebraic connectivity, for different graph constructions [35, 36].

- Consider the case $\lambda_2 = 0$: Here, beside the eigenvector (3.15), we can find at least one additional eigenvector

$$\widetilde{\mathbf{w}} = \text{stack}\left\{\mathbf{w}^{(i)}\right\}_{i=1}^{n} \text{ with } \mathbf{w}^{(i)} \neq \mathbf{w}^{(i')} \text{ for some } i, i' \in \mathcal{V}, \tag{3.16}$$

  of $\mathbf{L}^{(\mathcal{G})} \otimes \mathbf{I}$ with eigenvalue equal to 0. In this case, the graph $\mathcal{G}$ is not connected, i.e., we can find two subsets (components) of nodes that do not have any edge between them (see Fig. 3.4). For each connected component $\mathcal{C}$, we can construct the eigenvector by assigning the same (non-zero) vector $\mathbf{c} \in \mathbb{R}^d \setminus \{\mathbf{0}\}$ to all nodes $i \in \mathcal{C}$ and the zero vector $\mathbf{0}$ to the remaining nodes $i \in \mathcal{V} \setminus \mathcal{C}$.
- On the other hand, if $\lambda_2 > 0$ then $\mathcal{G}$ is connected. Moreover, the larger the value of λ_2, the stronger the connectivity between the nodes in $\mathcal{G}$. Indeed, adding edges to $\mathcal{G}$ can only increase the objective in (3.14) and, in turn, λ_2.

In what follows, we will make use of the lower bound [37]

$$\sum_{\{i,i'\} \in \mathcal{E}} A_{i,i'} \left\|\mathbf{w}^{(i)} - \mathbf{w}^{(i')}\right\|_2^2 \geq \lambda_2 \sum_{i=1}^{n} \left\|\mathbf{w}^{(i)} - \text{avg}\{\mathbf{w}^{(i)}\}\right\|_2^2. \tag{3.17}$$

[3] With slight abuse of language, we will sometimes speak about the eigenvalues of an FL network $\mathcal{G}$. However, we actually mean the eigenvalues of the Laplacian matrix (3.9) naturally associated with $\mathcal{G}$.

Here, $\text{avg}\{\mathbf{w}^{(i)}\} := (1/n)\sum_{i=1}^{n} \mathbf{w}^{(i)}$ is the average of all local model parameters. The bound (3.17) follows from (3.10) and the CFW for the eigenvalues of the matrix $\mathbf{L}^{(\mathcal{G})} \otimes \mathbf{I}$.

The quantity $\sum_{i=1}^{n} \left\|\mathbf{w}^{(i)} - \text{avg}\{\mathbf{w}^{(i)}\}_{i=1}^{n}\right\|_2^2$ on the right-hand side of (3.17) has an interesting geometric interpretation: It is the squared Euclidean norm of the projection of the stacked local model parameters

$$\mathbf{w} := \left(\left(\mathbf{w}^{(1)}\right)^T, \ldots, \left(\mathbf{w}^{(n)}\right)^T\right)^T$$

onto the orthogonal complement of the subspace

$$\mathcal{S} := \left\{\mathbf{1} \otimes \mathbf{a} : \mathbf{a} \in \mathbb{R}^d\right\} = \left\{\left(\mathbf{a}^T, \ldots, \mathbf{a}^T\right)^T, \text{ for some } \mathbf{a} \in \mathbb{R}^d\right\} \subseteq \mathbb{R}^{dn}. \tag{3.18}$$

The subspace $\mathcal{S}$ consists of stacked local model parameters $\mathbf{w}^{(i)}$ that are identical for all nodes $i = 1, \ldots, n$. Such a structure arises in certain FL settings where a single global model is shared among all devices. In this setting, the local model parameters satisfy $\mathbf{w}^{(i)} = \mathbf{a}$ for all $i = 1, \ldots, n$ and some common vector $\mathbf{a} \in \mathbb{R}^d$ (see Sect. 6.1). Equivalently, the condition

$$\left(\mathbf{w}^{(1)}, \ldots, \mathbf{w}^{(n)}\right)^T \in \mathcal{S}$$

characterizes this single-model setting as membership in the subspace $\mathcal{S}$.

The projection $\mathbf{P}_{\mathcal{S}}\mathbf{w}$ of $\mathbf{w} \in \mathbb{R}^{nd}$ on $\mathcal{S}$ is

$$\mathbf{P}_{\mathcal{S}}\mathbf{w} = \left(\mathbf{a}^T, \ldots, \mathbf{a}^T\right)^T, \text{ with } \mathbf{a} = \text{avg}\{\mathbf{w}^{(i)}\}_{i=1}^{n}. \tag{3.19}$$

The projection on the orthogonal complement $\mathcal{S}^{\perp}$, in turn, is

$$\mathbf{P}_{\mathcal{S}^{\perp}}\mathbf{w} = \mathbf{w} - \mathbf{P}_{\mathcal{S}}\mathbf{w} = \text{stack}\left\{\mathbf{w}^{(i)} - \text{avg}\{\mathbf{w}^{(i)}\}_{i=1}^{n}\right\}_{i=1}^{n}. \tag{3.20}$$

3.3 Generalized Total Variation Minimization

Consider some FL network $\mathcal{G}$ with nodes $i \in \mathcal{V}$ representing devices that learn personalized model parameters $\mathbf{w}^{(i)}$. The usefulness of a specific choice of the model parameters $\mathbf{w}^{(i)}$ is measured by a local loss function $L_i\left(\mathbf{w}^{(i)}\right)$. We are mainly interested in FL applications where the local loss functions do not provide enough information for learning accurate model parameters.[4] We therefore require learned model parameters to not only incur a small local loss but also to have a small GTV (3.5).

GTV minimization (GTVMin) optimally balances the (average) local loss and the GTV (3.5) of local model parameters $\mathbf{w}^{(i)}$:

$$\{\widehat{\mathbf{w}}^{(i)}\}_{i=1}^{n} \in \underset{\mathbf{w}^{(1)},\dots,\mathbf{w}^{(n)}}{\operatorname{argmin}} \sum_{i\in\mathcal{V}} L_i\left(\mathbf{w}^{(i)}\right) + \alpha \sum_{\{i,i'\}\in\mathcal{E}} A_{i,i'}\phi\big(\mathbf{w}^{(i)} - \mathbf{w}^{(i')}\big). \tag{3.21}$$

Note that (3.21) is parametrized by the choice for the penalty function $\phi(\cdot)$. We discuss the effect of different choices for $\phi(\cdot)$ in Sects. 3.3.1 and 3.3.2. Our main focus will be on the special case of (3.21), obtained with $\phi(\cdot) := \|\cdot\|_2^2$:

$$\{\widehat{\mathbf{w}}^{(i)}\}_{i=1}^{n} \in \underset{\mathbf{w}^{(1)},\dots,\mathbf{w}^{(n)}}{\operatorname{argmin}} \sum_{i\in\mathcal{V}} L_i\left(\mathbf{w}^{(i)}\right) + \alpha \sum_{\{i,i'\}\in\mathcal{E}} A_{i,i'} \left\|\mathbf{w}^{(i)} - \mathbf{w}^{(i')}\right\|_2^2. \tag{3.22}$$

The GTVMin parameter $\alpha > 0$ in (3.21) steers the preference for learning local model parameters $\mathbf{w}^{(i)}$ with small GTV versus incurring small local loss $\sum_{i\in\mathcal{V}} L_i\left(\mathbf{w}^{(i)}\right)$. For $\alpha = 0$, GTVMin decomposes into fully independent local ERM instances $\min_{\mathbf{w}^{(i)}} L_i(\cdot)$, for $i = 1, \dots, n$. On the other hand, increasing the value of α makes the solutions of (3.21) increasingly clustered: the local model parameters $\widehat{\mathbf{w}}^{(i)}$ become approximately constant over increasingly large subsets of nodes. This behavior is appealing for clustered FL which we discuss in Sect. 6.2.

Choosing α beyond a critical value—which depends on the shape of the local loss functions and the edges $\mathcal{E}$—results in $\widehat{\mathbf{w}}^{(i)}$ being (nearly) constant over all nodes $i \in \mathcal{V}$. In practice, the choice of α can be guided by validation [38] or by a probabilistic analysis of the solutions of (3.21). Section 3.3.2 presents an example of such an analysis.

Note that GTVMin (3.21) is an instance of RERM: The regularizer is the GTV of local model parameters over the weighted edges $A_{i,i'}$ of the FL network. Loosely speaking, GTVMin couples the training of local models by requiring them to be similar across the edges of the FL network. For the extreme case of an FL network without any edges, GTVMin decomposes into independent ERM instances

$$\underset{\mathbf{w}^{(i)}}{\operatorname{argmin}}\, L_i\left(\mathbf{w}^{(i)}\right), \text{ for each } i = 1, \dots, n.$$

The connectivity (i.e., the edges $\mathcal{E}$) of the FL network is an important design choice in GTVMin-based methods. This choice can be guided by computational aspects and statistical aspects of GTVMin-based FL systems. Some application

[4] For example, the local loss function can be obtained from the training error on a local dataset that is too small relative to the effective dimension of the local model. As discussed in Sect. 2.5, such a setting is prone to overfitting.

domains allow to leverage domain expertise to guess a useful choice for the FL network. If local datasets are generated at different geographic locations, we might use nearest-neighbor graphs based on geodesic distances between data generators (e.g., FMI weather stations). Chapter 7 discusses graph learning methods that determine the edge weights $A_{i,i'}$ in a data-driven fashion, i.e., directly from the local datasets $\mathcal{D}^{(i)}$, $\mathcal{D}^{(i')}$.

GTVMin for linear models Let us now consider the special case of GTVMin with local models being a linear model. For each node $i \in \mathcal{V}$ of the FL network, we want to learn the parameters $\mathbf{w}^{(i)}$ of a linear hypothesis $h^{(i)}(\mathbf{x}) := \left(\mathbf{w}^{(i)}\right)^T \mathbf{x}$. We measure the quality of the parameters via the average squared error loss

$$: L_i\left(\mathbf{w}^{(i)}\right) := (1/m_i) \sum_{r=1}^{m_i} \left(y^{(i,r)} - \left(\mathbf{w}^{(i)}\right)^T \mathbf{x}^{(i,r)}\right)^2$$

$$\stackrel{(3.2)}{=} (1/m_i) \left\| \mathbf{y}^{(i)} - \mathbf{X}^{(i)} \mathbf{w}^{(i)} \right\|_2^2. \tag{3.23}$$

Inserting (3.23) into (3.22) yields the following instance of GTVMin to train local linear models:

$$\left\{\widehat{\mathbf{w}}^{(i)}\right\} \in \underset{\{\mathbf{w}^{(i)}\}_{i=1}^n}{\operatorname{argmin}} \sum_{i \in \mathcal{V}} (1/m_i) \left\| \mathbf{y}^{(i)} - \mathbf{X}^{(i)} \mathbf{w}^{(i)} \right\|_2^2 + \alpha \sum_{\{i,i'\} \in \mathcal{E}} A_{i,i'} \left\| \mathbf{w}^{(i)} - \mathbf{w}^{(i')} \right\|_2^2. \tag{3.24}$$

The identity (3.10) allows to rewrite (3.24) using the Laplacian matrix $\mathbf{L}^{(\mathcal{G})}$ as

$$\widehat{\mathbf{w}}^{(i)} \in \underset{\mathbf{w} = \operatorname{stack}\left\{\mathbf{w}^{(i)}\right\}}{\operatorname{argmin}} \sum_{i \in \mathcal{V}} (1/m_i) \left\| \mathbf{y}^{(i)} - \mathbf{X}^{(i)} \mathbf{w}^{(i)} \right\|_2^2 + \alpha \mathbf{w}^T \left(\mathbf{L}^{(\mathcal{G})} \otimes \mathbf{I}_d\right) \mathbf{w}. \tag{3.25}$$

Let us rewrite the objective function in (3.25) as

$$\mathbf{w}^T \left(\begin{pmatrix} \mathbf{Q}^{(1)} & \cdots & 0 \\ \vdots & \ddots & \vdots \\ 0 & \cdots & \mathbf{Q}^{(n)} \end{pmatrix} + \alpha \mathbf{L}^{(\mathcal{G})} \otimes \mathbf{I} \right) \mathbf{w} + \left(\left(\mathbf{q}^{(1)}\right)^T, \ldots, \left(\mathbf{q}^{(n)}\right)^T\right) \mathbf{w} \tag{3.26}$$

$$\text{with } \mathbf{Q}^{(i)} = (1/m_i) \left(\mathbf{X}^{(i)}\right)^T \mathbf{X}^{(i)} \text{ and } \mathbf{q}^{(i)} := (-2/m_i) \left(\mathbf{X}^{(i)}\right)^T \mathbf{y}^{(i)}.$$

Thus, like linear regression (2.5) and ridge regression (2.23), GTVMin (3.25) (for local linear models $\mathcal{H}^{(i)}$) minimizes a convex quadratic function:

$$\left\{\widehat{\mathbf{w}}^{(i)}\right\}_{i=1}^n \in \underset{\mathbf{w} = \operatorname{stack}\left\{\mathbf{w}^{(i)}\right\}_{i=1}^n}{\operatorname{argmin}} \mathbf{w}^T \mathbf{Q} \mathbf{w} + \mathbf{q}^T \mathbf{w}. \tag{3.27}$$

Here, we used the psd matrix

$$\mathbf{Q}:=\begin{pmatrix} \mathbf{Q}^{(1)} & \cdots & \mathbf{0} \\ \vdots & \ddots & \vdots \\ \mathbf{0} & \cdots & \mathbf{Q}^{(n)} \end{pmatrix} + \alpha \mathbf{L}^{(\mathcal{G})} \otimes \mathbf{I} \text{ with } \mathbf{Q}^{(i)} := (1/m_i)\big(\mathbf{X}^{(i)}\big)^T \mathbf{X}^{(i)} \tag{3.28}$$

and the vector

$$\mathbf{q} := \big((\mathbf{q}^{(1)})^T, \ldots, (\mathbf{q}^{(n)})^T\big)^T, \text{ with } \mathbf{q}^{(i)} := (-2/m_i)\big(\mathbf{X}^{(i)}\big)^T \mathbf{y}^{(i)}. \tag{3.29}$$

3.3.1 Computational Aspects of GTVMin

Chapter 5 will apply optimization methods to solve GTVMin (3.21), resulting in practical FL algorithms. Different instances of GTVMin favor different classes of optimization methods. For example, using a differentiable loss function $L_i(\cdot)$ and penalty function $\phi(\cdot)$ allows to apply gradient-based methods (see Chap. 4) to solve GTVMin. Another important class of loss functions are those for which we can efficiently compute the proximal operator [39, 40]:

$$\mathbf{prox}_{L_i(\cdot),\rho}(\mathbf{w}) := \operatorname*{argmin}_{\mathbf{w}' \in \mathbb{R}^d} L_i\left(\mathbf{w}'\right) + (\rho/2)\left\|\mathbf{w} - \mathbf{w}'\right\|_2^2 \text{ for some } \rho > 0. \tag{3.30}$$

We refer to functions $L_i(\cdot)$ for which (3.30) can be computed easily as *simple* or proximable [41]. GTVMin (3.22) with proximable loss functions can be solved via proximal algorithms [40]. Besides influencing the choice of optimization method, the design choices underlying GTVMin also determine the amount of computation that is required by a given optimization method.

Chapter 5 discusses FL algorithms that are obtained by applying fixed-point iterations to solve GTVMin. These fixed-point iterations repeatedly apply a fixed-point operator which is determined by the FL network (including the choice for the local loss functions, local models, and edges in the FL network). The computational complexity of the resulting iterative method has two factors: (i) the amount of computation required by a single iteration (i.e., the per-iteration complexity) and (ii) the number of iterations required by the method to achieve a sufficiently accurate approximate solution of GTVMin.

The fixed-point iterations used in Chap. 5 to design FL algorithms can be implemented as message passing over the FL network. These algorithms require an amount of computation that is proportional to the number of edges of the FL network. Clearly, using an FL network with few edges (i.e., using a sparse graph) results in a smaller per-iteration complexity.

The number of iterations required by an FL algorithm employing a fixed-point operator $\mathcal{F}$ depends on the contraction properties of $\mathcal{F}$. These contraction properties can be influenced through design choices for the FL network, such as selecting local loss functions that are strongly convex. In addition to affecting the iteration count, the contraction properties of $\mathcal{F}$ also play a crucial role in determining whether the FL algorithm can tolerate asynchronous execution.

It is instructive to study the computational aspects of the special case of GTVMin (3.24) for local linear models. As discussed above, this instance is equivalent to solving (3.27). Any solution $\widehat{\mathbf{w}}$ of (3.27) (and, in turn, (3.24)) is characterized by the zero-gradient condition

$$\mathbf{Q}\widehat{\mathbf{w}} = -(1/2)\mathbf{q}, \tag{3.31}$$

with $\mathbf{Q}, \mathbf{q}$ as defined in (3.28) and (3.29). If the matrix $\mathbf{Q}$ in (3.31) is invertible, the solution to (3.31) and, in turn, to the GTVMin instance (3.24) is unique and given by $\widehat{\mathbf{w}} = (-1/2)\mathbf{Q}^{-1}\mathbf{q}$.

The size of the matrix $\mathbf{Q}$ (see (3.28)) is proportional to the number of nodes in the FL network $\mathcal{G}$ which might be in the order of millions (or even billions) for Internet-scale applications. For such large systems, we typically cannot use direct matrix inversion methods (such as Gaussian elimination) to compute $\mathbf{Q}^{-1}$.[5] Instead, we typically need to resort to iterative methods [42, 43].

One important family of such iterative methods are the gradient-based methods which we will discuss in Chap. 4. Starting from an initial choice of the local model parameters $\widehat{\mathbf{w}}_0 = \big(\widehat{\mathbf{w}}_0^{(1)}, \ldots, \widehat{\mathbf{w}}_0^{(n)}\big)$, these methods repeat variants of a gradient step:

$$\widehat{\mathbf{w}}_{k+1} := \widehat{\mathbf{w}}_k - \eta\big(2\mathbf{Q}\widehat{\mathbf{w}}_k + \mathbf{q}\big) \text{ for } k = 0, 1, \ldots.$$

The gradient step results in the updated local model parameters $\widehat{\mathbf{w}}^{(i)}$ which we stacked into

$$\widehat{\mathbf{w}}_{k+1} := \left(\left(\widehat{\mathbf{w}}^{(1)}\right)^T, \ldots, \left(\widehat{\mathbf{w}}^{(n)}\right)^T\right)^T.$$

We repeat the gradient step for a sufficient number of times, according to some stopping criterion (see Chap. 4).

3.3.2 *Statistical Aspects of GTVMin*

How useful are the solutions of GTVMin (3.22) as a choice for the local model parameters? To answer this question, we use—as for the statistical analysis of ERM in Chap. 2—a probabilistic model for the local datasets. In particular, we use a

[5] How many arithmetic operations (addition, multiplication) do you think are required to invert an arbitrary matrix $\mathbf{Q} \in \mathbb{R}^{d \times d}$?

variant of an i.i.d. assumption: Each local dataset $\mathcal{D}^{(i)}$ consists of data points whose features and labels are realizations of i.i.d. RVs:

$$\mathbf{y}^{(i)} = \underbrace{\left(\mathbf{x}^{(i,1)}, \ldots, \mathbf{x}^{(i,m_i)}\right)^T}_{\text{local feature matrix } \mathbf{X}^{(i)}} \overline{\mathbf{w}}^{(i)} + \boldsymbol{\varepsilon}^{(i)} \text{ with } \mathbf{x}^{(i,r)} \overset{\text{i.i.d.}}{\sim} \mathcal{N}(\mathbf{0}, \mathbf{I}), \boldsymbol{\varepsilon}^{(i)} \sim \mathcal{N}(\mathbf{0}, \sigma^2 \mathbf{I}). \tag{3.32}$$

In contrast to the probabilistic model (2.10) (which we already used for the analysis of ERM), the probabilistic model (3.32) allows for different node-specific parameters $\overline{\mathbf{w}}^{(i)}$, for $i \in \mathcal{V}$. In particular, the entire dataset obtained from pooling all local datasets does not conform to an i.i.d. assumption.

In what follows, we focus on the GTVMin instance (3.24) to learn the parameters $\mathbf{w}^{(i)}$ of a local linear model for each node $i \in \mathcal{V}$. For a reasonable choice of FL network, the parameters $\overline{\mathbf{w}}^{(i)}, \overline{\mathbf{w}}^{(i')}$ at connected nodes $\{i, i'\} \in \mathcal{E}$ should be similar. We cannot choose the edge weights based on parameters $\overline{\mathbf{w}}^{(i)}$ as they are unknown. However, we can still use estimates of $\overline{\mathbf{w}}^{(i)}$ that are computed from the available local datasets (see Chap. 7).

Consider an FL network with nodes carrying local datasets generated from the probabilistic model (3.32) with true model parameters $\overline{\mathbf{w}}^{(i)}$. For ease of exposition, we assume that

$$\overline{\mathbf{w}}^{(i)} = \overline{\mathbf{c}}, \text{ for some } \overline{\mathbf{c}} \in \mathbb{R}^d \text{ and all } i \in \mathcal{V}. \tag{3.33}$$

To study the deviation between the solutions $\widehat{\mathbf{w}}^{(i)}$ of (3.24) and the true underlying parameters $\overline{\mathbf{w}}^{(i)}$, we decompose it as

$$\widehat{\mathbf{w}}^{(i)} = \widetilde{\mathbf{w}}^{(i)} + \widehat{\mathbf{c}}, \text{ with } \widehat{\mathbf{c}} := (1/n) \sum_{i'=1}^{n} \widehat{\mathbf{w}}^{(i')}. \tag{3.34}$$

The component $\widehat{\mathbf{c}}$ is identical at all nodes $i \in \mathcal{V}$ and obtained as the orthogonal projection of $\widehat{\mathbf{w}} = \text{stack}\{\widehat{\mathbf{w}}^{(i)}\}_{i=1}^{n}$ on the subspace (3.18). The component $\widetilde{\mathbf{w}}^{(i)} := \widehat{\mathbf{w}}^{(i)} - (1/n) \sum_{i'=1}^{n} \widehat{\mathbf{w}}^{(i')}$ consists of the deviations, for each node i, between the GTVMin solution $\widehat{\mathbf{w}}^{(i)}$ and their average over all nodes. Trivially, the average of the deviations $\widetilde{\mathbf{w}}^{(i)}$ across all nodes is the zero vector, $(1/n) \sum_{i=1}^{n} \widetilde{\mathbf{w}}^{(i)} = \mathbf{0}$.

The decomposition (3.34) entails an analogous (orthogonal) decomposition of the error $\widehat{\mathbf{w}}^{(i)} - \overline{\mathbf{w}}^{(i)}$. Indeed, for identical true underlying model parameters (3.33) (which makes $\overline{\mathbf{w}}$ an element of the subspace (3.18)), we have

$$\sum_{i=1}^{n} \left\| \widehat{\mathbf{w}}^{(i)} - \overline{\mathbf{w}}^{(i)} \right\|_2^2 \overset{(3.33),(3.34)}{=} \underbrace{\sum_{i=1}^{n} \| \mathbf{c} - \widehat{\mathbf{c}} \|_2^2}_{n \| \mathbf{c} - \widehat{\mathbf{c}} \|_2^2} + \sum_{i=1}^{n} \left\| \widetilde{\mathbf{w}}^{(i)} \right\|_2^2. \tag{3.35}$$

The following proposition provides an upper bound on the second error component in (3.35).

Proposition 3.1 *Consider a connected FL network, i.e.,* $\lambda_2 > 0$ *(see* (3.12)*), and the solution* (3.34) *to GTVMin* (3.24) *for the local datasets* (3.32)*. If the true local model parameters in* (3.32) *are identical (see* (3.33)*), we can upper bound the deviation* $\widetilde{\mathbf{w}}^{(i)} := \widehat{\mathbf{w}}^{(i)} - (1/n)\sum_{i=1}^{n}\widehat{\mathbf{w}}^{(i)}$ *of learned model parameters* $\widehat{\mathbf{w}}^{(i)}$ *from their average, as*

$$\sum_{i=1}^{n}\left\|\widetilde{\mathbf{w}}^{(i)}\right\|_2^2 \leq \frac{1}{\lambda_2\alpha}\sum_{i=1}^{n}(1/m_i)\left\|\boldsymbol{\varepsilon}^{(i)}\right\|_2^2. \tag{3.36}$$

Proof See Sect. 3.7.1.

Note that Proposition 3.1 only applies to GTVMin over an FL network with a connected graph $\mathcal{G}$. A necessary and sufficient condition for $\mathcal{G}$ to be connected is that the second smallest eigenvalue is positive, $\lambda_2 > 0$. However, for an FL network with a graph $\mathcal{G}$ that is not connected, we can still apply Proposition 3.1 separately to each connected component of $\mathcal{G}$.

The upper bound (3.36) involves three components:

- the properties of local datasets, via the noise terms $\boldsymbol{\varepsilon}^{(i)}$ in (3.32),
- the FL network via the eigenvalue $\lambda_2(\mathbf{L}^{(\mathcal{G})})$ (see (3.12)),
- the GTVMin parameter α.

According to (3.36), we can ensure a small error component $\widetilde{\mathbf{w}}^{(i)}$ of the GTVMin solution by choosing a large value α. Thus, by (3.35), for sufficiently large α, the local model parameters $\widehat{\mathbf{w}}^{(i)}$ delivered by GTVMin are approximately identical for all nodes $i \in \mathcal{V}$ of a connected FL network (where $\lambda_2(\mathbf{L}^{(\mathcal{G})}) > 0$).

Enforcing identical local model parameters at all nodes of an FL network is desirable for FL applications that require to learn common (global) model parameters for all nodes [12]. However, some FL applications involve heterogeneous devices that generate local datasets with significantly different statistics [34]. For such applications, it is detrimental to enforce common model parameters at all nodes (see Chap. 6). Instead, we should enforce common model parameters only for nodes with local datasets having similar statistical properties. This is exactly the objective of clustered FL which we discuss in Sect. 6.2.

3.4 Non-parametric Models in FL Networks

In its basic form (3.22), GTVMin can only be applied to parametric local models with model parameters belonging to the same Euclidean space $\mathbb{R}^d$. Some FL applications involve non-parametric local models (such as decision trees) or parametric local models with varying parametrizations (e.g., nodes use different deep net architectures). Here, we cannot use the difference between model parameters as a measure for the discrepancy between $h^{(i)}$ and $h^{(i')}$ across an edge $\{i, i'\} \in \mathcal{E}$.

One simple approach to measuring the discrepancy between hypothesis maps $h^{(i)}, h^{(i')}$ is to compare their predictions on a dataset:

$$\mathcal{D}^{\{i,i'\}} = \left\{ \mathbf{x}^{(1)}, \ldots, \mathbf{x}^{(m')} \right\}.$$

For each edge $\{i, i'\}$, the connected nodes need to agree on dataset $\mathcal{D}^{(\{i,i'\})}$. Note that the dataset $\mathcal{D}^{\{i,i'\}}$ can be different for different edges. Examples for constructions of $\mathcal{D}^{\{i,i'\}}$ include i.i.d. realizations of some probability distribution or by using subsets of $\mathcal{D}^{(i)}$ and $\mathcal{D}^{(i')}$ (see Exercise 3.7).

We compare the predictions delivered by $h^{(i)}$ and $h^{(i')}$ on $\mathcal{D}^{\{i,i'\}}$ using some loss function L. In particular, we define the discrepancy measure

$$d^{(i,i')} := (1/m') \sum_{\mathbf{x} \in \mathcal{D}^{\{i,i'\}}} (1/2)\Big[L\left(\left(\mathbf{x}, h^{(i)}(\mathbf{x})\right), h^{(i')}\right) + L\left(\left(\mathbf{x}, h^{(i')}(\mathbf{x})\right), h^{(i)}\right)\Big]. \tag{3.37}$$

Different choices for the loss function in (3.37) result in different computational and statistical properties of the resulting FL algorithms. For real-valued predictions, we can use the squared error loss in (3.37), yielding

$$d^{(i,i')} := (1/m') \sum_{\mathbf{x} \in \mathcal{D}^{\{i,i'\}}} \left[h^{(i)}(\mathbf{x}) - h^{(i')}(\mathbf{x})\right]^2. \tag{3.38}$$

We can generalize GTVMin by replacing $\left\|\mathbf{w}^{(i)} - \mathbf{w}^{(i')}\right\|_2^2$ in (3.22) with the discrepancy $d^{(h^{(i)},h^{(i')})}$ (3.37) (or the special case (3.38)). This results in

$$\{\widehat{h}^{(i)}\}_{i=1}^{n} \in \underset{\substack{h^{(i)} \in \mathcal{H}^{(i)} \\ i \in \mathcal{V}}}{\operatorname{argmin}} \sum_{i \in \mathcal{V}} L_i\left(h^{(i)}\right) + \alpha \sum_{\{i,i'\} \in \mathcal{E}} A_{i,i'} d^{(h^{(i)},h^{(i')})}. \tag{3.39}$$

3.5 Interpretations

We next discuss some interpretations of GTVMin (3.21).

Empirical Risk Minimization GTVMin (3.22) is obtained as a special case of ERM (2.1) for specific choices for the model $\mathcal{H}$ and loss function L. The model (or hypothesis space) used by GTVMin is a product space generated by the local models at the nodes of an FL network. The loss function of GTVMin consists of two parts: the sum of loss functions at each node and a penalty term that measures the variation of local models across the edges of the FL network.

Generalized Convex Clustering One important special case of GTVMin (3.21) is convex clustering [44, 45]. Indeed, convex clustering is obtained from (3.21) using the local loss function

$$L_i\left(\mathbf{w}^{(i)}\right) = \|\mathbf{w}^{(i)} - \mathbf{a}^{(i)}\|^2, \text{ for all nodes } i \in \mathcal{V} \tag{3.40}$$

and the GTV penalty function $\phi(\mathbf{u}) = \|\mathbf{u}\|_p$ with some $p \geq 1$.[6] The vectors $\mathbf{a}^{(i)}$, for $i = 1, \ldots, n$, are the features of data points that we wish to cluster in (3.40). Thus, we can interpret GTVMin as a generalization of convex clustering: we replace the terms $\|\mathbf{w}^{(i)} - \mathbf{a}^{(i)}\|^2$ with a more general local loss function.

Dual of Minimum-Cost Flow Problem The optimization variables of GTVMin (3.21) are the local model parameters $\mathbf{w}^{(i)}$, for each node $i \in \mathcal{V}$ in an FL network $\mathcal{G}$. The optimization of node-wise variables $\mathbf{w}^{(i)}$, for $i = 1, \ldots, n$, is naturally associated with a dual problem [46]. This dual problem optimizes edge-wise variables $\mathbf{u}^{(\{i,i'\})}$, one for each edge $\{i, i'\} \in \mathcal{E}$ of $\mathcal{G}$:

$$\max_{\substack{\mathbf{u}^{(e)}, e \in \mathcal{E} \\ \mathbf{w}^{(i)}, i \in \mathcal{V}}} -\sum_{i \in \mathcal{V}} L_i^*\left(\mathbf{w}^{(i)}\right) - \alpha \sum_{e \in \mathcal{E}} A_e \phi^*\left(\mathbf{u}^{(e)}/(\alpha A_e)\right) \tag{3.41}$$

$$\text{subject to } -\mathbf{w}^{(i)} = \sum_{\substack{e \in \mathcal{E} \\ e_+ = i}} \mathbf{u}^{(e)} - \sum_{\substack{e \in \mathcal{E} \\ e_- = i}} \mathbf{u}^{(e)} \text{ for each } i \in \mathcal{V}. \tag{3.42}$$

Here, we have introduced an orientation for each edge $e := \{i, i'\}$, by defining the *head* $e_- := \min\{i, i'\}$ and the *tail* $e_+ := \max\{i, i'\}$.[7] Moreover, we used the convex conjugates $L_i^*(\cdot)$, ϕ^* of the local loss function $L_i(\cdot)$ and GTV penalty function ϕ (Fig. 3.5).[8]

The dual problem (3.41) generalizes the optimal flow problem [46, Sec. 1J] to vector-valued flows. The special case of (3.41), obtained when the GTV penalty function ϕ is a norm, is equivalent to a generalized minimum-cost flow problem

[6] Here, we used the p-norm $\|\mathbf{u}\|_p := \left(\sum_{j=1}^{d} |u_j|^p\right)^{1/p}$ of a vector $\mathbf{u} \in \mathbb{R}^d$.

[7] We use this orientation only for notational convenience to formulate the dual of GTVMin. The orientation of an edge (by choosing a head and tail) has no practical meaning in terms of GTVMin-based FL algorithms. After all, GTVMin (3.21) and its dual (3.41) are defined for an FL network with undirected edges $\mathcal{E}$.

[8] The convex conjugate of a function $f : \mathbb{R}^d \rightarrow \mathbb{R}$ is defined as [47]

$$f^*(\mathbf{x}) := \sup_{\mathbf{z} \in \mathbb{R}^d} \mathbf{x}^T \mathbf{z} - f(\mathbf{z}). \tag{3.43}$$

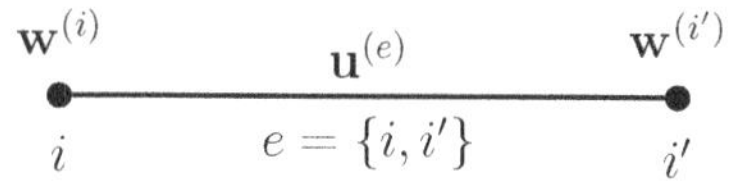

$\mathbf{w}^{(i)}$ $\mathbf{u}^{(e)}$ $\mathbf{w}^{(i')}$

i $e = \{i, i'\}$ i'

Fig. 3.5 Two nodes of an FL network that are connected by an edge $e = \{i, i'\}$. GTVMin (3.21) optimizes local model parameters $\mathbf{w}^{(i)}$ for each node $i \in \mathcal{V}$ in the FL network. The dual (3.41) of GTVMin optimizes local parameters $\mathbf{u}^{(e)}$ for each edge $e \in \mathcal{E}$ in the FL network

[48, Sec. 1.2.1]. Indeed, the maximization problem (3.41) is equivalent to the minimization

$$\min_{\substack{\mathbf{u}^{(e)}, e \in \mathcal{E} \\ \mathbf{w}^{(i)}, i \in \mathcal{V}}} \sum_{i \in \mathcal{V}} L_i^* \left(\mathbf{w}^{(i)}\right)$$

$$\text{subject to } -\mathbf{w}^{(i)} = \sum_{\substack{e \in \mathcal{E} \\ e_+ = i}} \mathbf{u}^{(e)} - \sum_{\substack{e \in \mathcal{E} \\ e_- = i}} \mathbf{u}^{(e)} \text{ for each node } i \in \mathcal{V}$$

$$\|\mathbf{u}^{(e)}\|_* \leq \alpha A_e \text{ for each edge } e \in \mathcal{E}. \tag{3.44}$$

The optimization problem (3.44) reduces to the minimum-cost flow problem [48, Eq. (1.3)–(1.5)] for scalar local model parameters $\mathbf{w}^{(i)} \in \mathbb{R}$.

Locally Weighted Learning The solution of GTVMin is local model parameters $\widehat{\mathbf{w}}^{(i)}$ that tend to be clustered: Each node $i \in \mathcal{V}$ belongs to a subset or cluster $\mathcal{C} \subseteq \mathcal{V}$. All the nodes in $\mathcal{C}$ have nearly identical local model parameters, $\widehat{\mathbf{w}}^{(i')} \approx \overline{\mathbf{w}}^{(\mathcal{C})}$ for all $i' \in \mathcal{C}$ [34]. The cluster-wise model parameters $\overline{\mathbf{w}}^{(\mathcal{C})}$ are the solutions of

$$\min_{\mathbf{w}} \sum_{i' \in \mathcal{C}} L_{i'}\left(\mathbf{w}\right), \tag{3.45}$$

which, in turn, is an instance of a locally weighted learning problem [49, Sec. 3.1.2]

$$\overline{\mathbf{w}}^{(\mathcal{C})} = \operatorname*{argmin}_{\mathbf{w} \in \mathbb{R}^d} \sum_{i' \in \mathcal{V}} \rho_{i'} L_{i'}\left(\mathbf{w}\right). \tag{3.46}$$

Indeed, we obtain (3.45) from (3.46) by setting the weights $\rho_{i'}$ equal to 1 if $i' \in \mathcal{C}$ and 0 otherwise.

3.6 Exercises

3.1 Spectral Radius of Laplacian Matrix. The spectral radius $\rho(\mathbf{Q})$ of a square matrix $\mathbf{Q}$ is the largest magnitude of an eigenvalue:

$$\rho(\mathbf{Q}) := \max\{|\lambda| : \lambda \text{ is an eigenvalue of } \mathbf{Q}\}.$$

Consider the Laplacian matrix $\mathbf{L}^{(\mathcal{G})}$ of an FL network with undirected graph $\mathcal{G}$. Show that $\rho\big(\mathbf{L}^{(\mathcal{G})}\big) = \lambda_n\big(\mathbf{L}^{(\mathcal{G})}\big)$ and verify the upper bound $\lambda_n\big(\mathbf{L}^{(\mathcal{G})}\big) \leq 2d^{(\mathcal{G})}_{\text{max}}$. Try to find a graph $\mathcal{G}$ such that $\lambda_n\big(\mathbf{L}^{(\mathcal{G})}\big) \approx 2d^{(\mathcal{G})}_{\text{max}}$.

3.2 Kernel of the Laplacian matrix. Consider an undirected weighted graph $\mathcal{G}$ with Laplacian matrix $\mathbf{L}^{(\mathcal{G})}$. A component of $\mathcal{G}$ is a subset $\mathcal{C} \subseteq \mathcal{V}$ of nodes that are connected, but there is no edge between $\mathcal{C}$ and the rest $\mathcal{V} \setminus \mathcal{C}$. The null-space (or *kernel*) of $\mathbf{L}^{(\mathcal{G})}$ is the subspace $\mathcal{K} \subseteq \mathbb{R}^n$ constituted by all vectors $\mathbf{v} \in \mathbb{R}^n$ such that $\mathbf{L}^{(\mathcal{G})}\mathbf{v} = \mathbf{0}$. Show that the dimension of $\mathcal{K}$ coincides with the number of components in $\mathcal{G}$.

3.3 Toy Example of Spectral Clustering. Consider the graph $\mathcal{G}$ depicted in Fig. 3.6. The Laplacian matrix has two zero eigenvalues $\lambda_1 = \lambda_2 = 0$.
Can you find corresponding orthonormal eigenvectors $\mathbf{u}^{(1)}, \mathbf{u}^{(2)}$? Are they unique?

3.4 Adding an Edge Increases Connectivity. Consider an undirected weighted graph $\mathcal{G}$ with Laplacian matrix $\mathbf{L}^{(\mathcal{G})}$. We construct a new graph $\mathcal{G}'$, with Laplacian matrix $\mathbf{L}^{(\mathcal{G}')}$, by adding a new edge to $\mathcal{G}$. Show that $\lambda_2(\mathcal{G}') \geq \lambda_2(\mathcal{G})$, i.e., the second smallest eigenvalue of $\mathbf{L}^{(\mathcal{G}')}$ is at least as large as the second smallest eigenvalue of $\mathbf{L}^{(\mathcal{G})}$.

3.5 Capacity of an FL network. Consider the FL network shown in Fig. 3.7. Each node holds a local dataset, with its size indicated by the adjacent numbers.

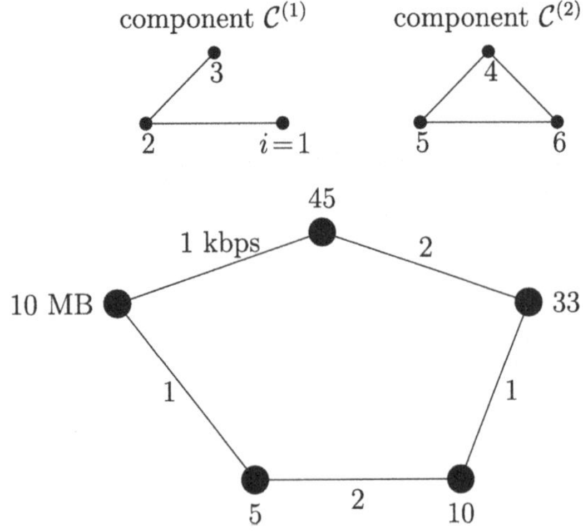

Fig. 3.6 An undirected graph $\mathcal{G}$ that consists of two connected components $\mathcal{C}^{(1)}, \mathcal{C}^{(2)}$

Fig. 3.7 An FL network whose nodes $i = 1, \ldots, 5$ represent devices that hold local datasets whose size is indicated next to each node

The devices (nodes) communicate over bi-directional links, whose capacities are specified by the numbers next to the edges. What is the minimum time required for the leftmost node to collect all local datasets from the other nodes?

3.6 Discrepancy Measures. Consider an FL network with nodes carrying parametric local models, each having model parameters $\mathbf{w}^{(i)} \in \mathbb{R}^d$. Is it possible to construct a dataset $\mathcal{D}^{\{i,i'\}}$ such that (3.38) coincides with $\left\|\mathbf{w}^{(i)} - \mathbf{w}^{(i')}\right\|_2^2$?

3.7 Privacy-Friendly Discrepancy Measures. The discrepancy measure (3.37) requires to choose a test-set $\mathcal{D}^{\{i,i'\}}$. One possible choice is to combine data points of the local datasets $\mathcal{D}^{(i)}$ and $\mathcal{D}^{(i')}$. However, sharing these data points can be harmful as they potentially leak sensitive information. Could you think of a simple message passing protocol between node i and i' that allows them to evaluate (3.37) only by sharing the predictions $h^{(i)}(\mathbf{x}), h^{(i')}(\mathbf{x})$ for $\mathbf{x} \in \mathcal{D}^{\{i,i'\}}$?

3.8 Structure of GTVMin. What are sufficient conditions for the local datasets and the edge weights used in GTVMin such that $\mathbf{Q}$ in (3.28) is invertible?

3.9 Existence and Uniqueness of GTVMin Solution. Consider the GTVMin instance (3.22), defined over an FL network with the weighted undirected graph $\mathcal{G}$.

1. **Existence.** Can you state a sufficient condition on the local loss functions and the weighted edges of $\mathcal{G}$ such that (3.22) has at least one solution?
2. **Uniqueness.** Then, try to find a condition that ensures that (3.22) has a unique solution.
3. Finally, try to find necessary conditions for the existence and uniqueness of solutions to (3.22).

3.10 Computing the Average. Consider an FL network with each node carrying a single model parameter $w^{(i)}$ and a local dataset, consisting of a single number $y^{(i)}$. Construct an instance of GTVMin such that its solutions are given by $\widehat{w}^{(i)} \approx (1/n)\sum_{i=1}^{n} y^{(i)}$ for all $i = 1, \ldots, n$.

3.11 Computing the Average over a Star. Consider the FL network depicted in Fig. 3.8, which consists of a center node i_0 which is connected to $n-1$ peripheral nodes $\mathcal{P} := \mathcal{V} \setminus \{i_0\}$. Each peripheral node $i \in \mathcal{P}$ carries a local dataset that consists of a single real-valued observation $y^{(i)} \in \mathbb{R}$. Construct an instance of GTVMin, using real-valued local model parameters $w^{(i)} \in \mathbb{R}$, such that the solution satisfies $\widehat{w}^{(i_0)} \approx (1/(n-1))\sum_{i \in \mathcal{P}} y^{(i)}$.

3.12 Fundamental Limits. Consider the FL network depicted in Fig. 3.9. Each node carries a local model with single parameter $w^{(i)}$ as well as a local dataset that consists of a single number $y^{(i)}$. We use a probabilistic model for the local datasets: $y^{(i)} = \bar{w} + n^{(i)}$. Here, $\bar{w}$ is some fixed but unknown number and $n^{(i)} \sim \mathcal{N}(0, 1)$ are i.i.d. Gaussian RVs. We use a message-passing FL algorithm to estimate c based on the local datasets. What is a fundamental limit on the accuracy of the estimate $\hat{c}^{(i)}$

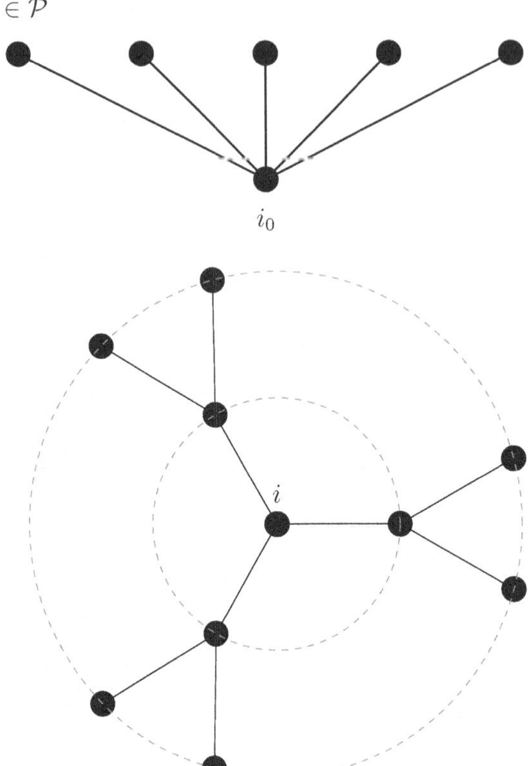

Fig. 3.8 An FL network that consists of a center node i_0 that is connected to several peripheral nodes $\mathcal{P} := \mathcal{V} \setminus \{i_0\}$

Fig. 3.9 An FL network containing a node i having degree $d^{(i)} = 3$ like all its neighbors $i' \in \mathcal{N}^{(i)}$. We use an FL algorithm to learn local model parameters $w^{(i)}$. If the algorithm employs message passing, the first iteration provides access only to the local datasets of the neighbors in $\mathcal{N}^{(i)}$ (located along the inner dashed circle). In the second iteration, the algorithm gains access to the local datasets of the neighbors $\mathcal{N}^{(i')}$ of each $i' \in \mathcal{N}^{(i)}$. These *second-hop* neighbors are located along the outer dashed circle

delivered at some fixed node i by such an algorithm after two iterations? Compare this limit with the risk $\mathbb{E}\{(\hat{w}^{(i)} - \bar{w})^2\}$ incurred by the estimate $\hat{w}^{(i)}$ delivered by running Algorithm 4 for two iterations.

3.13 Counting Number of Paths. Consider an undirected graph $\mathcal{G}$ with each edge $\{i, i'\} \in \mathcal{E}$ having unit edge weight $A_{i,i'} = 1$. A k-hop path, for some $k \in \{1, 2, \ldots .\}$, between two nodes $i, i' \in \mathcal{V}$ is a node sequence $i^{(1)}, \ldots, i^{(k+1)}$ such that $i^{(1)} = i$, $i^{(k+1)} = i'$, and $\{i^{(r)}, i^{(r+1)}\} \in \mathcal{E}$ for each $r = 1, \ldots, k$. Show that the number of k-hop paths between two nodes $i, i' \in \mathcal{V}$ is given by $(\mathbf{A}^k)_{i,i'}$.

3.14 Proximal operator of a quadratic function. Study the proximal operator (3.30) for a quadratic function

$$L_i\left(\mathbf{w}^{(i)}\right) = \left(\mathbf{w}^{(i)}\right)^T \mathbf{Q}\mathbf{w}^{(i)} + \mathbf{q}^T \mathbf{w}^{(i)} + q,$$

with some matrix $\mathbf{Q} \in \mathbb{R}^{d \times d}$, vector $\mathbf{q} \in \mathbb{R}^d$, and number $q \in \mathbb{R}$.

3.7 Proofs

3.7.1 *Proof of Proposition 3.1*

Let us introduce the shorthand $f\left(\mathbf{w}^{(i)}\right)$ for the objective function of the GTVMin instance (3.24). We verify the bound (3.36) by showing that if it does not hold, the choice of the local model parameters $\mathbf{w}^{(i)} := \overline{\mathbf{w}}^{(i)}$ (see (3.32)) results in a smaller objective function value, $f\left(\overline{\mathbf{w}}^{(i)}\right) < f\left(\widehat{\mathbf{w}}^{(i)}\right)$. This would contradict the fact that $\widehat{\mathbf{w}}^{(i)}$ is a solution to (3.24).

First, note that

$$\begin{aligned} f\left(\overline{\mathbf{w}}^{(i)}\right) &= \sum_{i \in \mathcal{V}} (1/m_i) \left\| \mathbf{y}^{(i)} - \mathbf{X}^{(i)} \overline{\mathbf{w}}^{(i)} \right\|_2^2 + \alpha \sum_{\{i,i'\} \in \mathcal{E}} A_{i,i'} \left\| \overline{\mathbf{w}}^{(i)} - \overline{\mathbf{w}}^{(i')} \right\|_2^2 \\ &\overset{(3.33)}{=} \sum_{i \in \mathcal{V}} (1/m_i) \left\| \mathbf{y}^{(i)} - \mathbf{X}^{(i)} \overline{\mathbf{w}}^{(i)} \right\|_2^2 \\ &\overset{(3.32)}{=} \sum_{i \in \mathcal{V}} (1/m_i) \left\| \mathbf{X}^{(i)} \overline{\mathbf{w}}^{(i)} + \boldsymbol{\varepsilon}^{(i)} - \mathbf{X}^{(i)} \overline{\mathbf{w}}^{(i)} \right\|_2^2 \\ &= \sum_{i \in \mathcal{V}} (1/m_i) \left\| \boldsymbol{\varepsilon}^{(i)} \right\|_2^2. \end{aligned} \tag{3.47}$$

Inserting (3.34) into (3.24),

$$\begin{aligned} f\left(\widehat{\mathbf{w}}^{(i)}\right) &= \underbrace{\sum_{i \in \mathcal{V}} (1/m_i) \left\| \mathbf{y}^{(i)} - \mathbf{X}^{(i)} \widehat{\mathbf{w}}^{(i)} \right\|_2^2}_{\geq 0} + \alpha \sum_{\{i,i'\} \in \mathcal{E}} A_{i,i'} \underbrace{\left\| \widehat{\mathbf{w}}^{(i)} - \widehat{\mathbf{w}}^{(i')} \right\|_2^2}_{\overset{(3.34)}{=} \left\| \widetilde{\mathbf{w}}^{(i)} - \widetilde{\mathbf{w}}^{(i')} \right\|_2^2} \\ &\geq \alpha \sum_{\{i,i'\} \in \mathcal{E}} A_{i,i'} \left\| \widetilde{\mathbf{w}}^{(i)} - \widetilde{\mathbf{w}}^{(i')} \right\|_2^2 \\ &\overset{(3.17)}{\geq} \alpha \lambda_2 \sum_{i=1}^{n} \left\| \widetilde{\mathbf{w}}^{(i)} \right\|_2^2. \end{aligned} \tag{3.48}$$

If the bound (3.36) would not hold, then by (3.48) and (3.47), we would obtain $f\left(\widehat{\mathbf{w}}^{(i)}\right) > f\left(\overline{\mathbf{w}}^{(i)}\right)$. This is a contradiction to the fact that $\widehat{\mathbf{w}}^{(i)}$ solves (3.24).

Chapter 4
Gradient Methods for Federated Optimization

Abstract This chapter focuses on gradient-based optimization methods for solving learning problems in federated learning systems. These methods update model parameters using local information about the gradient of the objective function. The chapter begins with a review of the basic gradient step and explains how to choose the learning rate and stopping criteria. It then discusses how to handle noisy or incomplete gradient information, which is common in distributed settings. To support constrained optimization tasks, the projected gradient method is introduced. The chapter also extends gradient methods to non-parametric models by using test datasets and replacing gradients with more general update rules. Finally, gradient-based methods are interpreted as fixed-point iterations, offering a unified view of many federated learning algorithms. Throughout, the emphasis is on practical implementation and theoretical guarantees. These techniques provide the foundation for efficient, scalable, and flexible optimization in federated learning, especially when working with large models or under limited communication and data-sharing constraints.

Chapter 3 introduced GTVMin as a central design principle for FL algorithms. Many important instances of GTVMin require the minimization of a smooth objective function $f(\mathbf{w})$ over a continuous parameter space. This chapter investigates how gradient-based methods—a broadly used family of iterative optimization methods—can be employed to solve such problems. These methods rely on local approximations of $f(\mathbf{w})$ using its gradient.

Section 4.1 introduces the basic gradient step and explains how it updates model parameters in the direction of steepest descent. Key considerations such as the choice of the learning rate are discussed in Sect. 4.2, along with stopping criteria in Sect. 4.3 that help determine when to terminate the optimization process. Section 4.4 studies how perturbations affect the convergence of gradient steps, which is particularly relevant in FL applications that involve unreliable communication or partial data access.

When optimization problems include explicit constraints on the model parameters, projected gradient descent (projected GD) presented in Sect. 4.5 provides a principled solution. Section 4.6 then extends gradient-based methods to non-

A. Jung, *Federated Learning*, https://doi.org/10.1007/978-981-95-1009-2_4

parametric models, using proximal operators and test datasets to generalize the notion of a gradient step. Finally, Sect. 4.7 interprets gradient-based methods as a special case of fixed-point iterations. This perspective allows for a unified understanding of FL algorithms as convergent processes driven by contraction operators.

4.1 Gradient Descent

Gradient-based methods are iterative algorithms for finding the minimum of a differentiable objective function $f(\mathbf{w})$ of a vector-valued argument $\mathbf{w}$. One example of such an optimization problem is the ERM instance (2.2). Unless stated otherwise, we consider an objective function of the form

$$f(\mathbf{w}) := \mathbf{w}^T \mathbf{Q} \mathbf{w} + \mathbf{q}^T \mathbf{w}. \tag{4.1}$$

Although restricting our discussion to objective functions of the form (4.1) may seem limiting, this formulation allows for a straightforward analysis and generalization to larger classes of differentiable functions. Moreover, we can use (4.1) also as an approximation for broader families of objective functions.

Note that (4.1) defines an entire family of convex quadratic functions $f(\mathbf{w})$. Each member of this family is specified by a psd matrix $\mathbf{Q} \in \mathbb{R}^{d \times d}$ and a vector $\mathbf{q} \in \mathbb{R}^d$. We have already encountered some ML and FL methods that minimize an objective function of the form (4.1): Linear regression (2.2) and ridge regression (2.23) in Chap. 2 as well as GTVMin (3.24) for local linear models in Chap. 3. Moreover, (4.1) is a useful approximation for the objective functions arising in other ML methods [50–52].

Given a current choice of model parameters $\mathbf{w}^{(k)}$, we want to update them toward a minimum of (4.1). To this end, we use the gradient $\nabla f(\mathbf{w}^{(k)})$ to locally approximate $f(\mathbf{w})$ (see Fig. 4.1). The gradient $\nabla f(\mathbf{w}^{(k)})$ indicates the direction in which the function $f(\mathbf{w})$ maximally increases. Therefore, it seems reasonable to

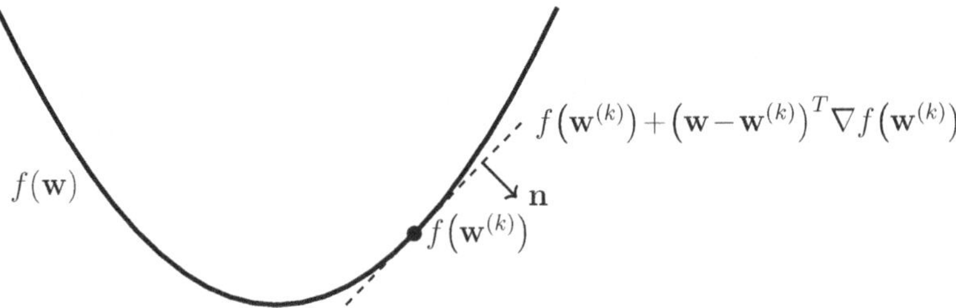

Fig. 4.1 We can approximate a differentiable function $f(\mathbf{w})$ locally around a point $\mathbf{w}^{(k)} \in \mathbb{R}^d$ using the linear function $f(\mathbf{w}^{(k)}) + (\mathbf{w} - \mathbf{w}^{(k)})^T \nabla f(\mathbf{w}^{(k)})$. Geometrically, we approximate the graph of $f(\mathbf{w})$ by a hyperplane with normal vector $\mathbf{n} = (\nabla f(\mathbf{w}^{(k)}), -1)^T \in \mathbb{R}^{d+1}$ of this approximating hyperplane is determined by the gradient $\nabla f(\mathbf{w}^{(k)})$ [2]

update $\mathbf{w}^{(k)}$ in the opposite direction of $\nabla f(\mathbf{w}^{(k)})$:

$$\mathbf{w}^{(k+1)} := \mathbf{w}^{(k)} - \eta \nabla f(\mathbf{w}^{(k)}) \stackrel{(4.1)}{=} \mathbf{w}^{(k)} - \eta(2\mathbf{Q}\mathbf{w}^{(k)} + \mathbf{q}). \quad (4.2)$$

The gradient step (4.2) involves the factor η which we refer to as step size or learning rate. Algorithm 2 summarizes the most basic variant of gradient-based methods, which simply iterates (4.2) until a predefined stopping criterion is met.

The usefulness of gradient-based methods crucially depends on the computational complexity of evaluating the gradient $\nabla f(\mathbf{w})$. Modern software libraries for automatic differentiation enable the efficient evaluation of the gradients arising in widely used ERM-based methods [53].

Besides the actual computation of the gradient, it might already be challenging to gather the required data points which define the objective function $f(\mathbf{w})$ (e.g., being the average loss over a large training set). Indeed, the matrix $\mathbf{Q}$ and vector $\mathbf{q}$ in (4.1) are constructed from the features and labels of data points in the training set. For example, the gradient of the objective function in ridge regression (2.23) is

$$\nabla f(\mathbf{w}) = -(2/m) \sum_{r=1}^{m} \mathbf{x}^{(r)} \big(y^{(r)} - \mathbf{w}^T \mathbf{x}^{(r)}\big) + 2\alpha \mathbf{w}.$$

Evaluating this gradient requires roughly $d \times m$ arithmetic operations such as adding and multiplying numbers.

Algorithm 2 A blueprint for gradient-based methods

Input: some objective function $f(\mathbf{w})$ (e.g., the average loss of a hypothesis $h^{(\mathbf{w})}$ on a training set); learning rate $\eta > 0$; some stopping criterion.
Initialize: set $\mathbf{w}^{(0)} := \mathbf{0}$; set iteration counter $k := 0$
1: **repeat**
2: $k := k+1$ (increase iteration counter)
3: $\mathbf{w}^{(k)} := \mathbf{w}^{(k-1)} - \eta \nabla f(\mathbf{w}^{(k-1)})$ (do a gradient step (4.2))
4: **until** stopping criterion is met
Output: learned model parameters $\widehat{\mathbf{w}} := \mathbf{w}^{(k)}$ (hopefully $f(\widehat{\mathbf{w}}) \approx \min_{\mathbf{w}} f(\mathbf{w})$)

Like most other gradient-based methods, Algorithm 2 involves two hyper-parameters: (i) the learning rate η used for the gradient step and (ii) a stopping criterion to decide when to stop repeating the gradient step. We next discuss how to choose these hyper-parameters.

Note that we can apply Algorithm 2 to find the minimum of any differentiable objective function $f(\mathbf{w})$. Indeed, Algorithm 2 only needs to be able to access the gradient $\nabla f(\mathbf{w}^{(k-1)})$. In particular, we apply Algorithm 2 to objective functions that do not belong to the family of convex quadratic functions (4.1).

4.2 How to Choose the Learning Rate

The learning rate must be chosen carefully: if it is too large, the gradient step may overshoot and diverge from the solution of (4.1); if it is too small, each step makes only negligible progress. Note that practical FL systems can only afford to compute a finite number of gradient steps. Therefore, we must ensure that each gradient step makes a sufficiently large progress toward the optimum of the objective function. Figure 4.2 illustrates both extremes.

One approach to choosing the learning rate is to start with some initial value (first guess) and monitor the decrease in the objective function. If this decrease does not agree with the decrease predicted by the (local linear approximation using the) gradient, we decrease the learning rate by a constant factor. After we decrease the learning rate, we re-consider the decrease in the objective function. We repeat this procedure until a sufficient decrease in the objective function is achieved [54, Sec 6.1].

Alternatively, we can use a prescribed sequence (schedule) η_k, for $k = 1, 2, \ldots,$ of learning rates that vary across successive gradient steps [55]. For example, we could require the learning rate η_k to satisfy a *diminishing step-size rule* [54, Sec. 6.1]:

$$\lim_{k \to \infty} \eta_k = 0, \sum_{k=1}^{\infty} \eta_k = \infty \text{ , and } \sum_{k=1}^{\infty} \eta_k^2 < \infty. \tag{4.3}$$

Running the gradient step (4.2) with a learning rate schedule η_k that satisfies (4.3) ensures convergence to a minimum of $f(\mathbf{w})$ if:

- the iterates $\left\|\mathbf{w}^{(k)}\right\|_2$ are bounded, i.e., $\sup_{k=1,\ldots} \left\|\mathbf{w}^{(k)}\right\|_2$ is finite, and
- the gradients $\left\|\nabla f\left(\mathbf{w}^{(k)}\right)\right\|_2$ are also bounded.

A detailed convergence proof can be found in [54, Sec. 3].

It is instructive to discuss the meanings of the individual conditions in (4.3). The first condition (4.3) requires that the learning rate eventually become sufficiently

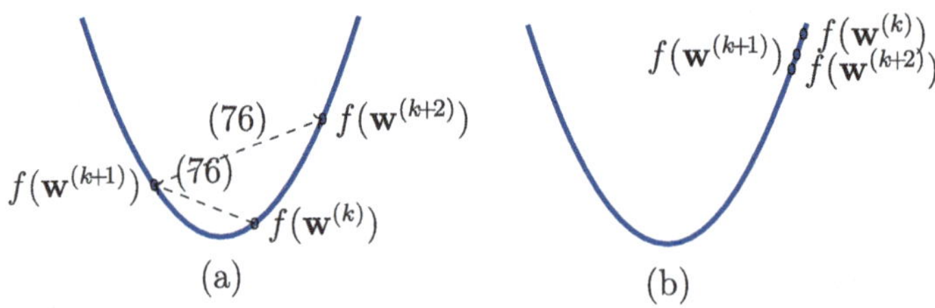

Fig. 4.2 Effect of inadequate learning rates η in the gradient step (4.2). (**a**) If η is too large, the gradient steps might "overshoot" such that the iterates $\mathbf{w}^{(k)}$ might diverge from the optimum, i.e., $f(\mathbf{w}^{(k+1)}) > f(\mathbf{w}^{(k)})$! (**b**) If η is too small, the gradient steps make very little progress toward the optimum or even fail to reach the optimum at all

small to avoid overshooting. The third condition (4.3) ensures that this required decay of the learning rate does not take "forever." Note that the first and third condition in (4.3) could be satisfied by the trivial learning rate schedule $\eta_k = 0$ which is clearly not useful as the gradient step has no effect.

The trivial schedule $\eta_k = 0$ is ruled out by the middle condition of (4.3). This middle condition ensures that the learning rate η_k is large enough such that the gradient steps make sufficient progress toward a minimizer of the objective function.

We emphasize that the conditions in (4.3) are independent of any properties of the matrix $\mathbf{Q}$ in (4.1). The matrix $\mathbf{Q}$ is determined by data points (see, e.g., (2.2)), whose statistical properties can typically be controlled only to a limited extent, such as through data normalization.

4.3 When to Stop?

For the stopping criterion, we may use a fixed number of iterations, $k_{\max}$. This hyperparameter can be determined by constraints on computational resources. We can optimize the number of iterations also via meta-learning, i.e., trying to predict the optimal $k_{\max}$ based on key characteristics (or features) of the objective function [56].

Another stopping criterion can be obtained by monitoring the decrease in the objective function $f\left(\mathbf{w}^{(k)}\right)$. Specifically, we stop repeating the gradient step (4.2) when $\left|f\left(\mathbf{w}^{(k)}\right) - f\left(\mathbf{w}^{(k+1)}\right)\right| \leq \varepsilon^{(\text{tol})}$ for a given tolerance $\varepsilon^{(\text{tol})}$. As before, we can optimize the tolerance level $\varepsilon^{(\text{tol})}$ via meta-learning techniques [56].

For an objective function of the form (4.1), we can use information about the psd matrix $\mathbf{Q}$ to construct a stopping criterion.[1] Indeed, the choice of the learning rate η and the stopping criterion can be guided by the eigenvalues:

$$0 \leq \lambda_1(\mathbf{Q}) \leq \ldots \leq \lambda_d(\mathbf{Q}).$$

Even if we do not know these eigenvalues precisely, we might know (or be able to ensure via feature learning) some upper and lower bounds:

$$0 \leq L \leq \lambda_1(\mathbf{Q}) \leq \ldots \leq \lambda_d(\mathbf{Q}) \leq U. \tag{4.4}$$

In what follows, we assume that $\mathbf{Q}$ is invertible and that we know some positive lower bound $L > 0$ on its eigenvalues (see (4.4)). The objective function (4.1) has then a unique solution $\widehat{\mathbf{w}}$. A gradient step (4.2) reduces the distance $\left\|\mathbf{w}^{(k)} - \widehat{\mathbf{w}}\right\|_2$ to

[1] For linear regression (2.5), the matrix $\mathbf{Q}$ is determined by the features of the data points in the training set. We can influence the properties of $\mathbf{Q}$ to some extent by feature transformation methods. One important example of such a transformation is the normalization of features.

$\widehat{\mathbf{w}}$ by a constant factor [54, Ch. 6]:

$$\left\| \mathbf{w}^{(k+1)} - \widehat{\mathbf{w}} \right\|_2 \leq \kappa^{(\eta_k)}(\mathbf{Q}) \left\| \mathbf{w}^{(k)} - \widehat{\mathbf{w}} \right\|_2. \tag{4.5}$$

Here, we used the contraction factor:

$$\kappa^{(\eta)}(\mathbf{Q}) := \max\{|1 - \eta 2\lambda_1|, |1 - \eta 2\lambda_d|\}. \tag{4.6}$$

The contraction factor depends on the learning rate η which is a hyper-parameter of gradient-based methods that we can control. However, the contraction factor also depends on the eigenvalues of the matrix $\mathbf{Q}$ in (4.1). In ML and FL applications, this matrix typically depends on data and can be controlled only to some extent, e.g., using feature transformation [23, Ch. 5]. To ensure $\kappa^{(\eta)}(\mathbf{Q}) < 1$, we require a positive learning rate satisfying $\eta_k < 1/U$.

Consider the gradient step (4.2) with fixed learning rate η and a contraction factor $\kappa^{(\eta)}(\mathbf{Q}) < 1$ (see (4.6)). We can then ensure an optimization error $\left\| \mathbf{w}^{(k)} - \widehat{\mathbf{w}} \right\|_2 \leq \varepsilon$ (see (4.5)) if the number k of gradient steps satisfies

$$k \geq \underbrace{\left\lceil \frac{\log\left(\left\| \mathbf{w}^{(0)} - \widehat{\mathbf{w}} \right\|_2 / \varepsilon \right)}{\log\left(1/\kappa^{(\eta)}(\mathbf{Q})\right)} \right\rceil}_{=:k^{(\varepsilon)}}. \tag{4.7}$$

According to (4.5), smaller values of the contraction factor $\kappa^{(\eta)}(\mathbf{Q})$ guarantee a faster convergence of (4.2) toward the solution of (4.1). Figure 4.3 illustrates the dependence of $\kappa^{(\eta)}(\mathbf{Q})$ on the learning rate η. Thus, choosing a small η (close to 0)

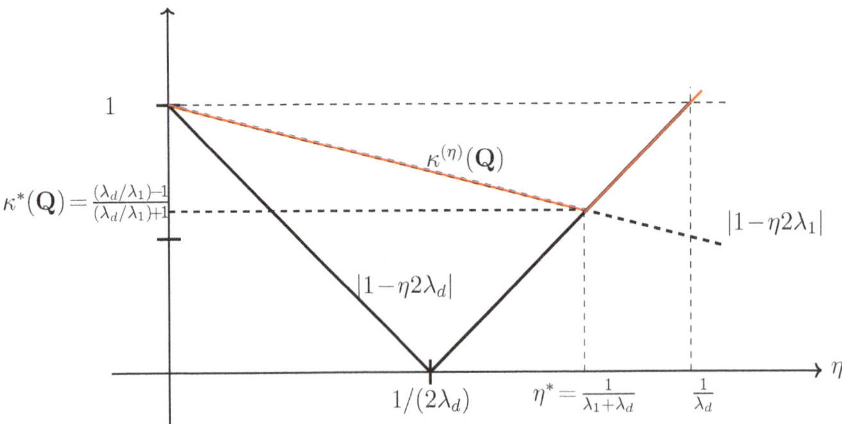

Fig. 4.3 The contraction factor $\kappa^{(\eta)}(\mathbf{Q})$ (4.6), used in the upper bound (4.5), as a function of the learning rate η. Note that $\kappa^{(\eta)}(\mathbf{Q})$ also depends on the eigenvalues of the matrix $\mathbf{Q}$ in (4.1)

will typically result in a larger $\kappa^{(\eta)}(\mathbf{Q})$ and, in turn, require more iterations to ensure optimization error level $\varepsilon^{(\mathrm{tol})}$ via (4.5).

We can minimize this contraction factor by choosing the learning rate (see Fig. 4.3):

$$\eta^{(*)} := \frac{1}{\lambda_1 + \lambda_d}. \tag{4.8}$$

[Note that evaluating (4.8) requires to know the extremal eigenvalues λ_1, λ_d of $\mathbf{Q}$.] Inserting the optimal learning rate (4.8) into (4.5),

$$\left\| \mathbf{w}^{(k+1)} - \widehat{\mathbf{w}} \right\|_2 \leq \underbrace{\frac{(\lambda_d/\lambda_1) - 1}{(\lambda_d/\lambda_1) + 1}}_{=:\kappa^*(\mathbf{Q})} \left\| \mathbf{w}^{(k)} - \widehat{\mathbf{w}} \right\|_2. \tag{4.9}$$

Carefully note that the formula (4.9) is valid only if the matrix $\mathbf{Q}$ in (4.1) is invertible, i.e., if $\lambda_1 > 0$. If the matrix $\mathbf{Q}$ is singular ($\lambda_1 = 0$), the convergence of (4.2) toward a solution of (4.1) is much slower than the decrease of the bound (4.9). However, we can still ensure the convergence of gradient steps $\mathbf{w}^{(k)}$ by using a fixed learning rate $\eta_k = \eta$ that satisfies [57, Thm. 2.1.14]

$$0 < \eta < 1/\lambda_d(\mathbf{Q}). \tag{4.10}$$

It is interesting to note that for linear regression, the matrix $\mathbf{Q}$ depends only on the features $\mathbf{x}^{(r)}$ of the data points in the training set (see (2.14)) but not on their labels $y^{(r)}$. Thus, the convergence of gradient steps is only affected by the features, whereas the labels are irrelevant. The same is true for ridge regression and GTVMin (using local linear models).

Note that both the optimal learning rate (4.8) and the optimal contraction factor

$$\kappa^*(\mathbf{Q}) := \frac{(\lambda_d/\lambda_1) - 1}{(\lambda_d/\lambda_1) + 1} \tag{4.11}$$

depend on the eigenvalues of the matrix $\mathbf{Q}$ in (4.1).

According to (4.9), the ideal case is when all eigenvalues are identical which leads, in turn, to a contraction factor $\kappa^*(\mathbf{Q}) = 0$. Here, a single gradient step arrives at the unique solution of (4.1).

In general, we do not have full control over the matrix $\mathbf{Q}$ and its eigenvalues. For example, the matrix $\mathbf{Q}$ arising in linear regression (2.5) is determined by the features of data points in the training set. These features might be obtained from sensing devices and therefore beyond our control. However, some applications might allow for some design freedom in the choice of feature vectors. We might also use feature transformations that nudge the resulting $\mathbf{Q}$ in (2.5) more toward a scaled identity matrix.

4.4 Perturbed Gradient Step

Consider the gradient step (4.2) used to find a minimum of (4.1). We again assume that the matrix $\mathbf{Q}$ in (4.1) is invertible ($\lambda_1(\mathbf{Q}) > 0$) and, in turn, (4.1) has a unique solution $\widehat{\mathbf{w}}$

In some applications, it is challenging to evaluate the gradient $\nabla f(\mathbf{w}) = 2\mathbf{Q}\mathbf{w}+\mathbf{q}$ of (4.1) exactly. For example, the evaluation could require to gather data points from distributed storage locations. These storage locations can become unavailable during the computation of $\nabla f(\mathbf{w})$ due to software or hardware failures (e.g., limited connectivity).

We can model imperfections during the computation of (4.2) as the perturbed gradient step:

$$\begin{aligned}\mathbf{w}^{(k+1)} &:= \mathbf{w}^{(k)} - \eta \nabla f\big(\mathbf{w}^{(k)}\big) + \boldsymbol{\varepsilon}^{(k)} \\ &\stackrel{(4.1)}{=} \mathbf{w}^{(k)} - \eta\big(2\mathbf{Q}\mathbf{w}^{(k)} + \mathbf{q}\big) + \boldsymbol{\varepsilon}^{(k)}, \text{ for } k = 0, 1, \ldots . \end{aligned} \tag{4.12}$$

We can use the contraction factor $\kappa := \kappa^{(\eta)}(\mathbf{Q})$ (4.6) to upper bound the deviation between $\mathbf{w}^{(k)}$ and the optimum $\widehat{\mathbf{w}}$ as (see (4.5))

$$\left\|\mathbf{w}^{(k)} - \widehat{\mathbf{w}}\right\|_2 \leq \kappa^k \left\|\mathbf{w}^{(0)} - \widehat{\mathbf{w}}\right\|_2 + \sum_{k'=1}^{k} \kappa^{k'} \left\|\boldsymbol{\varepsilon}^{(k-k')}\right\|_2 . \tag{4.13}$$

This bound applies for any number of iterations $k = 1, 2, \ldots$ of the perturbed gradient step (4.12).

The perturbed gradient step (4.12) could also be used as a tool to analyze the (exact) gradient step for an objective function $\tilde{f}(\mathbf{w})$ which does not belong to the family (4.1) of convex quadratic functions. Indeed, we can write the gradient step for minimizing $\tilde{f}(\mathbf{w})$ as

$$\begin{aligned}\mathbf{w}^{(k+1)} &:= \mathbf{w}^{(k)} - \eta \nabla \tilde{f}(\mathbf{w}) \\ &= \mathbf{w}^{(k)} - \eta \nabla f(\mathbf{w}) + \underbrace{\eta\big(\nabla f(\mathbf{w}) - \nabla \tilde{f}(\mathbf{w})\big)}_{:=\boldsymbol{\varepsilon}^{(k)}} .\end{aligned}$$

The last identity is valid for any choice of surrogate function $f(\mathbf{w})$. In particular, we can choose $f(\mathbf{w})$ as a convex quadratic function (4.1) that approximates $\tilde{f}(\mathbf{w})$. Note that the perturbation term $\boldsymbol{\varepsilon}^{(k)}$ is scaled by the learning rate η.

4.5 Handling Constraints: Projected Gradient Descent

Many important ML and FL methods amount to the minimization of an objective function of the form (4.1). The optimization variable $\mathbf{w}$ in (4.1) represents some model parameters.

Sometimes we might require the parameters $\mathbf{w}$ to belong to a subset $\mathcal{S} \subset \mathbb{R}^d$. One example is regularization via model pruning (see Chap. 2). Another example are FL methods that learn identical local model parameters $\mathbf{w}^{(i)}$ at all nodes $i \in \mathcal{V}$ of an FL network. This can be implemented by requiring the stacked local model parameters $\mathbf{w} = \big(\mathbf{w}^{(1)}, \ldots, \mathbf{w}^{(n)}\big)^T$ to belong to the subset

$$\mathcal{S} = \left\{ \big(\mathbf{w}^{(1)}, \ldots, \mathbf{w}^{(n)}\big)^T : \mathbf{w}^{(1)} = \ldots = \mathbf{w}^{(n)} \right\}.$$

Let us now show how to adapt the gradient step (4.2) to solve the constrained problem

$$f^* = \min_{\mathbf{w} \in \mathcal{S}} \mathbf{w}^T \mathbf{Q} \mathbf{w} + \mathbf{q}^T \mathbf{w}. \tag{4.14}$$

We assume that the constraint set $\mathcal{S} \subseteq \mathbb{R}^d$ is such that we can efficiently compute the projection

$$P_{\mathcal{S}}\big(\mathbf{w}\big) = \operatorname*{argmin}_{\mathbf{w}' \in \mathcal{S}} \big\|\mathbf{w} - \mathbf{w}'\big\|_2 \text{ for any } \mathbf{w} \in \mathbb{R}^d. \tag{4.15}$$

A suitable modification of the gradient step (4.2) to solve the constrained variant (4.14) is [54]

$$\begin{aligned} \mathbf{w}^{(k+1)} &:= P_{\mathcal{S}}\big(\mathbf{w}^{(k)} - \eta \nabla f\big(\mathbf{w}^{(k)}\big)\big) \\ &\overset{(4.1)}{=} P_{\mathcal{S}}\big(\mathbf{w}^{(k)} - \eta\big(2\mathbf{Q}\mathbf{w}^{(k)} + \mathbf{q}\big)\big). \end{aligned} \tag{4.16}$$

The projected GD step (4.16) consists of:

1. computing an ordinary gradient step $\mathbf{w}^{(k)} \mapsto \mathbf{w}^{(k)} - \eta \nabla f\big(\mathbf{w}^{(k)}\big)$ and then
2. projecting the result back to the constraint set $\mathcal{S}$.

Note that we re-obtain the basic gradient step (4.2) from the projected gradient step (4.16) for the trivial constraint set $\mathcal{S} = \mathbb{R}^d$.

The approaches for choosing the learning rate η and stopping criterion for basic gradient step (4.2) explained in Sects. 4.2 and 4.3 work also for the projected gradient step (4.16). In particular, the convergence speed of the projected gradient step is also characterized by (4.5) [54, Ch. 6]. This follows from the fact that the concatenation of a contraction (such as the gradient step (4.2) for sufficiently small η) and a projection (such as $P_{\mathcal{S}}(\cdot)$) results again in a contraction with the same contraction factor (Fig. 4.4).

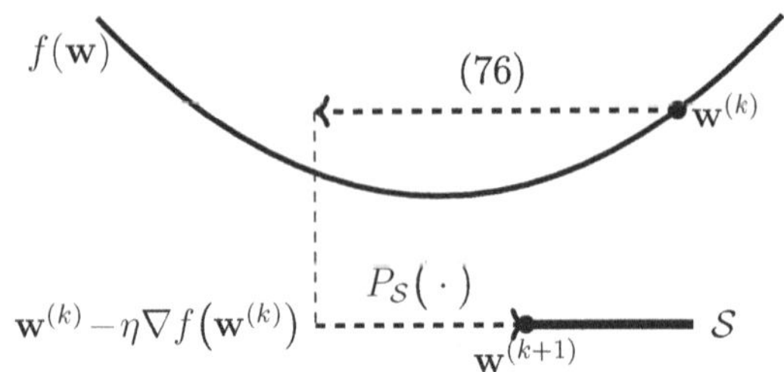

Fig. 4.4 Projected GD augments a basic gradient step with a projection back onto the constraint set $\mathcal{S}$

Thus, the convergence speed of projected GD, in terms of number of iterations required to ensure a given level of optimization error, is essentially the same as that of basic gradient descent (GD). However, the bound (4.5) is only telling about the number of projected gradient steps (4.16) required to achieve a guaranteed level of sub-optimality $\left|f\left(\mathbf{w}^{(k)}\right) - f^*\right|$. The iteration (4.16) of projected GD might require significantly more computation than the basic gradient step, as it requires to compute the projection (4.15).

4.6 Extended Gradient Methods for Federated Optimization

The gradient-based methods discussed so far can be used to learn a hypothesis from a parametric model. Let us now sketch one possible generalization of the gradient step (4.2) for a model $\mathcal{H}$ without a parametrization.

We start with rewriting the gradient step (4.2) as the optimization

$$\mathbf{w}^{(k+1)} = \operatorname*{argmin}_{\mathbf{w}\in\mathbb{R}^d} \left[(1/(2\eta)) \left\|\mathbf{w}-\mathbf{w}^{(k)}\right\|_2^2 + \underbrace{f\left(\mathbf{w}^{(k)}\right) + \left(\mathbf{w}-\mathbf{w}^{(k)}\right)^T \nabla f\left(\mathbf{w}^{(k)}\right)}_{\approx f(\mathbf{w})} \right]. \tag{4.17}$$

The objective function in (4.17) includes the first-order approximation

$$f(\mathbf{w}) \approx f\left(\mathbf{w}^{(k)}\right) + \left(\mathbf{w} - \mathbf{w}^{(k)}\right)^T \nabla f\left(\mathbf{w}^{(k)}\right)$$

of the function $f(\mathbf{w})$ around the location $\mathbf{w} = \mathbf{w}^{(k)}$ (see Fig. 4.1).

Let us modify (4.17) by using $f(\mathbf{w})$ itself (instead of an approximation),

$$\mathbf{w}^{(k+1)} = \operatorname*{argmin}_{\mathbf{w}\in\mathbb{R}^d} \left[f(\mathbf{w}) + (1/(2\eta)) \left\|\mathbf{w} - \mathbf{w}^{(k)}\right\|_2^2 \right]. \tag{4.18}$$

Like the gradient step, also (4.18) maps a given vector $\mathbf{w}^{(k)}$ to an updated vector $\mathbf{w}^{(k+1)}$. Note that (4.18) is nothing but the proximal operator of the function

$f(\mathbf{w})$ [40]. Similar to the role of the gradient step as the main building block of gradient-based methods, the proximal operator (4.18) is the main building block of proximal algorithms [40].

To obtain a version of (4.18) for a non-parametric model, we need to be able to evaluate its objective function directly in terms of a hypothesis h instead of its parameters $\mathbf{w}$. The objective function (4.18) consists of two components. The second component $f(\cdot)$, which is the function we want to minimize, is obtained from a training error incurred by a hypothesis, which might be parametric $h^{(\mathbf{w})}$. Thus, we can evaluate the function $f(h)$ by computing the training error for a given hypothesis.

The first component of the objective function in (4.18) uses $\left\|\mathbf{w}-\mathbf{w}^{(k)}\right\|_2^2$ to measure the difference between the hypothesis maps $h^{(\mathbf{w})}$ and $h^{(\mathbf{w}^{(k)})}$. Another measure for the difference between two hypothesis maps can be obtained by using some test dataset $\mathcal{D}' = \left\{\mathbf{x}^{(1)}, \ldots, \mathbf{x}^{(m')}\right\}$: The average squared difference between their predictions,

$$(1/m')\sum_{r=1}^{m'}\left(h\left(\mathbf{x}^{(r)}\right) - h^{(k)}\left(\mathbf{x}^{(r)}\right)\right)^2, \tag{4.19}$$

is a measure for the difference between h and $h^{(k)}$. Note that (4.19) only requires the predictions delivered by the hypothesis maps $h, h^{(k)}$ on $\mathcal{D}'$—no other information is needed about these maps.

It is interesting to note that (4.19) coincides with $\left\|\mathbf{w}-\mathbf{w}^{(k)}\right\|_2^2$ for the linear model $h^{(\mathbf{w})}(\mathbf{x}) := \mathbf{w}^T\mathbf{x}$ and a specific construction of the dataset $\mathcal{D}'$. This construction uses the realizations $\mathbf{x}^{(1)}, \mathbf{x}^{(2)}, \ldots$ of i.i.d. RVs with a common probability distribution $\mathbf{x} \sim \mathcal{N}(\mathbf{0}, \mathbf{I})$. Indeed, by the law of large numbers

$$\begin{aligned}
&\lim_{m'\to\infty}(1/m')\sum_{r=1}^{m'}\left(h^{(\mathbf{w})}\left(\mathbf{x}^{(r)}\right) - h^{(\mathbf{w}^{(k)})}\left(\mathbf{x}^{(r)}\right)\right)^2\\
&= \lim_{m'\to\infty}(1/m')\sum_{r=1}^{m'}\left(\left(\mathbf{w}-\mathbf{w}^{(k)}\right)^T\mathbf{x}^{(r)}\right)^2\\
&= \lim_{m'\to\infty}(1/m')\sum_{r=1}^{m'}\left(\mathbf{w}-\mathbf{w}^{(k)}\right)^T\mathbf{x}^{(r)}\left(\mathbf{x}^{(r)}\right)^T\left(\mathbf{w}-\mathbf{w}^{(k)}\right)\\
&= \left(\mathbf{w}-\mathbf{w}^{(k)}\right)^T\underbrace{\left[\lim_{m'\to\infty}(1/m')\sum_{r=1}^{m'}\mathbf{x}^{(r)}\left(\mathbf{x}^{(r)}\right)^T\right]}_{=\mathbf{I}}\left(\mathbf{w}-\mathbf{w}^{(k)}\right)\\
&= \left\|\mathbf{w}-\mathbf{w}^{(k)}\right\|_2^2.
\end{aligned} \tag{4.20}$$

Finally, we arrive at a generalized gradient step for the training of a non-parametric model $\mathcal{H}$ by replacing $\|\mathbf{w} - \mathbf{w}^{(k)}\|_2^2$ in (4.18) with (4.19). In other words,

$$h^{(k+1)} = \underset{h \in \mathcal{H}}{\operatorname{argmin}} \left[(1/(2\eta m')) \sum_{r=1}^{m'} \left(h(\mathbf{x}^{(r)}) - h^{(k)}(\mathbf{x}^{(r)}) \right)^2 + f(h) \right]. \tag{4.21}$$

We can turn gradient-based methods for the training of parametric models into corresponding training methods for non-parametric models by replacing the gradient step with the update (4.21). For example, we obtain Algorithm 3 from Algorithm 2 by modifying step 3 suitably.

Algorithm 3 A blueprint for generalized gradient-based methods

Input: some objective function $f : \mathcal{H} \rightarrow \mathbb{R}$ (e.g., the average loss of a hypothesis $h \in \mathcal{H}$ on a training set); learning rate $\eta > 0$; some stopping criterion; test dataset $\mathcal{D}' = \{\mathbf{x}^{(1)}, \ldots, \mathbf{x}^{(m')}\}$

Initialize: set $h^{(0)} := \mathbf{0}$; set iteration counter $k := 0$

1: **repeat**

2: $k := k+1$ (increase iteration counter)

3: do a generalized gradient step (4.21),

$$h^{(k)} = \underset{h \in \mathcal{H}}{\operatorname{argmin}} \left[(1/(2\eta m')) \sum_{r=1}^{m'} \left(h(\mathbf{x}^{(r)}) - h^{(k-1)}(\mathbf{x}^{(r)}) \right)^2 + f(h) \right]$$

4: **until** stopping criterion is met

Output: learned hypothesis $\widehat{h} := h^{(k)}$ (hopefully $f(\widehat{h}) \approx \min_{h \in \mathcal{H}} f(h)$)

4.7 Gradient Methods as Fixed-Point Iterations

The iterative optimization methods discussed in the previous sections are all special cases of a fixed-point iteration:

$$\mathbf{w}^{(k)} = \mathcal{F}\mathbf{w}^{(k-1)}, \text{ for } k = 1, 2, \ldots. \tag{4.22}$$

Different optimization methods use different choices for the operator $\mathcal{F}$ whose fixed points are solutions of the underlying optimization problem. For example, the gradient step (4.2) is obtained from (4.22) with the operator $\mathcal{F}^{(\mathrm{GD})} : \mathbf{w} \mapsto \mathbf{w} - \eta \nabla f(\mathbf{w})$. For a differentiable and convex objective function $f(\mathbf{w})$, every minimizer $\widehat{\mathbf{w}}$ is a fixed point of $\mathcal{F}^{(\mathrm{GD})}$.

The fixed-point iteration (4.22) will be the core computational step of every FL algorithm discussed in Chap. 5. These algorithms use (4.22) with an operator $\mathcal{F}$ determined by an instance of GTVMin. More precisely, any fixed point of $\mathcal{F}$ must be a GTVMin-solution $\widehat{\mathbf{w}} \in \mathbb{R}^{d \cdot n}$:

$$\mathcal{F}\widehat{\mathbf{w}} = \widehat{\mathbf{w}}. \tag{4.23}$$

Given an instance of GTVMin, there are many different operators $\mathcal{F}$ that satisfy (4.23). We obtain different FL algorithms by using different choices for $\mathcal{F}$ in (4.22). Clearly, we should use an operator $\mathcal{F}$ in (4.22) that reduces the distance to a solution:

$$\underbrace{\left\| \mathbf{w}^{(k+1)} - \widehat{\mathbf{w}} \right\|_2}_{\stackrel{(4.22),(4.23)}{=} \|\mathcal{F}\mathbf{w}^{(k)} - \mathcal{F}\widehat{\mathbf{w}}\|_2} \leq \left\| \mathbf{w}^{(k)} - \widehat{\mathbf{w}} \right\|_2 .$$

Thus, we require $\mathcal{F}$ to be at least non-expansive, i.e., the iteration (4.22) should not result in worse model parameters that have a larger distance to the GTVMin solution. Moreover, each iteration (4.22) should also make some progress, i.e., reduce the distance from a GTVMin solution. This requirement can be made precise using the notion of a contraction operator [58, 59].

The operator $\mathcal{F}$ is a contraction operator if, for some $\kappa \in [0, 1)$,

$$\left\| \mathcal{F}\mathbf{w} - \mathcal{F}\mathbf{w}' \right\|_2 \leq \kappa \left\| \mathbf{w} - \mathbf{w}' \right\|_2 \text{ holds for any } \mathbf{w}, \mathbf{w}' \in \mathbb{R}^{dn}.$$

For a contraction operator $\mathcal{F}$, the fixed-point iteration (4.22) generates a sequence $\mathbf{w}^{(k)}$ that converges to a GTVMin solution $\widehat{\mathbf{w}}$ quite rapidly. In particular [2, Theorem 9.23],

$$\left\| \mathbf{w}^{(k)} - \widehat{\mathbf{w}} \right\|_2 \leq \kappa^k \left\| \mathbf{w}^{(0)} - \widehat{\mathbf{w}} \right\|_2 .$$

Here, $\left\| \mathbf{w}^{(0)} - \widehat{\mathbf{w}} \right\|_2$ is the distance between the initialization $\mathbf{w}^{(0)}$ and the solution $\widehat{\mathbf{w}}$.

A well-known example of a fixed-point iteration (4.22) using a contraction operator is GD (4.2) for a smooth and strongly convex objective function $f(\mathbf{w})$.[2] In particular, (4.2) is obtained from (4.22) using $\mathcal{F} := \mathcal{G}^{(\eta)}$ with the "gradient step operator":

$$\mathcal{G}^{(\eta)} : \mathbf{w} \mapsto \mathbf{w} - \eta \nabla f(\mathbf{w}). \tag{4.24}$$

Note that the operator (4.24) is parametrized by the learning rate η.

[2] The objective function in (4.1) is convex and smooth for any choice of psd matrix $\mathbf{Q}$ and vector $\mathbf{q}$. Moreover, it is strongly convex whenever $\mathbf{Q}$ is invertible.

It is instructive to study the operator $\mathcal{G}^{(\eta)}$ for an objective function of the form (4.1). Here,

$$\mathcal{G}^{(\eta)} : \mathbf{w} \mapsto \mathbf{w} - \eta \underbrace{\left(2\mathbf{Q}\mathbf{w} + \mathbf{q}\right)}_{\overset{(4.1)}{=} \nabla f(\mathbf{w})}. \tag{4.25}$$

For $\eta := 1/(2\lambda_{\max}(\mathbf{Q}))$, the operator $\mathcal{G}^{(\eta)}$ is contractive with $\kappa = 1 - \lambda_{\min}(\mathbf{Q})/\lambda_{\max}(\mathbf{Q})$. Note that $\kappa < 1$ only when $\lambda_{\min}(\mathbf{Q}) > 0$, i.e., only when the matrix $\mathbf{Q}$ in (4.1) is invertible.

The gradient step operator (4.25) is not contractive for the objective function (4.1) with a singular matrix $\mathbf{Q}$ (for which $\lambda_{\min} = 0$). However, even then $\mathcal{G}^{(\eta)}$ is still firmly non-expansive [22]. We refer to an operator $\mathcal{F} : \mathbb{R}^{d \cdot n} \rightarrow \mathbb{R}^{d \cdot n}$ as firmly non-expansive if

$$\left\|\mathcal{F}\mathbf{w} - \mathcal{F}\mathbf{w}'\right\|_2^2 \leq \left(\mathcal{F}\mathbf{w} - \mathcal{F}\mathbf{w}'\right)^{?T} \left(\mathbf{w} - \mathbf{w}'\right), \text{ for any } \mathbf{w}, \mathbf{w}' \in \mathbb{R}^{d \cdot n}.$$

It turns out that a fixed-point iteration (4.22) with a firmly non-expansive operator $\mathcal{F}$ is guaranteed to converge to a fixed-point of $\mathcal{F}$ [58, Cor. 5.16]. Figure 4.5 depicts examples of a firmly non-expansive operator, a non-expansive operator, and a contraction operator. All these operators are defined on the one-dimensional space $\mathbb{R}$. Another example of a firmly non-expansive operator is the proximal operator (4.18) of a convex function [40, 58].

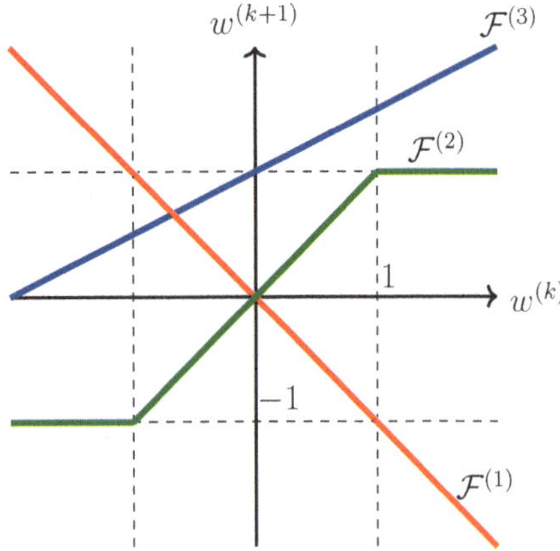

Fig. 4.5 Example of a non-expansive operator $\mathcal{F}^{(1)}$, a firmly non-expansive operator $\mathcal{F}^{(2)}$ and a contractive operator $\mathcal{F}^{(3)}$

4.8 Exercises

4.1 Learning Rate Schedule. Consider the gradient step method applied to a differentiable objective function $f(\mathbf{w})$

$$\mathbf{w}^{(k+1)} = \mathbf{w}^{(k)} - \eta_k \nabla f\big(\mathbf{w}^{(k)}\big), \quad \text{for } k = 1, 2, \ldots.$$

where the learning rate schedule is defined as $\eta_k := \frac{1}{k}$.

1. Verify that this learning rate schedule satisfies the standard conditions in (4.3).
2. Construct a differentiable, convex function $f(\mathbf{w})$ and an initialization $\mathbf{w}^{(0)}$ such that the gradient step iteration fails to converge to a minimizer of $f(\mathbf{w})$.

4.2 Online Gradient Descent. Linear regression methods learn model parameters of a linear model with minimum risk $\mathbb{E}\{(y - \mathbf{w}^T\mathbf{x})^2\}$ where $(\mathbf{x}, y)$ is a RV. In practice, we do not observe the RV $(\mathbf{x}, y)$ itself but a (realization of a) sequence of i.i.d. samples $\big(\mathbf{x}^{(t)}, y^{(t)}\big)$, for $t = 1, 2, \ldots$. Online GD is an online learning method that updates the current model parameters $\mathbf{w}^{(t)}$, after observing $\big(\mathbf{x}^{(t)}, y^{(t)}\big)$

$$\mathbf{w}^{(t+1)} := \mathbf{w}^{(t)} + 2\eta_t \mathbf{x}^{(t)}\big(y - \big(\mathbf{w}^{(t)}\big)^T \mathbf{x}^{(t)}\big) \text{ at time } t = 1, 2, \ldots.$$

Starting with initialization $\mathbf{w}^{(1)} := \mathbf{0}$, we run online gradient descent (online GD) for M time steps, resulting in the learned model parameters $\mathbf{w}^{(M+1)}$. Develop upper bounds on the risk $\mathbb{E}\{(y - (\mathbf{w}^{(M)})^T\mathbf{x})^2\}$ for two choices for the learning rate schedule: $\eta_t := 1/(t+5)$ or $\eta_t := 1/\sqrt{t+5}$.

4.3 Computing the Average—I. Consider an FL network with graph $\mathcal{G}$ and its Laplacian matrix $\mathbf{L}^{(\mathcal{G})}$. Each node carries a local dataset which consists of a single measurement $y^{(i)} \in \mathbb{R}$. To compute their average $(1/n)\sum_{i=1}^{n} y^{(i)}$, we try an iterative method that, starting from the initialization $\mathbf{u}^{(0)} := \big(y^{(1)}, \ldots, y^{(n)}\big)^T \in \mathbb{R}^n$, repeats the update

$$\mathbf{u}^{(k+1)} = \mathbf{u}^{(k)} - \eta \mathbf{L}^{(\mathcal{G})}\mathbf{u}^{(k)} \text{ for } k = 1, 2, \ldots. \tag{4.26}$$

Can you find a choice for η such that (4.26) becomes a fixed-point iteration (4.22) with a contractive operator $\mathcal{F}$. Given such a choice of η, how is the limit $\lim_{k\to\infty} \mathbf{u}^{(k+1)}$ related to the average $(1/n)\sum_{i=1}^{n} y^{(i)}$?

4.4 Computing the Average—II. Consider the FL network from Problem 4.3. Try to construct an instance of GTVMin for learning scalar local model parameters $w^{(i)}$ which coincide, for each node $i = 1, \ldots, n$ with the average $(1/n)\sum_{i'=1}^{n} y^{(i')}$. If you find such an instance of GTVMin, solve it using GD.

4.5 How to Quantize the Gradients? Any ML and FL application that uses a digital computer to implement a gradient step (4.2) must quantize the gradient $\nabla f(\mathbf{w})$ of the objective function $f(\mathbf{w})$. The quantization process introduces perturbations to the gradient step. Given a fixed total budget of bits available for quantization, a key question arises: Should we allocate more bits (reducing quantization noise) during the initial gradient steps or during the final gradient steps in gradient-based methods?
Hint: See Sect. 4.4.

4.6 When Is a Gradient Step (Firmly) Non-expansive? Consider the function $f(w) = (1/2)w^2$ and the associated gradient step $\mathcal{G}^{(\eta)} : w \mapsto w - \eta \nabla f(w)$. Discuss the value ranges for the learning rate η, for which the operator $\mathcal{G}^{(\eta)}$ is non-expansive or even firmly non-expansive.

Chapter 5
FL Algorithms

Chapter 3 introduced GTVMin as a flexible design principle for FL methods that arise from different design choices for the local models and edge weights of the FL network. The solutions of GTVMin are local model parameters that strike a balance between the loss incurred on local datasets and the GTV.

This chapter applies the gradient-based methods from Chapter 4 to solve GTVMin. We obtain FL algorithms by implementing these optimization methods as message passing across the edges of the FL network. These messages contain intermediate results of the computations carried out by FL algorithms. The details of how this message passing is implemented physically (e.g., via short-range wireless technology) are beyond the scope of this book.

Section 5.1 studies the gradient step for the GTVMin instance obtained for training local linear models. In particular, we show how the convergence rate of the gradient step can be characterized by the properties of the local datasets and their FL network.

Section 5.2 formulates the gradient step from Section 5.1 as a message passing across the edges of the FL network. This results in Algorithm 4 as a distributed FL method for parametric local models. Section 5.3 generalizes Algorithm 4 by replacing the exact gradient of local loss functions with some approximation. One possible approximation is to use a random subset (a batch) of a local dataset to estimate the gradient.

Section 5.4 discusses FL algorithms that train a single (global) model in a distributed fashion. We show how the widely-used FL algorithms FedAvg and FedProx are obtained from variations of projected GD, which we have discussed in Section 4.5.

Section 5.6 generalizes the gradient step, which is the core computation of FL algorithms for parametric models, to cope with non-parametric models. The idea is to compare the predictions of the local models at two nodes i and i' on a common test-set. By comparing their predictions, we can measure their variation across the edge $\{i, i'\}$.

A. Jung, *Federated Learning*, https://doi.org/10.1007/978-981-95-1009-2_5

Most of the algorithms discussed in this chapter operate in a synchronous manner: All devices must complete their local model updates (e.g., gradient steps) before exchanging updates simultaneously across the edges of the FL network. However, synchronous operation can be impractical or even infeasible for certain FL applications. Section 5.8 explores the design of FL algorithms that support asynchronous operation. These algorithms allow devices to update and communicate at different times within the FL system.

5.1 Gradient Descent for GTVMin

Consider a collection of n local datasets represented by the nodes $\mathcal{V} = \{1, \ldots, n\}$ of an FL network $\mathcal{G} = (\mathcal{V}, \mathcal{E})$. Each undirected edge $\{i, i'\} \in \mathcal{E}$ in the FL network $\mathcal{G}$ has a known edge weight $A_{i,i'}$. We want to learn local model parameters $\mathbf{w}^{(i)}$ of a personalized linear model for each node $i = 1, \ldots, n$. To this end, we solve the GTVMin instance

$$\{\widehat{\mathbf{w}}^{(i)}\}_{i=1}^{n} \in \underset{\{\mathbf{w}^{(i)}\}}{\operatorname{argmin}} \underbrace{\sum_{i \in \mathcal{V}} \overbrace{(1/m_i) \left\| \mathbf{y}^{(i)} - \mathbf{X}^{(i)} \mathbf{w}^{(i)} \right\|_2^2}^{\text{local loss } L_i(\mathbf{w}^{(i)})} + \alpha \sum_{\{i,i'\} \in \mathcal{E}} A_{i,i'} \left\| \mathbf{w}^{(i)} - \mathbf{w}^{(i')} \right\|_2^2}_{=: f(\mathbf{w})}. \tag{5.1}$$

As discussed in Chapter 3, the objective function in (5.1) - viewed as a function of the stacked local model parameters $\mathbf{w} := \text{stack}\{\mathbf{w}^{(i)}\}_{i=1}^{n}$ - is a quadratic function

$$\mathbf{w}^T \left(\begin{pmatrix} \mathbf{Q}^{(1)} & \cdots & \mathbf{0} \\ \vdots & \ddots & \vdots \\ \mathbf{0} & \cdots & \mathbf{Q}^{(n)} \end{pmatrix} + \alpha \mathbf{L}^{(\mathcal{G})} \otimes \mathbf{I} \right) \mathbf{w} + \left((\mathbf{q}^{(1)})^T, \ldots, (\mathbf{q}^{(n)})^T \right) \mathbf{w} \tag{5.2}$$

$$\text{with } \mathbf{Q}^{(i)} = (1/m_i) \big(\mathbf{X}^{(i)}\big)^T \mathbf{X}^{(i)} \text{ and } \mathbf{q}^{(i)} := (-2/m_i) \big(\mathbf{X}^{(i)}\big)^T \mathbf{y}^{(i)}.$$

Note that (5.2) is a special case of the generic quadratic function (4.1) studied in Chapter 4. Indeed, we obtain (5.2) from (4.1) for the choices

$$\mathbf{Q} := \left(\begin{pmatrix} \mathbf{Q}^{(1)} & \cdots & \mathbf{0} \\ \vdots & \ddots & \vdots \\ \mathbf{0} & \cdots & \mathbf{Q}^{(n)} \end{pmatrix} + \alpha \mathbf{L}^{(\mathcal{G})} \otimes \mathbf{I} \right), \text{ and } \mathbf{q} := \left((\mathbf{q}^{(1)})^T, \ldots, (\mathbf{q}^{(n)})^T \right)^T.$$

Therefore, the discussion and analysis of gradient-based methods from Chapter 4 also apply to GTVMin (5.1). In particular, we can use the gradient step

$$\begin{aligned}\mathbf{w}^{(k+1)} &:= \mathbf{w}^{(k)} - \eta \nabla f\big(\mathbf{w}^{(k)}\big) \\ &\overset{(5.2)}{=} \mathbf{w}^{(k)} - \eta\big(2\mathbf{Q}\mathbf{w}^{(k)} + \mathbf{q}\big)\end{aligned} \tag{5.3}$$

to iteratively compute an approximate solution $\widehat{\mathbf{w}}$ to (5.1). This solution consists of learned local model parameters $\widehat{\mathbf{w}}^{(i)}$, i.e., $\widehat{\mathbf{w}} = \mathrm{stack}\{\widehat{\mathbf{w}}^{(i)}\}$. Section 5.2 formulates the gradient step (5.3) directly in terms of local model parameters, resulting in a message passing over the FL network $\mathcal{G}$.

According to the convergence analysis in Chapter 4, the convergence rate of the iterations (5.3) is determined by the eigenvalues $\lambda_j(\mathbf{Q})$ of the matrix $\mathbf{Q}$ in (5.2). In general, these eigenvalues depend on the eigenvalues $\lambda_j\big(\mathbf{Q}^{(i)}\big)$ as well as the eigenvalues $\lambda_j\big(\mathbf{L}^{(\mathcal{G})}\big)$ of the Laplacian matrix $\mathbf{L}^{(\mathcal{G})}$. In particular, we will use the following two summary parameters

$$\lambda_{\max} := \max_{i=1,\ldots,n} \lambda_d\big(\mathbf{Q}^{(i)}\big), \text{ and } \bar{\lambda}_{\min} := \lambda_1\bigg((1/n)\sum_{i=1}^{n}\mathbf{Q}^{(i)}\bigg). \tag{5.4}$$

We first present an upper bound U (see (4.4)) on the eigenvalues of the matrix $\mathbf{Q}$ in (5.2).

Proposition 5.1 *The eigenvalues of* $\mathbf{Q}$ *in* (5.2) *are upper-bounded as*

$$\begin{aligned}\lambda_j(\mathbf{Q}) &\leq \lambda_{\max} + \alpha\lambda_n\big(\mathbf{L}^{(\mathcal{G})}\big) \\ &\leq \underbrace{\lambda_{\max} + 2\alpha d_{\max}^{(\mathcal{G})}}_{=:U}, \text{ for } j = 1, \ldots, dn.\end{aligned} \tag{5.5}$$

Proof See Section 5.10.1.

The next result offers a lower bound on the eigenvalues $\lambda_j(\mathbf{Q})$.

Proposition 5.2 *Consider the matrix* $\mathbf{Q}$ *in* (5.2). *If* $\lambda_2\big(\mathbf{L}^{(\mathcal{G})}\big) > 0$ *(i.e., the FL network in* (5.1) *is connected) and* $\bar{\lambda}_{\min} > 0$ *(i.e., the average of the matrices* $\mathbf{Q}^{(i)}$ *is non-singular), then the matrix* $\mathbf{Q}$ *is invertible and its smallest eigenvalue is lower bounded as*

$$\lambda_1(\mathbf{Q}) \geq \frac{1}{1+\rho^2}\min\{\lambda_2\big(\mathbf{L}^{(\mathcal{G})}\big)\alpha\rho^2, \bar{\lambda}_{\min}/2\}. \tag{5.6}$$

Here, we used the shorthand $\rho := \bar{\lambda}_{\min}/(4\lambda_{\max})$ *(see* (5.4)*).*

Proof See Section 5.10.2.

Proposition 5.1 and Proposition 5.2 provide some guidance for the design choices of GTVMin. According to the convergence analysis of gradient-based methods in Chapter 4, the eigenvalue $\lambda_1(\mathbf{Q})$ should be close to $\lambda_{dn}(\mathbf{Q})$ to ensure fast convergence. This suggests to favour FL networks $\mathcal{G}$ resulting in a small ratio between the upper bound (5.5) and the lower bound (5.6). A small ratio between these bounds, in turn, requires a large eigenvalue $\lambda_2(\mathbf{L}^{(\mathcal{G})})$ and a small node degree $d^{(\mathcal{G})}_{\max}$.[1]

The bounds in (5.5) and (5.6) also depend on the GTVMin parameter α. While these bounds might provide some guidance for the choice of α, the exact dependence of the convergence speed of (5.3) on α is complicated. For a fixed value of learning rate in (5.3), using larger values for α might slow down the convergence of (5.3) for some collection of local datasets but speed up the convergence of (5.3) for another collection of local datasets (see Exercise 5.1).

5.2 Message Passing Implementation

We now discuss in more detail the implementation of gradient-based methods to solve the GTVMin instances with a differentiable objective function $f(\mathbf{w})$. One such instance is GTVMin for local linear models (see (5.1)). The core of gradient-based methods is the gradient step

$$\mathbf{w}^{(k+1)} := \mathbf{w}^{(k)} - \eta \nabla f\big(\mathbf{w}^{(k)}\big). \tag{5.7}$$

The iterate $\mathbf{w}^{(k)}$ contains local model parameters $\mathbf{w}^{(i,k)}$,

$$\mathbf{w}^{(k)} =: \operatorname{stack}\big\{\mathbf{w}^{(i,k)}\big\}_{i=1}^{n}.$$

Inserting (5.1) into (5.7), we obtain the gradient step

$$\mathbf{w}^{(i,k+1)} := \mathbf{w}^{(i,k)} - \eta \Big[\underbrace{(2/m_i)\big(\mathbf{X}^{(i)}\big)^T \big(\mathbf{X}^{(i)}\mathbf{w}^{(i,k)} - \mathbf{y}^{(i)}\big)}_{\text{(I)}} + \underbrace{2\alpha \sum_{i' \in \mathcal{N}^{(i)}} A_{i,i'}\big(\mathbf{w}^{(i,k)} - \mathbf{w}^{(i',k)}\big)}_{\text{(II)}} \Big]. \tag{5.8}$$

[1] The are constructions of graphs with a prescribed value of $d^{(\mathcal{G})}_{\max}$ such that $\lambda_2(\mathbf{L}^{(\mathcal{G})})$ is maximal [60, 61].

We slightly modify this gradient step by using potentially different learning rates $\eta_{k,i}$ at different nodes i and iterations k,

$$\mathbf{w}^{(i,k+1)} := \mathbf{w}^{(i,k)} - \eta_{k,i}\Big[\underbrace{(2/m_i)(\mathbf{X}^{(i)})^T\big(\mathbf{X}^{(i)}\mathbf{w}^{(i,k)} - \mathbf{y}^{(i)}\big)}_{\text{(I)}} + \underbrace{2\alpha\sum_{i'\in\mathcal{N}^{(i)}} A_{i,i'}\big(\mathbf{w}^{(i,k)} - \mathbf{w}^{(i',k)}\big)}_{\text{(II)}}\Big]. \tag{5.9}$$

The update (5.9) consists of two components, denoted (I) and (II). The component (I) is the gradient $\nabla L_i\left(\mathbf{w}^{(i,k)}\right)$ of the local loss $L_i\left(\mathbf{w}^{(i)}\right) := (1/m_i)\left\|\mathbf{y}^{(i)} - \mathbf{X}^{(i)}\mathbf{w}^{(i)}\right\|_2^2$. Component (I) drives the updated local model parameters $\mathbf{w}^{(i,k+1)}$ towards the minimum of $L_i(\cdot)$, i.e., having a small deviation between labels $y^{(i,r)}$ and the predictions $\left(\mathbf{w}^{(i,k+1)}\right)^T\mathbf{x}^{(i,r)}$. Note that we can rewrite the component (I) in (5.9), as

$$(2/m_i)\sum_{r=1}^{m_i}\mathbf{x}^{(i,r)}\big(y^{(i,r)} - \big(\mathbf{x}^{(i,r)}\big)^T\mathbf{w}^{(i,k)}\big). \tag{5.10}$$

The purpose of component (II) in (5.9) is to force the local model parameters to be similar across an edge $\{i, i'\}$ with large weight $A_{i,i'}$. We control the relative importance of (II) and (I) using the GTVMin parameter α: Choosing a large value for α puts more emphasis on enforcing similar local model parameters across the edges. Using a smaller α puts more emphasis on learning local model parameters delivering accurate predictions (incurring a small loss) on the local dataset.

The execution of the gradient step (5.9) requires only local information at node i. Indeed, the update (5.9) at node i depends only on its current model parameters $\mathbf{w}^{(i,k)}$, the local loss function $L_i(\cdot)$, the neighbors' model parameters $\mathbf{w}^{(i',k)}$, for $i' \in \mathcal{N}^{(i)}$, and the corresponding edge weights $A_{i,i'}$ (see Figure 5.1). In particular,

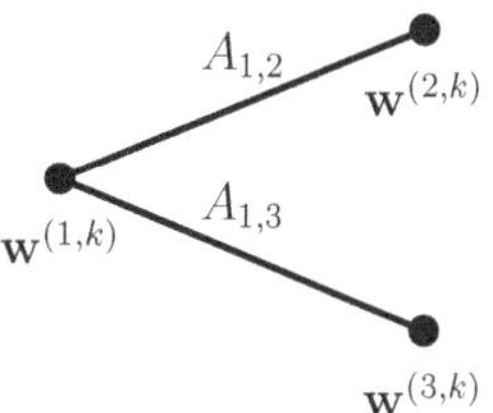

Fig. 5.1 At the beginning of iteration k, node $i = 1$ collects the current local model parameters $\mathbf{w}^{(2,k)}$ and $\mathbf{w}^{(3,k)}$ from its neighbors. Then, it computes the gradient step (5.9) to obtain the new local model parameters $\mathbf{w}^{(1,k+1)}$. These updated parameters are then used in the next iteration for the local updates at the neighbors $i = 2, 3$.

the update (5.9) does not depend on any properties k or edge weights) of the FL network beyond the neighbors $\mathcal{N}^{(i)}$.

We obtain Algorithm 4 by repeating the gradient step (5.9), simultaneously for each node $i \in \mathcal{V}$, until a stopping criterion is met. Algorithm 4 allows for potentially different learning rates $\eta_{k,i}$ at different nodes i and iterations k. It is important to note that Algorithm 4 requires a synchronous (simultaneous) execution of the updates (5.9) at all nodes $i \in \mathcal{V}$ [17, 18]. Loosely speaking, all nodes i rely on a single global clock that maintains the current iteration counter k [62].

Algorithm 4 FedGD for Local Linear Models

Input: FL network $\mathcal{G}$; GTV parameter α; learning rate $\eta_{k,i}$;
local dataset $\mathcal{D}^{(i)} = \left\{\left(\mathbf{x}^{(i,1)}, y^{(i,1)}\right); \ldots, \left(\mathbf{x}^{(i,m_i)}, y^{(i,m_i)}\right)\right\}$ for each i; some stopping criterion.
Output: linear model parameters $\widehat{\mathbf{w}}^{(i)}$ for each node $i \in \mathcal{V}$
Initialize: $k := 0$; $\mathbf{w}^{(i,0)} := \mathbf{0}$

while stopping criterion is not satisfied **do**
  for all nodes $i \in \mathcal{V}$ (simultaneously) **do**
    share local model parameters $\mathbf{w}^{(i,k)}$ with neighbors $i' \in \mathcal{N}^{(i)}$
    update local model parameters via (5.9)
  end for
  increment iteration counter: $k := k+1$
end while
$\widehat{\mathbf{w}}^{(i)} := \mathbf{w}^{(i,k)}$ for all nodes $i \in \mathcal{V}$

At the beginning of iteration k, each node $i \in \mathcal{V}$ sends its current model parameters $\mathbf{w}^{(i,k)}$ to their neighbors $i' \in \mathcal{N}^{(i)}$. Then, each node $i \in \mathcal{V}$ updates their model parameters according to (5.9), resulting in the updated model parameters $\mathbf{w}^{(i,k+1)}$. As soon as these local updates are completed, the global clock increments the counter $k \mapsto k+1$ and triggers the next iteration to be executed by all nodes. Figure 5.2 illustrates the alternating execution of message passing and local updates of Algorithm 4.

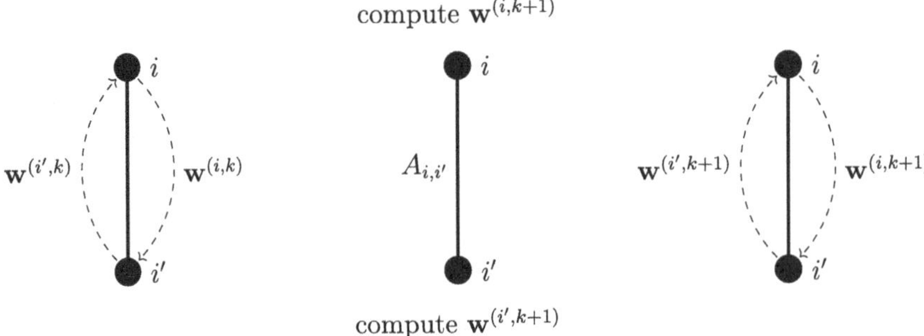

Fig. 5.2 Algorithm 4 alternates between message passing across the edges of the FL network (left and right) and updates of local model parameters (centre).

The implementation of Algorithm 4 in real-world computational infrastructures might incur deviations from the exact synchronous execution of (5.9) [63, Sec. 10]. This deviation can be modelled as a perturbation of the gradient step (5.7) and therefore analyzed using the concepts of Section 4.4 on perturbed GD. Section 8.2 will also discuss the effect of imperfect computation in the context of key requirements for trustworthy FL.

We close this section by generalizing Algorithm 4 which is limited FL networks using local linear models. This generalization, summarized in Algorithm 5, can be used to train parametric local models $\mathcal{H}^{(i)}$ with a differentiable loss function $L_i\left(\mathbf{w}^{(i)}\right)$, for $i = 1, \ldots, n$.

Algorithm 5 FedGD for Parametric Local Models

Input: FL network $\mathcal{G}$; GTV parameter α; learning rate $\eta_{k,i}$
local loss function $L_i\left(\mathbf{w}^{(i)}\right)$ for each $i = 1, \ldots, n$; some stopping criterion.
Output: linear model parameters $\widehat{\mathbf{w}}^{(i)}$ for each node $i \in \mathcal{V}$
Initialize: $k := 0$; $\mathbf{w}^{(i,0)} := \mathbf{0}$

while stopping criterion is not satisfied **do**
  for all nodes $i \in \mathcal{V}$ (simultaneously) **do**
    share local model parameters $\mathbf{w}^{(i,k)}$ with neighbors $i' \in \mathcal{N}^{(i)}$
    update local model parameters via

$$\mathbf{w}^{(i,k+1)} := \mathbf{w}^{(i,k)} - \eta_{k,i}\left[\nabla L_i\left(\mathbf{w}^{(i,k)}\right) + 2\alpha \sum_{i' \in \mathcal{N}^{(i)}} A_{i,i'}\left(\mathbf{w}^{(i,k)} - \mathbf{w}^{(i',k)}\right)\right].$$

  end for
  increment iteration counter: $k := k+1$
end while
$\widehat{\mathbf{w}}^{(i)} := \mathbf{w}^{(i,k)}$ for all nodes $i \in \mathcal{V}$

5.3 FedSGD

Consider Algorithm 4 for training local linear models $h^{(i)}(\mathbf{x}) = \mathbf{x}^T \mathbf{w}^{(i)}$ for each node $i = 1, \ldots, n$ of an FL network. Note that step 4 of Algorithm 4 requires to compute the sum (5.10). It might be infeasible to compute this sum exactly, e.g., when local datasets are generated by remote devices with limited connectivity. It is then useful to approximate the sum by

$$\underbrace{(2/B) \sum_{r \in \mathcal{B}} \mathbf{x}^{(i,r)} \left(y^{(i,r)} - \left(\mathbf{x}^{(i,r)}\right)^T \mathbf{w}^{(i,k)}\right)}_{\approx (5.10)}. \tag{5.11}$$

The approximation (5.11) uses a subset (so-called *batch*)

$$\mathcal{B} = \left\{ \left(\mathbf{x}^{(r_1)}, y^{(r_1)} \right), \ldots, \left(\mathbf{x}^{(r_B)}, y^{(r_B)} \right) \right\}$$

of B randomly chosen data points from $\mathcal{D}^{(i)}$. While (5.10) requires summing over m data points, the approximation requires to sum over B (typically $B \ll m$) data points.

Inserting the approximation (5.11) into the gradient step (5.9) yields the approximate gradient step

$$\mathbf{w}^{(i,k+1)} := \mathbf{w}^{(i,k)} - \eta_{k,i} \Bigg[\underbrace{(2/B) \sum_{r \in \mathcal{B}} \mathbf{x}^{(i,r)} \left(\left(\mathbf{x}^{(i,r)} \right)^T \mathbf{w}^{(i,k)} - y^{(i,r)} \right)}_{\approx (5.10)}$$

$$+ 2\alpha \sum_{i' \in \mathcal{N}^{(i)}} A_{i,i'} \left(\mathbf{w}^{(i,k)} - \mathbf{w}^{(i',k)} \right) \Bigg]. \tag{5.12}$$

We obtain Algorithm 6 from Algorithm 4 by replacing the gradient step (5.9) with the approximation (5.12).

Algorithm 6 FedSGD for Local Linear Models

Input: FL network $\mathcal{G}$; GTV parameter α; learning rate $\eta_{k,i}$;
local datasets $\mathcal{D}^{(i)} = \left\{ \left(\mathbf{x}^{(i,1)}, y^{(i,1)} \right), \ldots, \left(\mathbf{x}^{(i,m_i)}, y^{(i,m_i)} \right) \right\}$ for each node i; batch size B; some stopping criterion.
Output: linear model parameters $\widehat{\mathbf{w}}^{(i)}$ at each node $i \in \mathcal{V}$
Initialize: $k := 0$; $\mathbf{w}^{(i,0)} := \mathbf{0}$

while stopping criterion is not satisfied **do**
  for all nodes $i \in \mathcal{V}$ (simultaneously) **do**
    share local model parameters $\mathbf{w}^{(i,k)}$ with all neighbors $i' \in \mathcal{N}^{(i)}$
    draw fresh batch $\mathcal{B}^{(i)} := \{r_1, \ldots, r_B\}$
    update local model parameters via (5.12)
  end for
  increment iteration counter $k := k+1$
end while
$\widehat{\mathbf{w}}^{(i)} := \mathbf{w}^{(i,k)}$ for all nodes $i \in \mathcal{V}$

We close this section by generalizing Algorithm 6 which is limited to FL networks using local linear models. This generalization, summarized in Algorithm 7, can be used to train parametric local models $\mathcal{H}^{(i)}$ with a differentiable loss function $L_i \left(\mathbf{w}^{(i)} \right)$, for $i = 1, \ldots, n$. Algorithm 7 does not require these local loss function themselves, but only an oracle $\mathbf{g}^{(i)}(\cdot)$ for each node $i = 1, \ldots, n$. For a given vector $\mathbf{w}^{(i)}$, the oracle at node i delivers an approximate gradient (or estimate) $\mathbf{g}^{(i)}(\mathbf{w}^{(i)}) \approx \nabla L_i\left(\mathbf{w}^{(i)}\right)$. The analysis of Algorithm 7 can be facilitated by a probabilistic model which interprets the oracle output $\mathbf{g}^{(i)}(\mathbf{w}^{(i)})$ as the realization of a RV. Under such a probabilistic model, we refer to an oracle as unbiased if $\mathbb{E}\{\mathbf{g}^{(i)}(\mathbf{w}^{(i)})\} = \nabla L_i\left(\mathbf{w}^{(i)}\right)$.

Algorithm 7 FedSGD for Parametric Local Models

Input: FL network $\mathcal{G}$; GTV parameter α; learning rate $\eta_{k,i}$
gradient oracle $\mathbf{g}^{(i)}(\cdot)$ for each node $i = 1, \ldots, n$; some stopping criterion.
Output: linear model parameters $\widehat{\mathbf{w}}^{(i)}$ for each node $i \in \mathcal{V}$
Initialize: $k := 0$; $\mathbf{w}^{(i,0)} := \mathbf{0}$

while stopping criterion is not satisfied **do**
  for all nodes $i \in \mathcal{V}$ (simultaneously) **do**
    share local model parameters $\mathbf{w}^{(i,k)}$ with neighbors $i' \in \mathcal{N}^{(i)}$
    update local model parameters via

$$\mathbf{w}^{(i,k+1)} := \mathbf{w}^{(i,k)} - \eta_{k,i}\left[\mathbf{g}^{(i)}\left(\mathbf{w}^{(i,k)}\right) + 2\alpha \sum_{i' \in \mathcal{N}^{(i)}} A_{i,i'}\left(\mathbf{w}^{(i,k)} - \mathbf{w}^{(i',k)}\right)\right].$$

  end for
  increment iteration counter: $k := k+1$
end while
$\widehat{\mathbf{w}}^{(i)} := \mathbf{w}^{(i,k)}$ for all nodes $i \in \mathcal{V}$

5.4 FedAvg

Consider an FL method that learns model parameters $\widehat{\mathbf{w}} \in \mathbb{R}^d$ of a single (global) linear model from a decentralized collection local datasets $\mathcal{D}^{(i)}$, $i = 1, \ldots, n$.[2] How can we learn $\widehat{\mathbf{w}}$ without exchanging local datasets, but instead only exchanging updates for the model parameters?

One approach is to apply Algorithm 4 to GTVMin (5.1) with a sufficiently large α. According to our analysis in Chapter 3 (specifically Proposition 3.1), if α is sufficiently large, then the GTVMin solutions $\widehat{\mathbf{w}}^{(i)}$ are almost identical across all nodes $i \in \mathcal{V}$. We can interpret the local model parameters delivered by GTVMin as a local copy of the global model parameters.

Note that the bound in Proposition 3.1 only applies if the FL network (used in GTVMin) is connected. One example of a connected FL network is the star as depicted in Figure 5.3. Here, we choose one node $i = 1$ as a centre node that

[2] This setting if a special case of HFL which we discuss in Section 6.3.

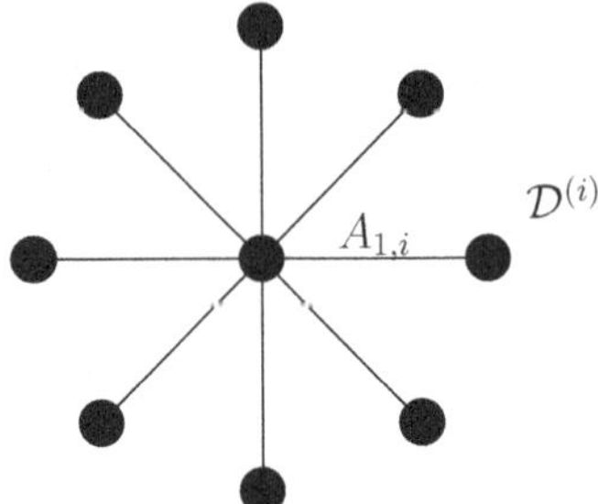

Fig. 5.3 Star-shaped graph $\mathcal{G}^{(\text{star})}$ with a centre node $i = 1$ representing a server that trains a (global) model which is shared with peripheral nodes. These peripheral nodes represent *clients* generating local datasets. The training process at the server is facilitated by receiving updates on the model parameters from the clients.

is connected by an edge with weight $A_{1,i}$ to the remaining nodes $i = 2, \ldots, n$. The star graph uses the minimum number of edges required to connect all n nodes [64].

Instead of using GTVMin with a connected FL network and a large value of α, we can also enforce identical local copies $\widehat{\mathbf{w}}^{(i)}$ via a constraint:

$$\widehat{\mathbf{w}} \in \underset{\mathbf{w} \in \mathcal{S}}{\arg\min} \sum_{i \in \mathcal{V}} (1/m_i) \left\| \mathbf{y}^{(i)} - \mathbf{X}^{(i)} \mathbf{w}^{(i)} \right\|_2^2$$
$$\text{with } \mathcal{S} = \left\{ \mathbf{w} = \text{stack}\{\mathbf{w}^{(i)}\}_{i=1}^{n} : \mathbf{w}^{(i)} = \mathbf{w}^{(i')} \text{ for any } i, i' \in \mathcal{V} \right\}. \tag{5.13}$$

Here, we use as constraint set the subspace $\mathcal{S}$ defined in (3.18). The projection of a given collection of local model parameters $\mathbf{w} = \text{stack}\{\mathbf{w}^{(i)}\}$ on $\mathcal{S}$ is given by

$$P_{\mathcal{S}}(\mathbf{w}) = \left(\mathbf{v}^T, \ldots, \mathbf{v}^T\right)^T \text{ with } \mathbf{v} := (1/n) \sum_{i \in \mathcal{V}} \mathbf{w}^{(i)}.$$

We can solve (5.13) using projected GD from Chapter 4. The resulting projected gradient step for solving (5.13) is

$$\widehat{\mathbf{w}}_{k+1/2}^{(i)} := \underbrace{\mathbf{w}^{(i,k)} - \eta_{i,k} (2/m_i) \left(\mathbf{X}^{(i)}\right)^T \left(\mathbf{X}^{(i)} \mathbf{w}^{(i,k)} - \mathbf{y}^{(i)}\right)}_{\text{(local gradient step)}} \tag{5.14}$$

$$\mathbf{w}^{(i,k+1)} := (1/n) \sum_{i' \in \mathcal{V}} \widehat{\mathbf{w}}_{k+1/2}^{(i')} \quad \text{(projection)}. \tag{5.15}$$

We can implement (5.15) conveniently in a server-client system with each node i being a client:

- First, each node computes the update (5.14), i.e., a gradient step towards a minimum of the local loss $L_i\left(\mathbf{w}^{(i)}\right) := \left\|\mathbf{y}^{(i)} - \mathbf{X}^{(i)}\mathbf{w}^{(i)}\right\|_2^2$.
- Second, each node i sends the result $\widehat{\mathbf{w}}_k^{(i)}$ of its local gradient step to a server.
- Finally, after receiving the updates $\widehat{\mathbf{w}}_k^{(i)}$ from all nodes $i \in \mathcal{V}$, the server computes the projection step (5.15). This projection results in the new local model parameters $\mathbf{w}^{(i,k+1)}$ that are sent back to each client i.

The averaging step (5.15) might take much longer to execute than the local update step (5.14). Indeed, (5.15) typically requires transmission of local model parameters from every client $i \in \mathcal{V}$ to a server or central computing unit. Thus, after the client $i \in \mathcal{V}$ has computed the local gradient step (5.14), it must wait until the server (i) has collected the updates $\widehat{\mathbf{w}}_k^{(i)}$ from all clients and (ii) sent back their average $\mathbf{w}^{(i,k+1)}$ to $i \in \mathcal{V}$.

Instead of using a single gradient step (5.14),[3] and then being forced to wait for receiving $\mathbf{w}^{(i,k+1)}$ back from the server, a client can make better use of its resources. For example, the device i could execute several local gradient steps (5.14) to make more progress towards the optimum,

$$
\begin{aligned}
\mathbf{v}^{(0)} &:= \widehat{\mathbf{w}}_k^{(i)} \\
\mathbf{v}^{(r)} &:= \mathbf{v}^{(r-1)} - \eta_{i,k}(2/m_i)\left(\mathbf{X}^{(i)}\right)^T\left(\mathbf{X}^{(i)}\mathbf{v}^{(r-1)} - \mathbf{y}^{(i)}\right), \text{ for } r = 1, \ldots, R \\
\widehat{\mathbf{w}}_{k+1/2}^{(i)} &:= \mathbf{v}^{(R)}.
\end{aligned}
\tag{5.16}
$$

We obtain Algorithm 8 by iterating the combination of (5.16) with the projection step (5.15).

One of the most popular server-based FL algorithms, referred to as FedAvg and summarized in Algorithm 9, is obtained by two modifications of Algorithm 8:

- replacing the updates in step 2 at the client in Algorithm 8 with $\mathbf{v}^{(r)} := \mathbf{v}^{(r-1)} - \eta_{i,k}\mathbf{g}\left(\mathbf{v}^{(r)}\right)$ using the gradient approximation $\mathbf{g}^{(i)}\left(\mathbf{v}^{(r)}\right) \approx \nabla L_i\left(\mathbf{v}^{(r)}\right)$,
- using a randomly selected subset $\mathcal{C}^{(k)}$ of clients during each global iteration k.

[3] For a large local dataset, the local gradient step (5.14) can become computationally too expensive and must be replaced by an approximation, e.g., using a stochastic gradient approximation (5.11).

Algorithm 8 Server-based FL for linear models

The Server.

Input. Some stopping criterion; list of clients $i = 1, \dots, n$, number R of local updates.

Output. Trained model parameters $\widehat{\mathbf{w}}^{(\text{global})}$

Initialize. $k := 0$; $\mathbf{w}^{(i,k)} = \mathbf{0}$ for all $i = 1, \dots, n$

while stopping criterion is not satisfied **do**

  Update the global model parameters

$$\widehat{\mathbf{w}}^{(k)} := (1/n) \sum_{i=1}^{n} \mathbf{w}^{(i,k)}.$$

  Send model parameters $\widehat{\mathbf{w}}^{(k)}$ (and k) to all clients. $i = 1, \dots, n$

  Gather updated local model parameters $\mathbf{w}^{(i,k+1)}$ from clients $i = 1, \dots, n$.

  Clock Tick. $k := k + 1$.

end while

The Client $i \in \{1, \dots, n\}$.

Input. Local dataset $\mathbf{X}^{(i)}, \mathbf{y}^{(i)}$, number of gradient steps R and learning rate (schedule) $\eta_{i,k}$.

Receive the current model parameters $\widehat{\mathbf{w}}^{(k)}$ from the server.

Update the local model parameters by R gradient steps

$$\begin{aligned}
\mathbf{v}^{(0)} &:= \widehat{\mathbf{w}}^{(\text{global})} \\
\mathbf{v}^{(r)} &:= \mathbf{v}^{(r-1)} - \eta_{i,k}(2/m_i)\big(\mathbf{X}^{(i)}\big)^T \big(\mathbf{X}^{(i)}\mathbf{v}^{(r-1)} - \mathbf{y}^{(i)}\big), \text{ for } r = 1, \dots, R \\
\mathbf{w}^{(i,k+1)} &:= \mathbf{v}^{(R)}.
\end{aligned}$$

Send the new local model parameters $\mathbf{w}^{(i,k+1)}$ back to server.

5.5 FedProx

A central challenge in FedAvg (Algorithm 9) is selecting an appropriate number of local updates, R, in (5.17). In each iteration, all clients perform exactly R approximate gradient steps. However, [65] argues that enforcing a uniform number R across clients can degrade performance in certain FL settings. To mitigate this, they propose an alternative to (5.17) for the local update step. This alternative is given by

$$\mathbf{w}^{(i)} := \operatorname*{argmin}_{\mathbf{v} \in \mathbb{R}^d} \left[L_i(\mathbf{v}) + (1/\eta) \left\| \mathbf{v} - \widehat{\mathbf{w}}^{(\text{global})} \right\|_2^2 \right]. \tag{5.18}$$

We have already encountered an update of the form (5.18) in Section 4.6. Indeed, (5.18) is the application of the proximal operator of $L_i(\mathbf{v})$ (see (4.18)) to the current model parameters. We obtain Algorithm 10 from Algorithm 9 by replacing the local update step (5.17) with (5.18). Empirical studies have shown that Algorithm 10 outperforms FedAvg (Algorithm 9) for FL applications with a high-

Algorithm 9 FedAvg [12]

The Server.
Input. List of clients $i = 1, \ldots, n$, number R of local updates
Output. Trained model parameters $\widehat{\mathbf{w}}^{(\text{global})}$
Initialize. $k := 0$; $\widehat{\mathbf{w}}^{(\text{global})} := \mathbf{0}$ for all $i = 1, \ldots, n$

while stopping criterion is not satisfied **do**
  randomly select a subset $\mathcal{C}^{(k)}$ of clients
  send $\widehat{\mathbf{w}}^{(\text{global})}$ to all clients $i \in \mathcal{C}^{(k)}$
  receive updated model parameters $\mathbf{w}^{(i)}$ from clients $i \in \mathcal{C}^{(k)}$
  update global model parameters

$$\widehat{\mathbf{w}}^{(\text{global})} := \left(1/\left|\mathcal{C}^{(k)}\right|\right) \sum_{i \in \mathcal{C}^{(k)}} \mathbf{w}^{(i)}.$$

  increase iteration counter $k := k+1$
end while

Client $i \in \{1, \ldots, n\}$, with local loss function $L_i(\cdot)$
receive global model parameters $\widehat{\mathbf{w}}^{(\text{global})}$ from server
update local model parameters by R approximate gradient steps

$$\begin{aligned} \mathbf{v}^{(0)} &:= \widehat{\mathbf{w}}^{(\text{global})} \\ \mathbf{v}^{(r)} &:= \mathbf{v}^{(r-1)} - \eta_{i,k} \underbrace{\mathbf{g}^{(i)}\left(\mathbf{v}^{(r-1)}\right)}_{\approx \nabla L_i\left(\mathbf{v}^{(r-1)}\right)}, \text{ for } r = 1, \ldots, R \\ \mathbf{w}^{(i)} &:= \mathbf{v}^{(R)}. \end{aligned} \tag{5.17}$$

return $\mathbf{w}^{(i)}$ back to server

level of heterogeneity among the computational capabilities of devices $i = 1, \ldots, n$ and the statistical properties of their local datasets $\mathcal{D}^{(i)}$ [65].

As the notation in (5.18) indicates, the parameter η plays a role similar to the learning rate of a gradient step (4.2). It controls the size of the neighborhood of $\mathbf{w}^{(i,k)}$ over which (5.18) optimizes the local loss function $L_i(\cdot)$. Choosing a small η forces the update (5.18) to not move too far from the current model parameters $\mathbf{w}^{(i,k)}$.

The core computation (5.19) of FedProx Algorithm 10 can be interpreted as form of regularization. Indeed, we obtain (5.19) from (2.21) by

- replacing the average squared error loss with the local loss function $L_i(\mathbf{v})$,
- using the regularizer

$$\mathcal{R}\{\mathbf{v}\} := \left\|\mathbf{v} - \widehat{\mathbf{w}}^{(\text{global})}\right\|_2^2, \tag{5.20}$$

- and the regularization parameter $\alpha := 1/\eta$.

Algorithm 10 FedProx [65]

The Server.
Input. List of clients $i = 1, \ldots, n$
Output. Trained model parameters $\widehat{\mathbf{w}}^{(\mathrm{global})}$
Initialize. $k := 0$; $\widehat{\mathbf{w}}^{(\mathrm{global})} := \mathbf{0}$ for all $i = 1, \ldots, n$

1: **while** stopping criterion is not satisfied **do**
2: randomly select a subset $\mathcal{C}^{(k)}$ of clients
3: send $\widehat{\mathbf{w}}^{(\mathrm{global})}$ to all clients $i \in \mathcal{C}^{(k)}$
4: receive updated model parameters $\mathbf{w}^{(i)}$ from clients $i \in \mathcal{C}^{(k)}$
5: update global model parameters

$$\widehat{\mathbf{w}}^{(\mathrm{global})} := \left(1/\left|\mathcal{C}^{(k)}\right|\right) \sum_{i \in \mathcal{C}^{(k)}} \mathbf{w}^{(i)}.$$

6: increase iteration counter $k := k+1$
7: **end while**

Client $i \in \{1, \ldots, n\}$, with local loss function $L_i(\cdot)$
1: receive global model parameters $\widehat{\mathbf{w}}^{(\mathrm{global})}$ from server
2: update local model parameters by

$$\mathbf{w}^{(i)} := \underset{\mathbf{v} \in \mathbb{R}^d}{\mathrm{argmin}} \left[L_i(\mathbf{v}) + (1/\eta) \left\| \mathbf{v} - \widehat{\mathbf{w}}^{(\mathrm{global})} \right\|_2^2 \right] \tag{5.19}$$

3: return $\mathbf{w}^{(i)}$ back to server

Note that Algorithms 10 and 9 provide only an abstract description of a practical FL system. The details of their actual implementation, such as the synchronization between the server and all clients (see steps 4 and 3 in Algorithm 10) is beyond the scope of this book. Instead, we refer the reader to relevant literature on the implementation of distributed computing systems [18, 66].

5.6 FedRelax

We now apply a simple block-coordinate minimization method [17] to solve GTVMin (3.22). To this end, we rewrite (3.22) as

$$\widehat{\mathbf{w}} \in \underset{\mathbf{w} \in \mathbb{R}^{d \cdot n}}{\arg\min} \underbrace{\sum_{i \in \mathcal{V}} f^{(i)}(\mathbf{w})}_{=: f^{(\mathrm{GTV})}(\mathbf{w})}$$

$$\text{with } f^{(i)}(\mathbf{w}) := L_i\left(\mathbf{w}^{(i)}\right) + (\alpha/2) \sum_{i' \in \mathcal{N}^{(i)}} A_{i,i'} \left\| \mathbf{w}^{(i)} - \mathbf{w}^{(i')} \right\|_2^2,$$

$$\text{and the stacked model parameters } \mathbf{w} = \left(\mathbf{w}^{(1)}, \ldots, \mathbf{w}^{(n)}\right)^T. \tag{5.21}$$

According to (5.21), the objective function of (3.22) decomposes into components $f^{(i)}(\mathbf{w})$, one for each node $\mathcal{V}$ of the FL network. Moreover, the local model parameters $\mathbf{w}^{(i)}$ influence the objective function only via the components at the nodes $i \cup \mathcal{N}^{(i)}$. We exploit this structure of (5.21) to decouple the optimization of the local model parameters $\{\widehat{\mathbf{w}}^{(i)}\}_{i\in\mathcal{V}}$ as described next.

Consider some local model parameters $\mathbf{w}^{(i,k)}$, for $i = 1, \ldots, n$, at time k. We then update (in parallel) each $\mathbf{w}^{(i,k)}$ by minimizing $f^{(\mathrm{GTV})}(\cdot)$ along $\mathbf{w}^{(i)}$ with the other local model parameters $\mathbf{w}^{(i')} := \mathbf{w}^{(i',k)}$ held fixed for all $i' \neq i$,

$$\begin{aligned}
\mathbf{w}^{(i,k+1)} &\in \operatorname*{argmin}_{\mathbf{w}^{(i)}\in\mathbb{R}^d} f^{(\mathrm{GTV})}\left(\mathbf{w}^{(1,k)}, \ldots, \mathbf{w}^{(i-1,k)}, \mathbf{w}^{(i)}, \mathbf{w}^{(i+1,k)}, \ldots\right) \\
&\overset{(5.21)}{=} \operatorname*{argmin}_{\mathbf{w}^{(i)}\in\mathbb{R}^d} f^{(i)}\left(\mathbf{w}^{(1,k)}, \ldots, \mathbf{w}^{(i-1,k)}, \mathbf{w}^{(i)}, \mathbf{w}^{(i+1,k)}, \ldots\right) \\
&\overset{(5.21)}{=} \operatorname*{argmin}_{\mathbf{w}^{(i)}\in\mathbb{R}^d} L_i\left(\mathbf{w}^{(i)}\right) + \alpha \sum_{i'\in\mathcal{N}^{(i)}} A_{i,i'} \left\|\mathbf{w}^{(i)} - \mathbf{w}^{(i',k)}\right\|_2^2 .
\end{aligned} \tag{5.22}$$

The update rule in (5.22) can be viewed as a non-linear Jacobi method applied to (5.21) [17, Sec. 3.2.4]. It also admits an interpretation as a form of block-coordinate optimization [67]. By iterating this update sufficiently many times, we arrive at Algorithm 11. There is an interesting connection between the update (5.22) and the basic gradient steps used by FedGD and FedSGD (see Algorithm 5 and 7). Indeed, we obtain step 4 in Algorithm 5 from (5.22) by replacing the loss function $L_i\left(\mathbf{w}^{(i)}\right)$ with the approximation

$$L_i\left(\mathbf{w}^{(i,k)}\right) + \left(\nabla L_i\left(\mathbf{w}^{(i,k)}\right)\right)\left(\mathbf{w}^{(i)} - \mathbf{w}^{(i,k)}\right) + (1/(2\eta)) \left\|\mathbf{w}^{(i)} - \mathbf{w}^{(i,k)}\right\|_2^2 .$$

Algorithm 11 FedRelax for Parametric Models

Input: FL network $\mathcal{G}$ with local loss functions $L_i(\cdot)$, GTV parameter α
Initialize: $k := 0$; $\mathbf{w}^{(i,0)} := \mathbf{0}$

- **while** stopping criterion is not satisfied **do**
  - **for** all nodes $i \in \mathcal{V}$ in parallel **do**
    - compute $\mathbf{w}^{(i,k+1)}$ via (5.22)
    - share $\mathbf{w}^{(i,k+1)}$ with neighbors $\mathcal{N}^{(i)}$
  - **end for**
  - $k := k+1$
- **end while**

A Model-Agnostic Method. The applicability of Algorithm 11 is limited to FL networks with parametric local models (such as linear regression or ANNs with a common structure). We can generalize Algorithm 11 to non-parametric local models by applying the non-linear Jacobi method to the GTVMin variant (3.39). This results in the update

$$\widehat{h}^{(i)}_{k+1} \in \underset{h^{(i)} \in \mathcal{H}^{(i)}}{\operatorname{argmin}} \, L_i\left(h^{(i)}\right) + \alpha \sum_{i' \in \mathcal{N}^{(i)}} A_{i,i'} \underbrace{d^{(h^{(i)},\widehat{h}^{(i')}_k)}}_{\text{see (3.37)}}. \tag{5.23}$$

We obtain Algorithm 12 as a model-agnostic variant of Algorithm 11 by replacing the update (5.22) in its step 3 with the update (5.23).

Algorithm 12 is model-agnostic as it allows devices of an FL network to train different types of local models. The only restriction for the local models is that the update (5.23) can be computed efficiently. For some choices of local models and loss function, the update (5.23) can be implemented by basic data augmentation (see Exercise 5.3).

Algorithm 12 Model Agnostic FedRelax

Input: FL network with $\mathcal{G}$, local models $\mathcal{H}^{(i)}$, loss functions $L_i\,(\cdot)$, GTV parameter α, loss $L\,(\cdot,\cdot)$ used in (3.37).

Initialize: $k := 0$; $\widehat{h}^{(i)}_0 := \mathbf{0}$

while stopping criterion is not satisfied **do**
  for all nodes $i \in \mathcal{V}$ in parallel **do**
    compute $\widehat{h}^{(i)}_{k+1}$ via (5.23)
  end for
  $k := k+1$
end while

5.7 A Unified Formulation

The previous sections have presented some widely-used FL algorithms. These algorithms are obtained by applying distributed optimization methods to solve GTVMin. Despite their different formulations they share a common underlying structure. In particular, they can all be expressed as synchronous fixed-point iterations:

$$\widehat{h}^{(i)}_{k+1} = \mathcal{F}^{(i)}\big(\widehat{h}^{(1)}_k, \ldots, \widehat{h}^{(n)}_k\big), \text{ for } i = 1, \ldots, n. \tag{5.24}$$

Each operator $\mathcal{F}^{(i)} : \mathcal{H}^{(1)} \times \ldots \times \mathcal{H}^{(n)} \rightarrow \mathcal{H}^{(i)}$ represents a local update rule at the $i = 1, \ldots, n$ (see Figure 5.4). Some algorithms use time-varying update rules,

$$\widehat{h}^{(i)}_{k+1} = \mathcal{F}^{(i)}\big(\widehat{h}^{(1)}_k, \ldots, \widehat{h}^{(n)}_k\big). \tag{5.25}$$

with operators $\mathcal{F}^{(i,k)}$ that can vary across nodes $i = 1, \ldots, n$ and time instants $k = 1, 2, \ldots$. One example of (5.25) is used in Algorithm 5 for a time-varying learning rate.

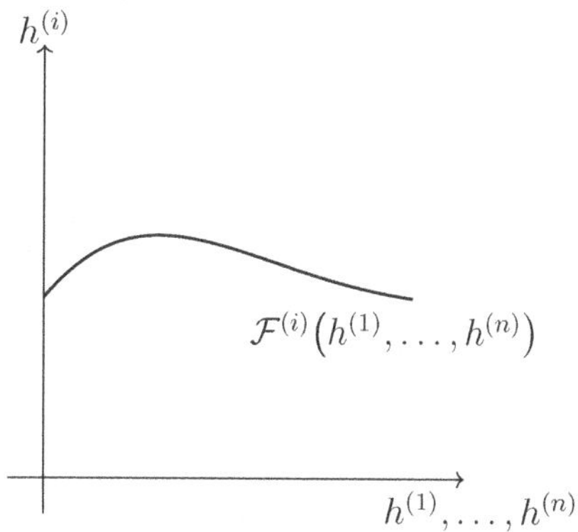

Fig. 5.4 A key computational step in many FL algorithms is the evaluation of an operator $\mathcal{F}^{(i)}$ at each node $i = 1, \ldots, n$ of the FL network.

Clearly, any FL algorithm of the form 5.24 is fully specified by the operators $\mathcal{F}^{(1)}, \ldots, \mathcal{F}^{(n)}$. This - rather trivial - observation implies that we can study the behaviour of FL algorithms via analyzing the properties of the operators $\mathcal{F}^{(i)}$, for $i = 1, \ldots, n$. In particular, the robustness of FL algorithms crucially depends on the shape of $\mathcal{F}^{(i)}$.

For parametric local models, we can re-formulate the fixed-point iteration (5.24) directly in terms of the model parameters

$$\mathbf{w}^{(i,k+1)} = \mathcal{F}^{(i)}\big(\mathbf{w}^{(1,k)}, \ldots, \mathbf{w}^{(n,k)}\big), \text{ for } k = 0, 1, \ldots, \tag{5.26}$$

with operators $\mathcal{F}^{(i)} : \mathbb{R}^{nd} \rightarrow \mathbb{R}^{d}$, for $i = 1, \ldots, n$. One example of (5.26) is the update 5.22 used by FedRelax (see Algorithm 11).

5.8 Asynchronous FL Algorithms

The FL algorithms presented so far rely on synchronous coordination among devices $i = 1, \ldots, n$ within an FL network [18, Ch. 6]. A new iteration is only initiated once all devices have completed their local updates (5.24) and communicated them to their neighbors [68, Sec. 10], [17, Sec. 1.4].

The implementation of synchronous FL algorithms can be difficult (or impossible) in practice. As highlighted in Chapter 8, trustworthy FL systems should tolerate unreliable or failing devices. Synchronous methods lack this robustness–any device failure or dropout can cause the entire algorithm execution to stall. Moreover, synchronous execution is inefficient in heterogeneous FL systems. Devices often vary in computational power or communication bandwidth, leading to the *straggler problem*: faster devices are forced to wait idly for slower ones [69, 70]. Having devices to wait idly for slower devices results in a waste of their computational resources.

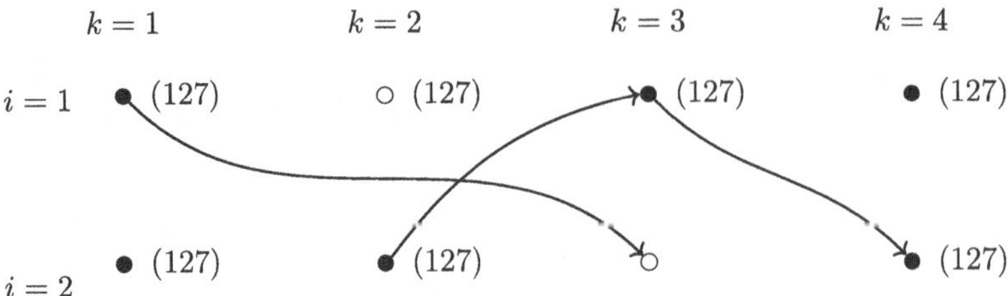

Fig. 5.5 The execution of an asynchronous FL algorithm consists of a sequence of update events, indexed by $k = 0, 1, 2, \ldots$. During each event k, the active nodes $i \in \mathcal{A}^{(k)} \subseteq \mathcal{V}$ of an FL network update their local model parameters $\mathbf{w}^{(i)}$ by computing (5.27). Active nodes are depicted as filled circles.

To address the limitations of synchronous FL algorithms, we now show how to build asynchronous variants of the FL algorithms discussed in Section 5.7. We focus here on parametric local models, each represented by their own model parameters $\mathbf{w}^{(i)}$. The basic idea is to let each device $i = 1, \ldots, n$ execute the update (5.26) independently, using potentially outdated updates from its neighbors $\mathcal{N}^{(i)}$.

An asynchronous FL algorithm consists of a sequence of update events, which we index by $k = 0, 1, 2, \ldots$ (see Figure 5.5). During each event k, a subset $\mathcal{A}^{(k)} \subseteq \mathcal{V}$ of devices performs updates:

$$\mathbf{w}^{(i,k+1)} = \mathcal{F}^{(i)}\big(\mathbf{w}^{(1,k_{i,1})}, \ldots, \mathbf{w}^{(n,k_{i,n})}\big). \tag{5.27}$$

Here, $k_{i,i'} \leq k$ is event index of the latest available model parameters of device i' at device i.

The set of nodes performing the update (5.27) during event k is denoted as the active set $\mathcal{A}^{(k)} \subseteq \mathcal{V}$. It is convenient to summarize the resulting asynchronous algorithm as

$$\mathbf{w}^{(i,k+1)} = \begin{cases} \mathcal{F}^{(i)}\big(\mathbf{w}^{(1,k_{i,1})}, \ldots, \mathbf{w}^{(n,k_{i,n})}\big) & \text{for } k \in T^{(i)} \\ \mathbf{w}^{(i,k)} & \text{otherwise.} \end{cases} \tag{5.28}$$

Here, we used the set

$$T^{(i)} := \big\{k \in \{0, 1, \ldots,\} : i \in \mathcal{A}^{(k)}\big\},$$

which consists, for each $i = 1, \ldots, n$, of those clock ticks during which node i is active. Note that (5.28) reduces to the synchronous algorithm (5.26) for the extreme case when $T^{(i)} = 0, 1, 2, \ldots,$ for all $i = 1, \ldots, n$.

Like the synchronous algorithm (5.26), also the asynchronous variant 5.28 uses an iteration counter k. However, the practical meaning of k in the asynchronous variant is fundamentally different: Instead of representing a global clock tick (or wall-clock time), the counter k in (5.28) indexes some update event during which at least one node is active and computes a local update. We denote the set of active nodes (or devices) during event k by $\mathcal{A}^{(k)} \subseteq \mathcal{V}$. The inactive nodes $i \notin \mathcal{A}^{(k)}$ leave their current model parameters unchanged, i.e., $\mathbf{w}^{(i,k+1)} = \mathbf{w}^{(i,k)}$.

For each active node $i \in \mathcal{A}^{(k)}$, the local update (5.28) uses potentially outdated model parameters $\mathbf{w}^{(i',k_{i,i'})}$ from its neighbors $i' \in \mathcal{N}^{(i)}$. Indeed, some of the neighbors might have not been in the active sets $\mathcal{A}^{(k-1)}, \mathcal{A}^{(k-2)}, \ldots$ of the most recent iterations. In this case, the update (5.28) does not have access to $\mathbf{w}^{(i',k)}$. Instead, we can only use $\mathbf{w}^{(i',k_{i,i'})}$ that has been obtained during some previous iteration $k_{i,i'} < k$.

The update (5.27) involves an operator $\mathcal{F}^{(i)} : \mathbb{R}^{dn} \rightarrow \mathbb{R}^{d}$ that determines the resulting FL algorithm. We can interpret (5.27) as an asynchronous variant of the synchronous algorithm (5.26) obtained for the same $\mathcal{F}^{(i)}$. For example, an asynchronous variant of Algorithm 5 (with a fixed learning rate) can be obtained for the choice

$$\mathcal{F}^{(i)}\left(\mathbf{w}^{(1)}, \ldots, \mathbf{w}^{(n)}\right) = \mathbf{w}^{(i)} - \eta\left(\nabla L_i\left(\mathbf{w}^{(i)}\right) + \sum_{i' \in \mathcal{N}^{(i)}} 2A_{i,i'}\left(\mathbf{w}^{(i)} - \mathbf{w}^{(i')}\right)\right). \tag{5.29}$$

Note that the choice (5.29) involves the local loss functions and the weighted edges of an FL network.

The update (5.27), at an active node $i \in \mathcal{A}^{(k)}$, involves potentially outdated local model parameters $\mathbf{w}^{(i',k_{i,i'})}$, with $k_{i,i'} \leq k$, for $i' = 1, \ldots, n$. The quantity $k_{i,i'}$ represents the most recent update event during which node i' has shared its updated local model parameters with node i. We can, in turn, interpret the difference $k - k_{i,i'}$ as a measure of the communication delay between node i' and node i.

Depending on the extent of the delays $k - k_{i,i'}$ in the update (5.27), we distinguish between [17]

- **Totally asynchronous algorithms.** These are algorithms of the form (5.28) with unbounded delays $k - k_{i,i'}$, i.e., they can can become arbitrarily large. Moreover, we require that no device stops updating, i.e., the set $T^{(i)}$ is infinite for each $i = 1, \ldots, n$.
- **Partially asynchronous algorithms.** These are algorithms of the form (5.28) with bounded delays $k - k_{i,i'} \leq B$, with some fixed (but possibly unknown) maximum delay $B \in \mathbb{N}$. Moreover, each device updates at least once during B consecutive clock ticks, i.e., $T^{(i)} \cap \{t, t+1, t+B-1\} \neq \emptyset$ for each $t = 1, 2, \ldots,$ and $i = 1, \ldots, n$.

For some choices of $\mathcal{F}^{(i)}$ in (5.27), a partially asynchronous algorithm can converge for any value of B. However, there are also choices of $\mathcal{F}^{(i)}$ for which a partially asynchronous algorithm will only converge if B is sufficiently small [17, Ch. 7].

Convergence Guarantees. There is an elegant characterization of the convergence of totally and partially asynchronous FL algorithms of the form (5.28). This characterization applies whenever the operators $\mathcal{F}^{(i)}$, for $i = 1, \ldots, n$, in (5.28) form a pseudo-contraction [71]

$$\max_{i=1,\ldots,n} \left\| \mathcal{F}^{(i)}\left(\mathbf{w}^{(1)}, \ldots, \mathbf{w}^{(n)}\right) - \mathcal{F}^{(i)}\left(\widehat{\mathbf{w}}^{(1)}, \ldots, \widehat{\mathbf{w}}^{(n)}\right) \right\| \leq \kappa \cdot \max_{i=1,\ldots,n} \left\| \mathbf{w}^{(i)} - \widehat{\mathbf{w}}^{(i)} \right\|, \tag{5.30}$$

with some contraction rate $\kappa \in [0, 1)$ and some fixed-point $\widehat{\mathbf{w}}^{(1)}, \ldots, \widehat{\mathbf{w}}^{(n)}$.

The operators $\mathcal{F}^{(i)}$, for $i = 1, \ldots, n$, underlying GTVMin-based algorithms are determined by the design choices of the GTVMin building blocks. These include the choices of local loss functions $L_i(\cdot)$, for $i = 1, \ldots, n$ and edge weights $A_{i,i'}$, for $\{i, i'\} \in \mathcal{E}$. Let us next discuss specific design choices which yield operators that form a pseudo-contraction (5.30).

Consider the operator $\mathcal{F}^{(i)}$ defined by the update (5.22) of FedRelax (see Algorithm 11). If the local loss functions $L_i(\cdot)$ are strongly convex,[4] we can decompose $\mathcal{F}^{(i)}$ as

$$\mathcal{F}^{(i)} = \mathbf{prox}_{L_i(\cdot), 2\alpha d^{(i)}}(\cdot) \circ \mathcal{T}^{(i)}. \tag{5.31}$$

Here, we used the proximal operator as defined in (3.30) as well as the averaging-neighbors-operator

$$\mathcal{T}^{(i)} : \underbrace{\mathbb{R}^d \times \ldots \times \mathbb{R}^d}_{n \text{ times}} \rightarrow \mathbb{R}^d : \mathbf{w}^{(1)}, \ldots, \mathbf{w}^{(n)} \mapsto (1/d^{(i)}) \sum_{i' \in \mathcal{N}^{(i)}} A_{i,i'} \mathbf{w}^{(i')}.$$

It can be easily verified that the operators $\mathcal{T}^{(1)}, \ldots, \mathcal{T}^{(n)}$ are non-expansive. Moreover, by the basic properties of proximal operators (see, e.g., [72, Sec. 6]), the operators

$$\mathbf{prox}_{L_1(\cdot), 2\alpha d^{(1)}}(\cdot), \ldots, \mathbf{prox}_{L_n(\cdot), 2\alpha d^{(n)}}(\cdot)$$

also form a pseudo-contraction with $\kappa = \frac{1}{1+(\sigma/(2\alpha d^{(i)}))}$. Combining these facts with (5.31) yields that the operators $\mathcal{F}^{(i)}$, for $i = 1, \ldots, n$, form a pseudo-contraction with

$$\kappa = \frac{1}{1 + (\sigma/(2\alpha d^{(i)}))}. \tag{5.32}$$

[4] Strictly speaking, we also need to require that the epigraph of each $L_i(\cdot)$, for $i = 1, \ldots, n$ is non-empty and closed [40].

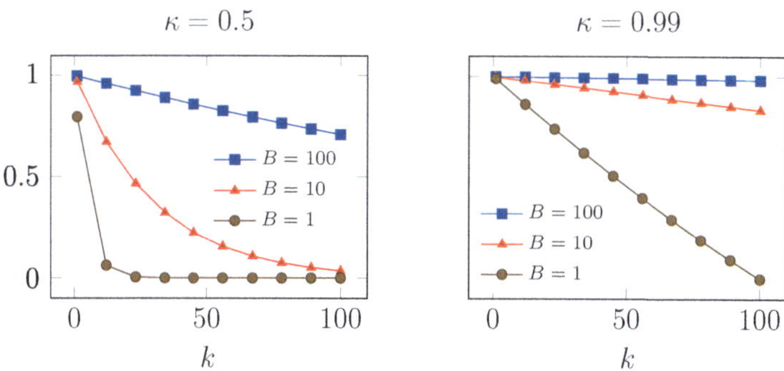

Fig. 5.6 Illustration of the factor $\kappa^{k/(2B+1)}$ in the convergence bound (5.33) for a partially asynchronous FL algorithm (5.28) using a pseudo-contraction (see (5.30)).

For any FL algorithm (5.28) such that (5.30) is satisfied, the following holds:

- A totally asynchronous algorithm of the form (5.28) converges to $\widehat{\mathbf{w}}^{(1)}, \ldots, \widehat{\mathbf{w}}^{(n)}$ [71, Thm. 23].
- In the partially asynchronous case with maximum delay B [71, Thm. 24],

$$\max_{i=1,\ldots,n} \left\| \mathbf{w}^{(i,k)} - \widehat{\mathbf{w}}^{(i)} \right\| \leq \kappa^{k/(2B+1)} \cdot \max_{i=1,\ldots,n} \left\| \mathbf{w}^{(i,0)} - \widehat{\mathbf{w}}^{(i)} \right\|. \tag{5.33}$$

The bound (5.33) is quite intuitive: smaller contraction factors κ and smaller delay bounds B lead to faster convergence of the algorithm (5.28). Figure 5.6 illustrates the factor $\kappa^{k/(2B+1)}$ for different values of κ and maximum delay B.

The contraction factor κ of the operators $\mathcal{F}^{(i)}$, for $i = 1, \ldots, n$, arising in GTVMin-based methods depends on the properties of local loss functions and the connectivity of the FL network. According to (5.32), the operators underlying FedRelax (see (5.22) and Algorithm 11), have a small contraction factor if we use

- local loss functions that are strongly convex with large coefficient σ,
- an FL network with small weighted node degrees $d^{(i)}$, for $i = 1, \ldots, n$.

Moreover, the contraction factor (5.32) decreases with decreasing GTVMin parameter α. In the extreme case of $\alpha = 0$ - where GTVMin decomposes into fully independent local instances of ERM $\min_{\mathbf{w}^{(i)}} L_i(\cdot)$ - the contraction factor becomes $\kappa = 0$. This makes sense as in this extreme case, there is no information sharing required among the nodes of an FL network. Clearly, the delays $k - k_{i,i'}$ are then irrelevant for the performance of FL algorithms.

5.9 Exercises

5.1 (The convergence speed of gradient-based methods.) Study the convergence speed of (5.3) for two different collections of local datasets assigned to the nodes of the FL network $\mathcal{G}$ with nodes $\mathcal{V} = \{1, 2\}$ and (unit weight) edges $\mathcal{E} = \{\{1, 2\}\}$. The first collection of local datasets results in the local loss functions $L_1(w) := (w+5)^2$ and $L_2(w) := 1000(w+5)^2$. The second collection of local datasets results in the local loss functions $L_1(w) := 1000(w+5)^2$ and $L_2(w) := 1000(w-5)^2$. Use a fixed learning rate $\eta := 0.5 \cdot 10^{-3}$ for the iteration (5.3).

5.2 (Convergence speed for homogeneous data.) Study the convergence speed of (5.3) when applied to GTVMin (5.1) with the following FL network $\mathcal{G}$: Each node $i = 1, \ldots, n$ carries a simple local model with single parameter $w^{(i)}$ and the local loss function $L_i(w) := \left(y^{(i)} - x^{(i)} w^{(i)}\right)^2$. The local dataset consists of a constant $x^{(i)} := 1$ and some $y^{(i)} \in \mathbb{R}$. The edges $\mathcal{E}$ are obtained by connecting each node i with 4 other randomly chosen nodes. We learn model parameters $\widehat{w}^{(i)}$ by repeating (5.3), starting with the initializations $w^{(i,0)} := y^{(i)}$. Study the dependence of the convergence speed of (5.3) (towards a solution of (5.1)) on the value of α in (5.1).

5.3 (Implementing FedRelax via data augmentation.) Consider the application of Algorithm 12 to an FL network whose nodes carry regression tasks. In particular, each device $i = 1, \ldots, n$ learns a hypothesis $h^{(i)}$ to predict the numeric label $y \in \mathbb{R}$ of a data point with feature vector $\mathbf{x}$. The usefulness of a hypothesis is measured by the average squared error loss incurred on a labelled local dataset

$$\mathcal{D}^{(i)} := \left\{ \left(\mathbf{x}^{(1)}, y^{(1)}\right), \ldots, \left(\mathbf{x}^{(m_i)}, \mathbf{x}^{(m_i)}\right) \right\}.$$

To compare the learned hypothesis maps at the nodes of an edge $\{i, i'\}$, we use (3.37) with the squared error loss. Show that the update (5.23) is equivalent to plain ERM (2.1) using a dataset $\mathcal{D}$ that is obtained by a specific augmentation of $\mathcal{D}^{(i)}$.

5.4 (FedAvg as fixed-point iteration.) Consider Algorithm 8 for training the model parameters $\mathbf{w}^{(i)}$ of local linear models for each $i = 1, \ldots, n$ of an FL network. Each client uses a constant learning rate schedule $\eta_{i,k} := \eta_i$. Try to find a collection of operators $\mathcal{F}^{(i)} : \mathbb{R}^{nd} \rightarrow \mathbb{R}^d$, for each node $i = 1, \ldots, n$, such that Algorithm 8 is equivalent to the fixed-point iteration

$$\mathbf{w}^{(i,k+1)} = \mathcal{F}^{(i)}\left(\mathbf{w}^{(1,k)}, \ldots, \mathbf{w}^{(n,k)}\right).$$

5.5 (Fixed-Points of a pseudo-contraction.) Show that a pseudo-contraction cannot have more than one fixed-point.

5.6 (FedRelax update.) Show that the update (5.22) of FedRelax for parametric local models can be rewritten as

$$\operatorname*{argmin}_{\mathbf{w}^{(i)} \in \mathbb{R}^d} L_i\left(\mathbf{w}^{(i)}\right) + \alpha d^{(i)} \left\| \mathbf{w}^{(i)} - \widehat{\mathbf{w}}^{(\mathcal{N}^{(i)})} \right\|_2^2.$$

Here, we used $\widehat{\mathbf{w}}^{(\mathcal{N}^{(i)})} := (1/d^{(i)}) \sum_{i' \in \mathcal{N}^{(i)}} A_{i,i'} \mathbf{w}^{(i',k)}$ and the weighted node degree $d^{(i)} = \sum_{i' \in \mathcal{N}^{(i)}} A_{i,i'}$ (see (3.7)).

5.7 (FedRelax vs. FedSGD) Show that the update in step (4) of Algorithm 5 is obtained from the update (5.22) of FedRelax by replacing the local loss function $L_i\left(\mathbf{w}^{(i)}\right)$ with a local approximation by a quadratic function, centred around $\mathbf{w}^{(i,k)}$.

5.8 (FedSGD as fixed-point iteration.) Show that Algorithm 6 can be written as the distributed fixed-point iteration (5.26). Try to find an elegant characterization of the resulting operators $\mathcal{F}^{(i)}$, for $i = 1, \ldots, n$.

5.9 (FedRelax as fixed-point iteration.) Show that Algorithm 12 can be written as the distributed fixed-point iteration (5.24). Try to find an elegant characterization of the resulting operators $\mathcal{F}^{(i)}$, for $i = 1, \ldots, n$..

5.10 Proofs

5.10.1 Proof of Proposition 5.1

The first inequality in (5.5) follows from well-known results on the eigenvalues of a sum of symmetric matrices (see, e.g., [3, Thm 8.1.5]). In particular,

$$\lambda_{\max}\big(\mathbf{Q}\big) \leq \max\big\{ \underbrace{\max_{i=1,\ldots,n} \lambda_d\big(\mathbf{Q}^{(i)}\big)}_{\stackrel{(5.4)}{=} \lambda_{\max}}, \lambda_{\max}\big(\alpha \mathbf{L}^{(\mathcal{G})} \otimes \mathbf{I}\big) \big\}. \tag{5.34}$$

The second inequality in (5.5) uses the following upper bound on the maximum eigenvalue $\lambda_n\big(\mathbf{L}^{(\mathcal{G})}\big)$ of the Laplacian matrix:

$$\begin{aligned}
\lambda_n\big(\mathbf{L}^{(\mathcal{G})}\big) &\stackrel{(a)}{=} \max_{\mathbf{v} \in \mathbb{S}^{(n-1)}} \mathbf{v}^T \mathbf{L}^{(\mathcal{G})} \mathbf{v} \\
&\stackrel{(3.10)}{=} \max_{\mathbf{v} \in \mathbb{S}^{(n-1)}} \sum_{\{i,i'\} \in \mathcal{E}} A_{i,i'} \big(v_i - v_{i'}\big)^2 \\
&\stackrel{(b)}{\leq} \max_{\mathbf{v} \in \mathbb{S}^{(n-1)}} \sum_{\{i,i'\} \in \mathcal{E}} 2A_{i,i'} \big(v_i^2 + v_{i'}^2\big)
\end{aligned}$$

Fig. 5.7 Illustration of step (c) in (5.35).

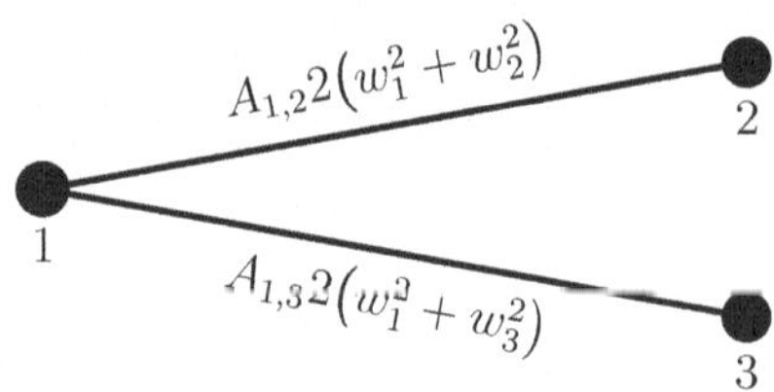

$$\begin{aligned} &\stackrel{(c)}{=} \max_{\mathbf{v}\in\mathbb{S}^{(n-1)}} \sum_{i\in\mathcal{V}} 2v_i^2 \sum_{i'\in\mathcal{N}^{(i)}} A_{i,i'} \\ &\stackrel{(3.8)}{\leq} \max_{\mathbf{v}\in\mathbb{S}^{(n-1)}} \sum_{i\in\mathcal{V}} 2v_i^2 d_{\max}^{(\mathcal{G})} \\ &= 2d_{\max}^{(\mathcal{G})}. \end{aligned} \tag{5.35}$$

Here, step (a) uses the CFW of eigenvalues [3, Thm. 8.1.2.] and step (b) uses the inequality $(u+v)^2 \leq 2(u^2+v^2)$ for any $u, v \in \mathbb{R}$. For step (c) we use the identity $\sum_{i\in\mathcal{V}}\sum_{i'\in\mathcal{N}^{(i)}} f(i,i') = \sum_{\{i,i'\}} \big(f(i,i') + f(i',i)\big)$ (see Figure 5.7). The bound (5.35) is essentially tight.[5]

5.10.2 *Proof of Proposition 5.2*

Similar to the upper bound (5.35) we also start with the CFW for the eigenvalues of $\mathbf{Q}$ in (5.2). In particular,

$$\lambda_1 = \min_{\|\mathbf{w}\|_2^2=1} \mathbf{w}^T\mathbf{Q}\mathbf{w}. \tag{5.36}$$

We next analyze the right-hand side of (5.36) by partitioning the constraint set $\{\mathbf{w} : \|\mathbf{w}\|_2^2 = 1\}$ of (5.36) into two complementary regimes for the optimization variable $\mathbf{w} = \text{stack}\{\mathbf{w}^{(i)}\}$. To define these two regimes, we use the orthogonal decomposition

$$\mathbf{w} = \underbrace{\mathbf{P}_{\mathcal{S}}\mathbf{w}}_{=:\overline{\mathbf{w}}} + \underbrace{\mathbf{P}_{\mathcal{S}^\perp}\mathbf{w}}_{=:\widetilde{\mathbf{w}}} \text{ for subspace } \mathcal{S} \text{ in (3.18).} \tag{5.37}$$

[5] Consider an FL network being a chain (or path).

Explicit expressions for the orthogonal components $\overline{\mathbf{w}}$, $\widetilde{\mathbf{w}}$ are given by (3.19) and (3.20). In particular, the component $\overline{\mathbf{w}}$ satisfies

$$\overline{\mathbf{w}} = \left((\mathbf{c})^T, \ldots, (\mathbf{c})^T\right)^T \text{ with } \mathbf{c} := \operatorname{avg}\{\mathbf{w}^{(i)}\}_{i=1}^{n}.$$

Note that

$$\|\mathbf{w}\|_2^2 = \|\overline{\mathbf{w}}\|_2^2 + \|\widetilde{\mathbf{w}}\|_2^2. \tag{5.38}$$

Regime I. This regime is obtained for $\|\widetilde{\mathbf{w}}\|_2 \geq \rho \|\overline{\mathbf{w}}\|_2$. Since $\|\mathbf{w}\|_2^2 = 1$, and due to (5.38), we have

$$\|\widetilde{\mathbf{w}}\|_2^2 \geq \rho^2/(1+\rho^2). \tag{5.39}$$

This implies, in turn, via (3.17) that

$$\begin{aligned}
\mathbf{w}^T \mathbf{Q} \mathbf{w} &\overset{(5.2)}{\geq} \alpha \mathbf{w}^T \left(\mathbf{L}^{(\mathcal{G})} \otimes \mathbf{I}\right) \mathbf{w} \\
&\overset{(3.10),(3.17)}{\geq} \alpha \lambda_2\left(\mathbf{L}^{(\mathcal{G})}\right) \|\widetilde{\mathbf{w}}\|_2^2 \\
&\overset{(5.39)}{\geq} \alpha \lambda_2\left(\mathbf{L}^{(\mathcal{G})}\right) \rho^2/(1+\rho^2).
\end{aligned} \tag{5.40}$$

Regime II. This regime is obtained for $\|\widetilde{\mathbf{w}}\|_2 < \rho \|\overline{\mathbf{w}}\|_2$. Here we have $\|\overline{\mathbf{w}}\|_2^2 > (1/\rho^2)\left(1 - \|\overline{\mathbf{w}}\|_2^2\right)$ and, in turn,

$$n \|\mathbf{c}\|_2^2 = \|\overline{\mathbf{w}}\|_2^2 > 1/(1+\rho^2). \tag{5.41}$$

We next develop the right-hand side of (5.36) according to

$$\begin{aligned}
\mathbf{w}^T \mathbf{Q} \mathbf{w} &\overset{(5.2)}{\geq} \sum_{i=1}^{n} \left(\mathbf{w}^{(i)}\right)^T \mathbf{Q}^{(i)} \mathbf{w}^{(i)} \\
&\overset{(5.37)}{\geq} \sum_{i=1}^{n} \left(\mathbf{c} + \widetilde{\mathbf{w}}^{(i)}\right)^T \mathbf{Q}^{(i)} \left(\mathbf{c} + \widetilde{\mathbf{w}}^{(i)}\right) \\
&\overset{(5.41)}{\geq} \|\overline{\mathbf{w}}\|_2^2 \underbrace{\lambda_1\left((1/n) \sum_{i=1}^{n} \mathbf{Q}^{(i)}\right)}_{\bar{\lambda}_{\min}} + \sum_{i=1}^{n} \Big[2\left(\widetilde{\mathbf{w}}^{(i)}\right)^T \mathbf{Q}^{(i)} \mathbf{c} + \underbrace{\left(\widetilde{\mathbf{w}}^{(i)}\right)^T \mathbf{Q}^{(i)} \widetilde{\mathbf{w}}^{(i)}}_{\geq 0}\Big] \\
&\geq \|\overline{\mathbf{w}}\|_2^2 \bar{\lambda}_{\min} + \sum_{i=1}^{n} 2\left(\widetilde{\mathbf{w}}^{(i)}\right)^T \mathbf{Q}^{(i)} \mathbf{c}.
\end{aligned} \tag{5.42}$$

To develop (5.42) further, we note that

$$\left|\sum_{i=1}^{n} 2\big(\widetilde{\mathbf{w}}^{(i)}\big)^{T}\mathbf{Q}^{(i)}\mathbf{c}\right| \overset{(a)}{\leq} 2\lambda_{\max}\,\|\widetilde{\mathbf{w}}\|_2\,\|\overline{\mathbf{w}}\|_2$$

$$\overset{\|\widetilde{\mathbf{w}}\|_2<\rho\|\overline{\mathbf{w}}\|_2}{\leq} 2\lambda_{\max}\rho\,\|\overline{\mathbf{w}}\|_2^2. \tag{5.43}$$

Here, step (a) follows from $\max_{\|\mathbf{y}\|_2=1,\|\mathbf{x}\|_2=1}\mathbf{y}^T\mathbf{Q}\mathbf{x} = \lambda_{\max}$. Inserting (5.43) into (5.42) for $\rho = \bar{\lambda}_{\min}/(4\lambda_{\max})$,

$$\mathbf{w}^T\mathbf{Q}\mathbf{w} \geq \|\overline{\mathbf{w}}\|_2^2\,\bar{\lambda}_{\min}/2 \overset{(5.41)}{\geq} (1/(1+\rho^2))\bar{\lambda}_{\min}/2 \tag{5.44}$$

For each $\mathbf{w}$ with $\|\mathbf{w}\|_2^2 = 1$, either (5.40) or (5.44) must hold.

Chapter 6
Key Variants of Federated Learning

Abstract This chapter explores major variants of federated learning (FL) that arise from different structures of local data and network connectivity. It shows how each variant can be seen as a special case of the GTVMin framework introduced earlier. The chapter begins with single-model FL, where all devices collaboratively train a shared model. Next, it covers clustered FL, which groups devices into clusters based on data similarity, allowing each cluster to train its own model. Horizontal FL is discussed as a setup where devices hold different samples of a shared dataset, while vertical FL covers cases where devices share the same data points but observe different features. The chapter then introduces personalized FL, where only parts of a model are shared across devices, allowing customization for local data. Finally, it presents few-shot learning as a setting where FL helps train models across categories with limited data. Each variant reflects different assumptions about data distribution and model structure. The chapter explains how to implement these variants using the same core optimization tools, making GTVMin a flexible framework for diverse FL applications.

Chapter 3 discussed GTVMin as a main design principle for FL algorithms. GTVMin learns local model parameters that optimally balance the individual local loss with their variation across the edges of an FL network. Chapter 5 discussed how to obtain practical FL algorithms. These algorithms solve GTVMin using distributed optimization methods, such as those from Chap. 4.

This chapter discusses important special cases (or "main flavors") of GTVMin obtained for specific construction of local datasets, choices of local models, measures for their variation, and the weighted edges of the FL network. We next briefly summarize the resulting main flavors of FL discussed in the following sections.

Section 6.1 discusses single-model FL that learns model parameters of a single (global) model from local datasets. This single-model flavor can be obtained from GTVMin using a connected FL network with large edge weights or, equivalently, a sufficient large value for the GTVMin parameter.

A. Jung, *Federated Learning*, https://doi.org/10.1007/978-981-95-1009-2_6

Section 6.2 discusses how clustered federated learning (CFL) is obtained from GTVMin over FL networks with a cluster structure. CFL exploits the presence of clusters (subsets of local datasets) which can be approximated using an i.i.d. assumption. GTVMin captures these clusters if they are well-connected by many (large weight) edges of the FL network.

Section 6.3 discusses horizontal FL (HFL) which is obtained from GTVMin over an FL network whose nodes carry different subsets of a single underlying global dataset. Loosely speaking, HFL involves local datasets characterized by the same set of features but obtained from different data points from an underlying dataset.

Section 6.4 discusses vertical federated learning (VFL) which is obtained from GTVMin over an FL network whose nodes share the same data points but hold different features. As an example, consider the local datasets at different public institutions including tax authorities social insurance institutes or public healthcare providers. These organizations hold different information about the same underlying population such as those with a Finnish social security number.

Section 6.5 shows how personalized FL can be obtained from GTVMin by using specific measures for the GTV of local model parameters. For example, using deep ANNs as local models, we might only use the model parameters corresponding to the first few input layers to define the GTV.

6.1 Single-Model FL

Some FL use cases require to train a single (global) model $\mathcal{H}$ from a decentralized collection of local datasets $\mathcal{D}^{(i)}, i = 1, \ldots, n$ [13, 73]. In what follows we assume that the model $\mathcal{H}$ is parametrized by a vector $\mathbf{w} \in \mathbb{R}^d$. Figure 6.1 depicts a server-client architecture for an iterative FL algorithm that generates a sequence of (global) model parameters $\mathbf{w}^{(k)}, k = 1, \ldots$.

After computing the new model parameters $\mathbf{w}^{(k+1)}$, the server broadcasts it to the devices $i = 1, \ldots, i$ and increments the clock $k := k + 1$. In the next iteration, each device i uses the current global model parameters $\mathbf{w}^{(k)}$ to compute a local update $\mathbf{w}^{(i,k)}$ based on its local dataset $\mathcal{D}^{(i)}$. The precise implementation of this local update step depends on the choice of the global model $\mathcal{H}$ (trained by the server). One example of such a local update has been discussed in Chap. 5 (see (5.18)).

Chapter 5 already hinted at an alternative to the server-based system in Fig. 6.1. Indeed, we might learn local model parameters $\mathbf{w}^{(i)}$ for each client i using a distributed optimization of GTVMin. We can force the resulting model parameters $\mathbf{w}^{(i)}$ to be (approximately) identical by using a connected FL network and a sufficiently large GTVMin parameter α.

To minimize the computational complexity of the resulting single-model FL system, we prefer FL networks with a small number of edges such as the star graph in Fig. 5.3 [64]. However, to increase the robustness against node/link failures, we should use an FL network with more edges. This redundancy helps to ensure that the FL network is connected even after removing some of its edges [74].

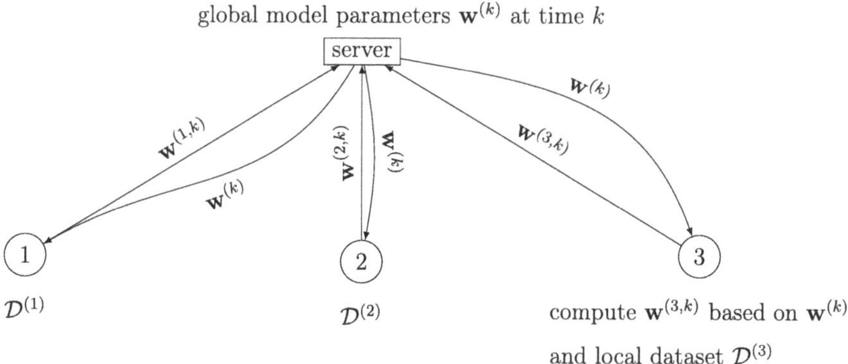

Fig. 6.1 The operation of a server-based (or centralized) FL system during iteration k. First, the server broadcasts the current global model parameters $\mathbf{w}^{(k)}$ to each client $i \in \mathcal{V}$. Each device i then computes the update $\mathbf{w}^{(i,k)}$ by combining the previous model parameters $\mathbf{w}^{(k)}$ (received from the server) and its local dataset $\mathcal{D}^{(i)}$. The updates $\mathbf{w}^{(i,k)}$ are then sent back to the server who aggregates them to obtain the updated global model parameters $\mathbf{w}^{(k+1)}$

Much like the server-based system from Fig. 6.1, GTVMin-based methods using a star graph offers a single point of failure which is the server in Fig. 6.1 or the center node in Fig. 5.3. Chapter 8 will discuss the robustness of GTVMin-based FL systems in slightly more detail.

6.2 Clustered FL

Single-model FL systems require the local datasets to be well approximated as i.i.d. realizations from a common underlying probability distribution. However, requiring homogeneous local datasets, generated from the same probability distribution, might be overly restrictive. Indeed, the local datasets might be heterogeneous and need to be modeled using different probability distribution [16, 34].

CFL relaxes the requirement of a common probability distribution underlying all local datasets. Instead, we approximate subsets of local datasets as i.i.d. realizations from a common probability distribution. In other words, CFL assumes that local datasets form clusters. Each cluster $\mathcal{C} \subseteq \mathcal{V}$ has a cluster-specific probability distribution $p^{(\mathcal{C})}$.

The idea of CFL is to pool the local datasets $\mathcal{D}^{(i)}$ in the same cluster $\mathcal{C}$ to obtain a training set to learn cluster-specific $\widehat{\mathbf{w}}^{(\mathcal{C})}$. Each node $i \in \mathcal{C}$ then uses these learned model parameters $\widehat{\mathbf{w}}^{(\mathcal{C})}$. A main challenge in CFL is that the cluster assignments of the local datasets are unknown in general.

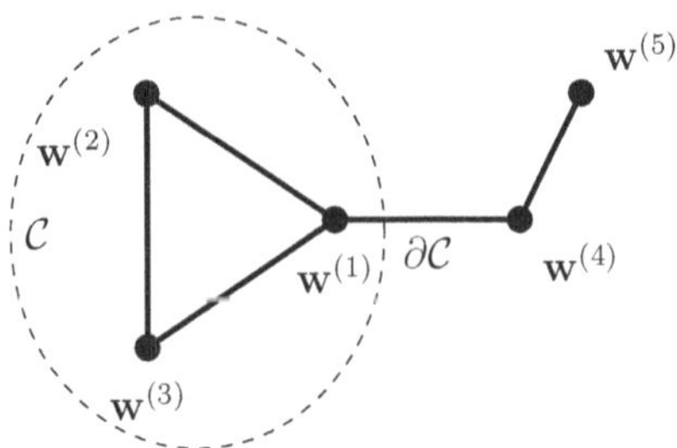

Fig. 6.2 The solution of GTVMin (3.22) are local model parameters that are approximately identical for all nodes in a tight-knit cluster $\mathcal{C}$

To determine a cluster $\mathcal{C}$, we could apply basic clustering methods, such as k-Means or Gaussian mixture model (GMM) to vector representations for local datasets [23, Ch. 5]. We can obtain a vector representation for local dataset $\mathcal{D}^{(i)}$ via the learned model parameters $\widehat{\mathbf{w}}$ of some parametric ML model that is trained on $\mathcal{D}^{(i)}$.

We can also implement CFL via GTVMin with a suitably chosen FL network. In particular, the FL network should contain many edges (with large weight) between nodes in the same cluster and few edges (with a small weight) between nodes in different clusters. To fix ideas, consider the FL network in Fig. 6.2, which contains a cluster $\mathcal{C} = \{1, 2, 3\}$.

Chapter 3 discussed how the eigenvalues of the Laplacian matrix can be used to measure the connectivity of $\mathcal{G}$. Similarly, we can measure the connectivity of a cluster $\mathcal{C}$ via the eigenvalue $\lambda_2\big(\mathbf{L}^{(\mathcal{C})}\big)$ of the Laplacian matrix $\mathbf{L}^{(\mathcal{C})}$ of the induced sub-graph $\mathcal{G}^{(\mathcal{C})}$:[1]

The larger $\lambda_2\big(\mathbf{L}^{(\mathcal{C})}\big)$, the better the connectivity among the nodes in $\mathcal{C}$. While $\lambda_2\big(\mathbf{L}^{(\mathcal{C})}\big)$ describes the intrinsic connectivity of a cluster $\mathcal{C}$, we also need to characterize its connectivity with the other nodes in the FL network. To this end, we use the cluster boundary:

$$|\partial\mathcal{C}| := \sum_{\{i,i'\}\in\partial\mathcal{C}} A_{i,i'} \text{ with } \partial\mathcal{C} := \big\{\{i,i'\} \in \mathcal{E} : i \in \mathcal{C}, i' \notin \mathcal{C}\big\}.$$

Note that for a single-node cluster $\mathcal{C} = \{i\}$, the cluster boundary coincides with the node degree, $|\partial\mathcal{C}| = d^{(i)}$ (see (3.7)).

Intuitively, GTVMin tends to deliver (approximately) identical model parameters $\mathbf{w}^{(i)}$ for nodes $i \in \mathcal{C}$ if $\lambda_2\big(\mathbf{L}^{(\mathcal{C})}\big)$ is large and the cluster boundary $|\partial\mathcal{C}|$ is small. The following result makes this intuition more precise for the special case of GTVMin (5.1) for local linear models.

[1] The graph $\mathcal{G}^{(\mathcal{C})}$ consists of the nodes in $\mathcal{C}$ and the edges $\{i, i'\} \in \mathcal{E}$ for $i, i' \in \mathcal{C}$.

Proposition 6.1 *Consider an FL network $\mathcal{G}$ which contains a cluster $\mathcal{C}$ of local datasets with labels $\mathbf{y}^{(i)}$ and feature matrix $\mathbf{X}^{(i)}$ related via*

$$\mathbf{y}^{(i)} = \mathbf{X}^{(i)} \overline{\mathbf{w}}^{(\mathcal{C})} + \boldsymbol{\varepsilon}^{(i)}, \text{ for all } i \in \mathcal{C}. \tag{6.1}$$

We learn local model parameters $\widehat{\mathbf{w}}^{(i)}$ via solving GTVMin (5.1). If the cluster is connected, the error component

$$\widetilde{\mathbf{w}}^{(i)} := \widehat{\mathbf{w}}^{(i)} - (1/|\mathcal{C}|) \sum_{i \in \mathcal{C}} \widehat{\mathbf{w}}^{(i)} \tag{6.2}$$

is upper bounded as

$$\sum_{i \in \mathcal{C}} \left\| \widetilde{\mathbf{w}}^{(i)} \right\|_2^2 \leq \frac{1}{\alpha \lambda_2\big(\mathbf{L}^{(\mathcal{C})}\big)} \left[\sum_{i \in \mathcal{C}} \frac{1}{m_i} \left\| \boldsymbol{\varepsilon}^{(i)} \right\|_2^2 + \alpha \, |\partial \mathcal{C}| \, 2 \Big(\left\| \overline{\mathbf{w}}^{(\mathcal{C})} \right\|_2^2 + R^2 \Big) \right]. \tag{6.3}$$

Here, we used $R := \max_{i' \in \mathcal{V} \setminus \mathcal{C}} \left\| \widehat{\mathbf{w}}^{(i')} \right\|_2$.

Proof See Sect. 6.8.1. □

The bound (6.3) depends on the cluster $\mathcal{C}$ (via the eigenvalue $\lambda_2\big(\mathbf{L}^{(\mathcal{C})}\big)$ and the boundary $|\partial \mathcal{C}|$) and the GTVMin parameter α. Using a larger $\mathcal{C}$ typically results in a decreased eigenvalue $\lambda_2\big(\mathbf{L}^{(\mathcal{C})}\big)$.[2] According to (6.3), we should then increase α to maintain a small deviation $\widetilde{\mathbf{w}}^{(i)}$ of the learned local model parameters from their cluster-wise average. Thus, increasing α in (3.22) enforces its solutions to be approximately constant over increasingly larger subsets (clusters) of nodes (see Fig. 6.3).

For a connected FL network $\mathcal{G}$ and a sufficiently large α, the solution of GTVMin consists of learned model parameters $\mathbf{w}^{(i)}$ that are approximately identical for all

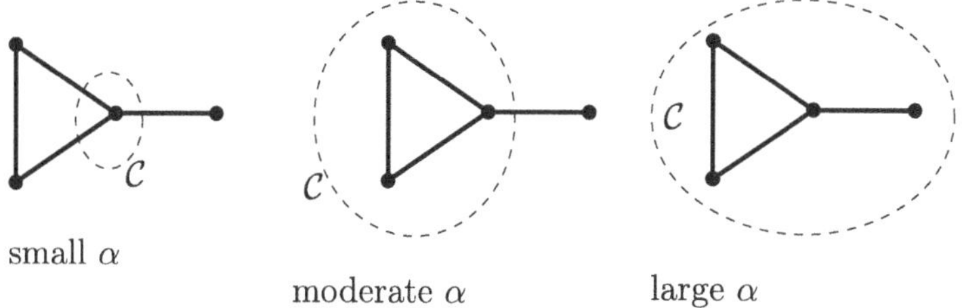

Fig. 6.3 As the regularization parameter α increases, the solutions of the GTVMin (3.22) become approximately constant over larger subsets of nodes, i.e., they exhibit stronger clustering

[2] Consider an FL network (with uniform edge weights) that contains a fully connected cluster $\mathcal{C}$ which is connected via a single edge with another node $i' \in \mathcal{V} \setminus \mathcal{C}$ (see Fig. 6.2). Compare the corresponding eigenvalues $\lambda_2\big(\mathbf{L}^{(\mathcal{C})}\big)$ and $\lambda_2\big(\mathbf{L}^{(\mathcal{C}')}\big)$ of $\mathcal{C}$ and the enlarged cluster $\mathcal{C}' := \mathcal{C} \cup \{i'\}$.

$\mathcal{V} = 1, \ldots, n$. The resulting approximation error is quantified by Proposition 6.1 for the extreme case where the entire FL network forms a single cluster, i.e., $\mathcal{C} = \mathcal{V}$. Trivially, the cluster boundary is then equal to 0 and the bound (6.3) specializes to (3.36).

We hasten to add that the bound (6.3) only applies for local datasets that conform with the probabilistic model (6.1). In particular, it assumes that all cluster nodes $i \in \mathcal{C}$ have identical model parameters $\overline{\mathbf{w}}^{(\mathcal{C})}$. Trivially, this is no restriction if we allow for arbitrary error terms $\boldsymbol{\varepsilon}^{(i)}$ in the probabilistic model (6.3). However, as soon as we place additional assumptions on these error terms (such as being realizations of i.i.d. Gaussian RVs), we should verify their validity using principled statistical tests [32, 75]. Finally, we might replace $\left\|\overline{\mathbf{w}}^{(\mathcal{C})}\right\|_2^2$ in (6.3) with an upper bound for this quantity.

6.3 Horizontal FL

HFL uses local datasets $\mathcal{D}^{(i)}$, for $i \in \mathcal{V}$, that contain data points characterized by the same features [76]. As illustrated in Fig. 6.4, we can think of each local dataset $\mathcal{D}^{(i)}$ as being a subset (or batch) of an underlying global dataset:

$$\mathcal{D}^{(\text{global})} := \left\{ \left(\mathbf{x}^{(1)}, y^{(1)} \right), \ldots, \left(\mathbf{x}^{(m)}, y^{(m)} \right) \right\}.$$

In particular, local dataset $\mathcal{D}^{(i)}$ is constituted by the data points of $\mathcal{D}^{(\text{global})}$ with indices in $\{r_1, \ldots, r_{m_i}\}$:

$$\mathcal{D}^{(i)} := \left\{ \left(\mathbf{x}^{(r_1)}, y^{(r_1)} \right), \ldots, \left(\mathbf{x}^{(r_{m_i})}, y^{(r_{m_i})} \right) \right\}.$$

We can interpret HFL as a generalization of semi-supervised learning (SSL) [77]: For some local datasets $i \in \mathcal{U}$, we might not have access to the label values of data points. Still, we can use the features of the data points to construct (the weighted edges of) the FL network. To implement SSL, we can solve GTVMin using a trivial

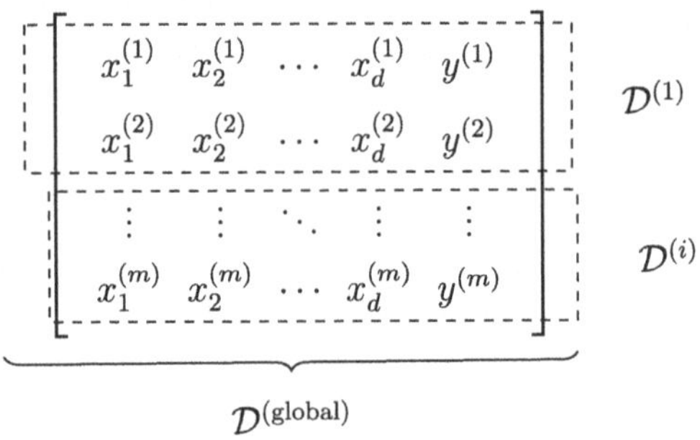

$$\begin{bmatrix} x_1^{(1)} & x_2^{(1)} & \cdots & x_d^{(1)} & y^{(1)} \\ x_1^{(2)} & x_2^{(2)} & \cdots & x_d^{(2)} & y^{(2)} \\ \vdots & \vdots & \ddots & \vdots & \vdots \\ x_1^{(m)} & x_2^{(m)} & \cdots & x_d^{(m)} & y^{(m)} \end{bmatrix}$$

$\mathcal{D}^{(1)}$ $\mathcal{D}^{(i)}$ $\mathcal{D}^{(\text{global})}$

Fig. 6.4 HFL uses the same features to characterize data points in different local datasets. Different local datasets are constituted by different subsets of data points out of an underlying global dataset

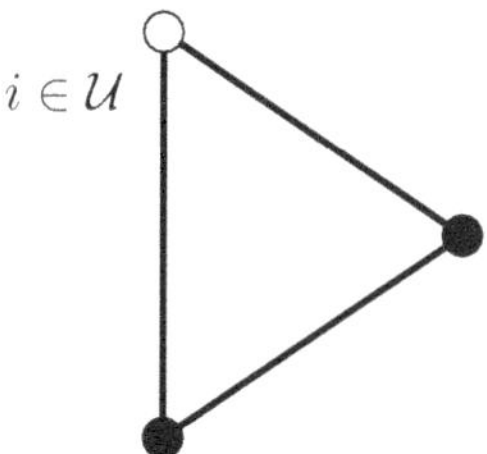

Fig. 6.5 HFL includes SSL as a special case. SSL involves a subset of nodes $\mathcal{U}$, for which the local datasets do not contain labels. We can take this into account by using the trivial loss function $L_i(\cdot) = 0$ for each node $i \in \mathcal{U}$. However, we can still use the features in $\mathcal{D}^{(i)}$ to construct an FL network $\mathcal{G}$

loss function $L_i\left(\mathbf{w}^{(i)}\right) = 0$ for each unlabeled node $i \in \mathcal{U}$. Solving GTVMin delivers model parameters $\mathbf{w}^{(i)}$ for all nodes i (including the unlabeled ones $\mathcal{U}$). GTVMin-based methods combine the information in the labeled local datasets $\mathcal{D}^{(i)}$, for $i \in \mathcal{V} \setminus \mathcal{U}$ and their connections (via the edges of $\mathcal{G}$) with nodes in $\mathcal{U}$ (see Fig. 6.5).

6.4 Vertical FL

VFL uses local datasets that are constituted by the same (identical) data points. However, each local dataset uses a different choice of features to characterize these data points [78]. Formally, VFL applications revolve around an underlying global dataset:

$$\mathcal{D}^{(\text{global})} := \left\{\left(\mathbf{x}^{(1)}, y^{(1)}\right), \ldots, \left(\mathbf{x}^{(m)}, y^{(m)}\right)\right\}.$$

Each data point in the global dataset is characterized by d' features $\mathbf{x}^{(r)} = \left(x_1^{(r)}, \ldots, x_{d'}^{(r)}\right)^T$. The global dataset can only be accessed indirectly via local datasets that use different subsets of the feature vectors $\mathbf{x}^{(r)}$ (see Fig. 6.6).

For example, the local dataset $\mathcal{D}^{(i)}$ consists of feature vectors

$$\mathbf{x}^{(i,r)} = \left(x_{j_1}^{(r)}, \ldots, x_{j_d}^{(r)}\right)^T.$$

Here, we used a subset $\mathcal{F}^{(i)} := \{j_1, \ldots, j_d\}$ of the original d' features in $\mathbf{x}^{(r)}$. There is typically one node i' with a a local datasets that contains the label values $y^{(1)}, \ldots, y^{(m)}$.

A potential toy application for vertical FL is a national social insurance system. The global dataset comprises data points representing individuals enrolled in the system. Each individual is characterized by multiple sets of features sourced from

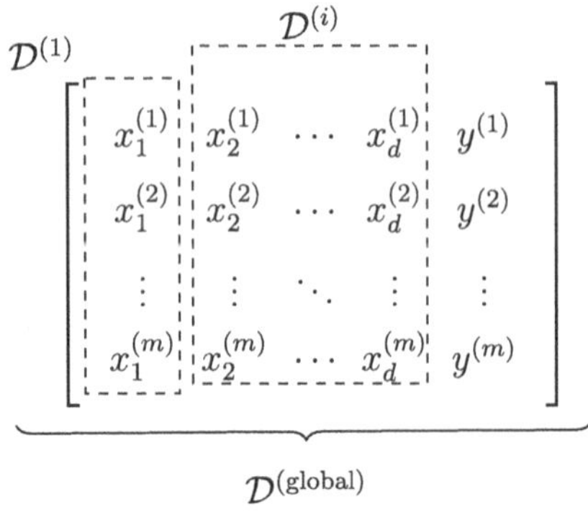

Fig. 6.6 VFL uses local datasets that are derived from the same data points. The local datasets differ in the choice of features used to characterize the common data points

different institutions. Healthcare providers contribute medical records, offering health-related features. Financial service providers, such as banks, supply financial features. Some individuals participate in retailer loyalty programs, which generate consumer behavior features. Additionally, social network accounts can provide real-time data on user activities and mobility patterns, further enriching the available features. Since these diverse data sources belong to separate entities, VFL enables collaborative learning while preserving data privacy.

6.5 Personalized Federated Learning

Consider GTVMin (3.22) for learning local model parameters $\widehat{\mathbf{w}}^{(i)}$ for each local dataset $\mathcal{D}^{(i)}$. If the value of α in (3.22) is not too large, the local model parameters $\widehat{\mathbf{w}}^{(i)}$ can be different for each $i \in \mathcal{V}$. However, the local model parameters are still coupled via the GTV term in (3.22).

For some FL use cases, we should use different coupling strengths for different components of the local model parameters. For example, if local models are deep ANNs, we might enforce the parameters of input layers to be identical, while the parameters of the deeper layers might be different for each local dataset.

The partial parameter sharing for local models can be implemented in many different ways [79, Sec. 4.3.]:

- One way is to use a choice of the GTV penalty that is different from $\phi = \left\|\mathbf{w}^{(i)} - \mathbf{w}^{(i')}\right\|_2^2$. In particular, we could construct the penalty function as a combination of two terms:

$$\phi\left(\mathbf{w}^{(i)} - \mathbf{w}^{(i')}\right) := \alpha^{(1)}\phi^{(1)}\left(\mathbf{w}^{(i)} - \mathbf{w}^{(i')}\right) + \alpha^{(2)}\phi^{(2)}\left(\mathbf{w}^{(i)} - \mathbf{w}^{(i')}\right). \tag{6.4}$$

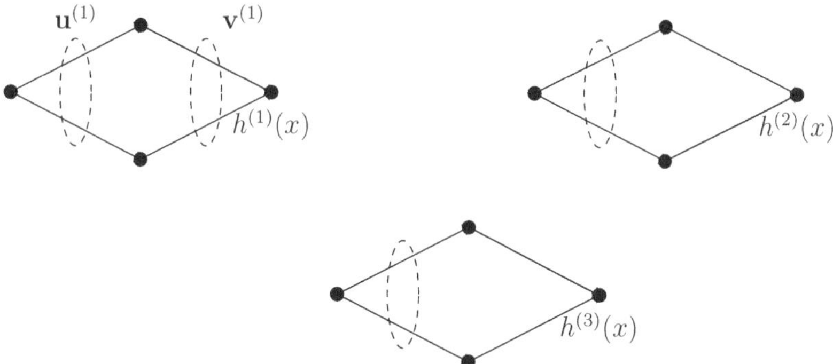

Fig. 6.7 Personalized FL with local models being ANNs with a single hidden layer. The ANN $h^{(i)}$ is parametrized by the vector $\mathbf{w}^{(i)} = \left(\left(\mathbf{u}^{(i)}\right)^T, \left(\mathbf{v}^{(i)}\right)^T \right)^T$, with parameters $\mathbf{u}^{(i)}$ of the hidden layer and the parameters $\mathbf{v}^{(i)}$ of the output layer. We couple the training of $\mathbf{u}^{(i)}$ via GTVMin using the discrepancy measure $\phi = \left\| \mathbf{u}^{(i)} - \mathbf{u}^{(i')} \right\|_2^2$

The functions $\phi^{(1)}$ and $\phi^{(2)}$ measure different components of the variation $\mathbf{w}^{(i)} - \mathbf{w}^{(i')}$ across the edge $\{i, i'\} \in \mathcal{E}$. For example, we might construct $\phi^{(1)}$ and $\phi^{(2)}$ by (3.37) with different choices for the dataset $\mathcal{D}^{\{i,i'\}}$.

- Moreover, we might use different regularization strengths $\alpha^{(1)}$ and $\alpha^{(2)}$ for different penalty components in (6.4) to enforce different subsets of the model parameters to be clustered with different granularity, i.e., enforcing some of the model parameters to be constant across larger subsets of nodes.
- For local models being deep ANNs, we enforce identical model parameters for the layers closer to the input. In contrast, we allow the layers that closer to the output to have different model parameters at different devices. Figure 6.7 illustrates this idea for local models constituted by ANNs with a single hidden layer.
- Yet another technique for partial sharing of model parameters is to train a hyper-model which, in turn, is used to initialize the training of local models [80].

6.6 Few-Shot Learning

Some ML applications involve data points belonging to a large number of different categories. A prime example is the detection of a specific object in a given image [81, 82]. Here, the object category is the label $y \in \mathcal{Y}$ of a data point (image). The label space $\mathcal{Y}$ is constituted by the possible object categories and, in turn, can be quite large. Moreover, for some categories, we might only have a few example images in the training set.

Few-shot learning exploits structural similarities between object categories to accurately detect objects with limited (or even no) training examples. A principled approach to few-shot learning is GTVMin, which leverages relational information between categories. To formalize this approach, we define an FL network $\mathcal{G} = (\mathcal{V}, \mathcal{E}, \mathbf{A})$, where each node $i \in \mathcal{V}$ corresponds to an element of the label space $\mathcal{Y}$. The edge weights $\mathbf{A}$ encode prior knowledge about category relationships, providing a structured way to propagate information between object categories.

Each node i in $\mathcal{G}$ represents a distinct object category and corresponding object detector. Solving GTVMin yields model parameters $\widehat{\mathbf{w}}^{(i)}$ for each of these specialized object detectors. The coupling of these tailored object detectors via GTVMin enables knowledge transfer across categories, improving detection performance even in low-data regimes.

6.7 Exercises

6.1 Horizontal FL of a Linear Model [68, Sec. 8.2]**.** Linear regression learns the model parameters of a linear model by minimizing the average squared error loss on a given dataset $\mathcal{D}$. Consider an application where the data points are gathered by different devices. We can model such an application using an FL network with nodes i carrying different subsets of $\mathcal{D}$. Construct an instance of GTVMin such that its solutions coincide (approximately) with the solution of plain vanilla linear regression.

6.2 Vertical FL of a Linear Model [68, Sec. 8.3]. Linear regression learns the model parameters of a linear model by minimizing the average squared error loss on a given dataset $\mathcal{D}$. Consider an application where the features of a data point are measured by different devices. We can model such an application using an FL network with nodes i carrying different features of the same dataset $\mathcal{D}$. In particular, node i carries the features x_j with $j \in \mathcal{F}^{(i)}$. Construct an instance of GTVMin such that its solutions coincide (approximately) with the solution of plain vanilla linear regression.

6.8 Proofs

6.8.1 *Proof of Proposition 6.1*

To verify (6.3), we follow a similar argument as used in the proof of Proposition 3.1.

First, we decompose the objective function $f(\mathbf{w})$ in (5.1) as follows:

$$f(\mathbf{w}) = \underbrace{\sum_{i\in\mathcal{C}}(1/m_i)\left\|\mathbf{y}^{(i)}-\mathbf{X}^{(i)}\mathbf{w}^{(i)}\right\|_2^2+\alpha\left[\sum_{i,i'\in\mathcal{C}}A_{i,i'}\left\|\mathbf{w}^{(i)}-\mathbf{w}^{(i')}\right\|_2^2+\sum_{\{i,i'\}\in\partial\mathcal{C}}A_{i,i'}\left\|\mathbf{w}^{(i)}-\mathbf{w}^{(i')}\right\|_2^2\right]}_{=:f'(\mathbf{w})} + f''(\mathbf{w}). \tag{6.5}$$

Note that only the first component f' depends on the local model parameters $\mathbf{w}^{(i)}$ of cluster nodes $i \in \mathcal{C}$. Let us introduce the shorthand $f'\left(\mathbf{w}^{(i)}\right)$ for the function obtained from $f'(\mathbf{w})$ for varying $\mathbf{w}^{(i)}$, $i \in \mathcal{C}$, but fixing $\mathbf{w}^{(i')} := \widehat{\mathbf{w}}^{(i')}$ for $i' \notin \mathcal{C}$.

We obtain the bound (6.3) via a proof by contradiction: If (6.3) does not hold, the local model parameters $\overline{\mathbf{w}}^{(i)} := \overline{\mathbf{w}}^{(\mathcal{C})}$, for $i \in \mathcal{C}$, result in a smaller value $f'\left(\overline{\mathbf{w}}^{(i)}\right) < f'\left(\widehat{\mathbf{w}}^{(i)}\right)$ than the choice $\widehat{\mathbf{w}}^{(i)}$, for $i \in \mathcal{C}$. This would contradict the fact that $\widehat{\mathbf{w}}^{(i)}$ is a solution to (5.1).

First, note that

$$\begin{aligned}
f'\left(\overline{\mathbf{w}}^{(i)}\right) &= \sum_{i\in\mathcal{C}}(1/m_i)\left\|\mathbf{y}^{(i)}-\mathbf{X}^{(i)}\overline{\mathbf{w}}^{(\mathcal{C})}\right\|_2^2 \\
&\quad +\alpha\left[\sum_{\substack{\{i,i'\}\in\mathcal{E}\\ i,i'\in\mathcal{C}}}A_{i,i'}\left\|\overline{\mathbf{w}}^{(\mathcal{C})}-\overline{\mathbf{w}}^{(\mathcal{C})}\right\|_2^2+\sum_{\substack{\{i,i'\}\in\mathcal{E}\\ i\in\mathcal{C},i'\notin\mathcal{C}}}A_{i,i'}\left\|\overline{\mathbf{w}}^{(\mathcal{C})}-\widehat{\mathbf{w}}^{(i')}\right\|_2^2\right] \\
&\overset{(6.1)}{=} \sum_{i\in\mathcal{C}}(1/m_i)\left\|\boldsymbol{\varepsilon}^{(i)}\right\|_2^2+\alpha\sum_{\substack{\{i,i'\}\in\mathcal{E}\\ i\in\mathcal{C},i'\notin\mathcal{C}}}A_{i,i'}\left\|\overline{\mathbf{w}}^{(\mathcal{C})}-\widehat{\mathbf{w}}^{(i')}\right\|_2^2 \\
&\overset{(a)}{\leq} \sum_{i\in\mathcal{C}}(1/m_i)\left\|\boldsymbol{\varepsilon}^{(i)}\right\|_2^2+\alpha\sum_{\substack{\{i,i'\}\in\mathcal{E}\\ i\in\mathcal{C},i'\notin\mathcal{C}}}2A_{i,i'}\left(\left\|\overline{\mathbf{w}}^{(\mathcal{C})}\right\|_2^2+\left\|\widehat{\mathbf{w}}^{(i')}\right\|_2^2\right) \\
&\leq \sum_{i\in\mathcal{C}}(1/m_i)\left\|\boldsymbol{\varepsilon}^{(i)}\right\|_2^2+\alpha\,|\partial\mathcal{C}|\,2\left(\left\|\overline{\mathbf{w}}^{(\mathcal{C})}\right\|_2^2+R^2\right).
\end{aligned} \tag{6.6}$$

Step (a) uses the inequality $\|\mathbf{u}+\mathbf{v}\|_2^2 \leq 2\left(\|\mathbf{u}\|_2^2+\|\mathbf{v}\|_2^2\right)$ which is valid for any two vectors $\mathbf{u}, \mathbf{v} \in \mathbb{R}^d$.

On the other hand,

$$f'\left(\widehat{\mathbf{w}}^{(i)}\right) \geq \alpha \sum_{i,i' \in \mathcal{C}} A_{i,i'} \underbrace{\left\| \widehat{\mathbf{w}}^{(i)} - \widehat{\mathbf{w}}^{(i')} \right\|_2^2}_{\stackrel{(6.2)}{=} \left\| \widetilde{\mathbf{w}}^{(i)} - \widetilde{\mathbf{w}}^{(i')} \right\|_2^2}$$

$$\stackrel{(3.17)}{\geq} \alpha \lambda_2\left(\mathbf{L}^{(\mathcal{C})}\right) \sum_{i \in \mathcal{C}} \left\| \widetilde{\mathbf{w}}^{(i)} \right\|_2^2. \tag{6.7}$$

If the bound (6.3) would not hold, then by (6.7) and (6.6), we would obtain $f'\left(\widehat{\mathbf{w}}^{(i)}\right) > f'\left(\overline{\mathbf{w}}^{(i)}\right)$, which contradicts the fact that $\widehat{\mathbf{w}}^{(i)}$ solves (5.1).

Chapter 7
Graph Learning for FL Networks

Abstract This chapter focuses on how to construct the communication network that underlies a federated learning (FL) system. It shows how the structure of the FL network affects both the statistical accuracy and computational efficiency of FL algorithms. The chapter begins by analyzing how network properties—such as node degrees and eigenvalues of the Laplacian matrix—influence convergence speed and per-iteration cost. It then discusses how to measure similarities and differences between local datasets, using tools like probabilistic modeling, gradient analysis, and neural embeddings. These measures form the basis for graph learning methods, where the FL network is inferred directly from data. The final section formulates graph construction as an optimization problem that selects edge weights to minimize dataset discrepancy while meeting connectivity and complexity constraints. Together, these sections provide a principled framework for designing or learning FL networks that balance accuracy, efficiency, and communication overhead.

Chapter 3 introduced GTVMin as a flexible design principle for FL algorithms. Chapter 5 explores how algorithms can be obtained by applying optimization methods—such as the gradient-based methods from Chap. 4—to solve GTVMin instances.

The computational and statistical properties of such algorithms depend crucially on the structure of the underlying FL network. For example, both the computational and communication costs of FL systems typically increase with the number of edges in the FL network. Moreover, the graph topology governs how local datasets are pooled into clusters with shared model parameters.

In some settings, domain expertise can guide the construction of the FL network. For instance, in healthcare, known clinical similarities between disease types are used to define edges connecting patients or diseases [83]. In sensor networks, physical proximity and hardware connectivity constraints naturally shape the graph structure [84, 85]. However, other applications lack strong prior structure and require to learn the graph from data [86–89]. This chapter presents techniques to infer FL networks from local datasets and associated local loss functions.

This chapter is organized as follows. Section 7.1 discusses how the analysis of FL algorithms can inform the design of the FL network. Section 7.2 presents methods to

A. Jung, *Federated Learning*, https://doi.org/10.1007/978-981-95-1009-2_7

quantify discrepancies between local datasets. Section 7.3 formulates graph learning as an optimization problem that minimizes the discrepancy between datasets stored at nodes that are connected by an edge. The structure of the resulting graph can be influenced by imposing connectivity constraints, such as a minimum required node degree.

7.1 Edges as Design Choice

Consider the GTVMin instance (3.24), which aims to learn local model parameters for each linear model associated with a local dataset $\mathcal{D}^{(i)}$. To solve (3.24), we use Algorithm 4, which implements the gradient step (5.3) in a message-passing fashion.

The GTVMin formulation (3.24) is defined for a fixed FL network $\mathcal{G}$. Hence, the structure of $\mathcal{G}$ significantly impacts both the statistical and computational properties of Algorithm 4.

Statistical Properties These can be assessed using a probabilistic model for the local datasets. An important example is the clustering assumption (6.1), discussed in the context of CFL in Sect. 3.3.1. Under the CFL assumption, nodes in the same cluster should learn similar model parameters.

According to Proposition 6.1, the solution to GTVMin will be approximately constant across a cluster $\mathcal{C}$ if the second smallest eigenvalue $\lambda_2\big(\mathbf{L}^{(\mathcal{C})}\big)$ is large and the cluster boundary $|\partial\mathcal{C}|$ is small. Here, $\lambda_2\big(\mathbf{L}^{(\mathcal{C})}\big)$ refers to the smallest nonzero eigenvalue of the Laplacian matrix of the induced subgraph $\mathcal{G}^{(\mathcal{C})}$.

Intuitively, $\lambda_2\big(\mathbf{L}^{(\mathcal{C})}\big)$ increases with the number of internal edges in $\mathcal{C}$. This can be made precise via Cheeger's inequality [36, Ch. 21]. Alternatively, we can approximate $\mathcal{G}^{(\mathcal{C})}$ as a realization of an Erdős-Rényi (ER) graph, a useful assumption especially if $\mathcal{G}$ itself resembles a typical ER graph instance.

In an ER graph over $\mathcal{C}$, each pair of nodes $i, i' \in \mathcal{C}$ is connected independently with probability p_e. The presence of each edge is governed by realizations $b^{(i,i')}$ of i.i.d. RVs, one for each pair of different nodes $i, i' \in \mathcal{V}$. As a result, edge occurrences between different pairs of nodes are mutually independent.

This independence greatly simplifies analysis. For instance, the Laplacian matrix $\mathbf{L}^{(ER)}$ of an ER graph can be expressed as a sum of independent random matrices:

$$\mathbf{L}^{(\mathrm{ER})} = \sum_{\{i,i'\}} b^{(i,i')}\mathbf{T}^{(i,i')}.$$

This decomposition involves, for each pair of different nodes $i, i' \in \mathcal{V}$, the deterministic matrix $\mathbf{T}^{(i,i')}$. The decomposition is useful for the analysis of the eigenvalues of $\mathbf{L}^{(\mathrm{ER})}$, e.g., via matrix concentration inequalities [90, 91].

Interpreting a graph $\mathcal{G}$ as (the realization of) an ER graph turns quantities such as node degrees $d^{(i)}$ and eigenvalues like $\lambda_2(\mathbf{L}^{(\mathcal{C})})$ into (realizations of) RVs. The expected node degree is

$$\mathbb{E}\{d^{(i)}\} = p_e(|\mathcal{C}| - 1).$$

With high probability,

$$d_{\max}^{(\mathcal{G})} \approx p_e(|\mathcal{C}| - 1). \tag{7.1}$$

Increasing p_e results in a larger expected node degree and, thus, a higher connectivity of $\mathcal{G}^{(\mathcal{C})}$.

We can approximate $\lambda_2(\mathbf{L}^{(\mathcal{C})})$ by the second smallest eigenvalue of the expected Laplacian matrix:

$$\overline{\mathbf{L}} := \mathbb{E}\{\mathbf{L}^{(\mathcal{C})}\} = |\mathcal{C}|p_e\mathbf{I} - p_e\mathbf{1}\mathbf{1}^T.$$

A straightforward calculation yields

$$\lambda_2(\overline{\mathbf{L}}) = |\mathcal{C}|p_e.$$

Thus, we arrive at the approximation

$$\lambda_2(\mathbf{L}^{(\mathcal{C})}) \approx \lambda_2(\overline{\mathbf{L}}) = |\mathcal{C}|p_e \overset{(7.1)}{\approx} d_{\max}^{(\mathcal{G})}. \tag{7.2}$$

The precise quantification of the approximation error in (7.2) is beyond our scope. We refer interested readers to [90, 92] for further analysis of random graphs.

Computational Properties The computational complexity of Algorithm 4 depends on the amount of computation required by a single iteration of its steps (3) and (4). Clearly, the *per-iteration* complexity of Algorithm 4 increases with increasing node degrees $d^{(i)}$. Indeed, step (3) requires to communicate local model parameters across each edge of the FL network. This communication can be implemented using different physical channels, such as short-range wireless links or optical fiber connections [93, 94].

To summarize, using an FL network with smaller $d^{(i)}$ results in less computation and communication per iteration of Algorithm 4. Trivially, the lowest per-iteration cost occurs when $d^{(i)} = 0$, i.e., an empty FL network with $\mathcal{E} = \emptyset$. However, the overall computational cost also depends on the number of iterations required to approximate the GTVMin solution (3.24).

According to (4.5), the convergence speed of the gradient steps (5.9) used in Algorithm 4 depends on the condition number of the matrix $\mathbf{Q}$ in (5.2):

$$\text{condition number} = \frac{\lambda_{nd}(\mathbf{Q})}{\lambda_1(\mathbf{Q})}.$$

Faster convergence is achieved when this ratio is close to one (see (4.9)).

The condition number of $\mathbf{Q}$ tends to be smaller when the ratio between the maximum node degree $d_{\max}^{(\mathcal{G})}$ and the second smallest eigenvalue $\lambda_2(\mathbf{L}^{(\mathcal{G})})$ is small (see (5.5) and (5.6)).

Thus, for a given maximum node degree $d_{\max}^{(\mathcal{G})}$, we should place the edges of an FL network so that $\lambda_2(\mathbf{L}^{(\mathcal{G})})$ is large—leading to faster convergence of Algorithm 4 without increasing per-iteration complexity.

Spectral graph theory also provides upper bounds on $\lambda_2(\mathbf{L}^{(\mathcal{G})})$ in terms of the node degrees [36, 37, 95]. These upper bounds can serve as a baseline for evaluating practical constructions of the FL network: If the resulting value $\lambda_2(\mathbf{L}^{(\mathcal{G})})$ is close to its upper bound, then further attempts to improve connectivity (in terms of spectral properties) are unlikely to yield significant gains.

The next result provides an example of such an upper bound.

Proposition 7.1 *Consider an FL network $\mathcal{G}$ with $n > 1$ nodes and associated Laplacian matrix $\mathbf{L}^{(\mathcal{G})}$. Then, $\lambda_2(\mathbf{L}^{(\mathcal{G})})$ cannot exceed the node degree $d^{(i)}$ of any node by more than a factor $n/(n-1)$. In other words,*

$$\lambda_2\big(\mathbf{L}^{(\mathcal{G})}\big) \leq \frac{n}{n-1} d^{(i)}, \quad \text{for every } i = 1, \ldots, n. \tag{7.3}$$

Proof The bound (7.3) follows from the variational characterization (3.14) by evaluating the quadratic form $\mathbf{w}^T \mathbf{L}^{(\mathcal{G})} \mathbf{w}$ for the specific vector

$$\widetilde{\mathbf{w}} = \sqrt{\frac{n}{n-1}} \Big(-\frac{1}{n}, \ldots, \underbrace{1-\frac{1}{n}}_{\tilde{w}^{(i)}}, \ldots, -\frac{1}{n} \Big)^T.$$

This "test" vector is tailored to a particular node $i \in \mathcal{V}$; its only positive entry is $\tilde{w}^{(i)} = 1-(1/n)$. It satisfies $\|\widetilde{\mathbf{w}}\| = 1$ and $\widetilde{\mathbf{w}}^T \mathbf{1} = 0$, making it a feasible vector for the optimization in (3.14). □

Alternative—and potentially tighter—upper bounds for $\lambda_2(\mathbf{L}^{(\mathcal{G})})$ can be found in the graph theory literature [35, 36, 92, 96].

The per-iteration complexity of FL algorithms increases with the node degrees $d^{(i)}$ (and thus the total number of edges) in the FL network $\mathcal{G}$. On the other hand, the number of iterations required by Algorithm 4 typically decreases as the second smallest eigenvalue $\lambda_2(\mathbf{L}^{(\mathcal{G})})$ increases.

According to the upper bound in (7.3), a large value of $\lambda_2(\mathbf{L}^{(\mathcal{G})})$ is only possible if the node degrees $d^{(i)}$—and hence the total number of edges—are sufficiently large. Recent work has focused on constructing graphs that maximize $\lambda_2(\mathbf{L}^{(\mathcal{G})})$ given a fixed maximum node degree $d_{\max}^{(\mathcal{G})} = \max_{i \in \mathcal{V}} d^{(i)}$ [60, 97].

Figure 7.1 illustrates this trade-off between per-iteration complexity and the number of iterations required by FL algorithms.

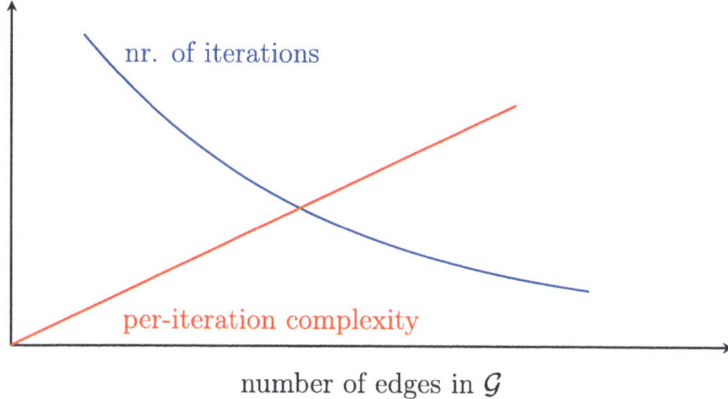

Fig. 7.1 Computational trade-off in GTVMin-based methods such as Algorithm 4: Increasing the number of edges in the FL network $\mathcal{G}$ raises the per-iteration complexity, but typically reduces the total number of iterations required for convergence

7.2 Measuring (Dis-)Similarity Between Datasets

The main idea behind GTVMin is to enforce similar model parameters at two different nodes i and i' that are connected by an edge $\{i, i'\}$ with (relatively) large edge weight $A_{i,i'}$. In general, the edges (and their weights) of the FL network are design choices. Placing an edge between two nodes i, i' is typically only useful if the local datasets $\mathcal{D}^{(i)}, \mathcal{D}^{(i')}$ (generated by devices i, i') have similar statistical properties. We next discuss different approaches for measuring the similarity—or, equivalently, the discrepancy (i.e., the lack of similarity)—between two local datasets.

The first approach is based on a probabilistic model, i.e., we interpret the local dataset $\mathcal{D}^{(i)}$ as realizations of RVs with some parametric probability distribution $p^{(i)}\big(\mathcal{D}^{(i)}; \mathbf{w}^{(i)}\big)$. We can then measure the discrepancy between $\mathcal{D}^{(i)}$ and $\mathcal{D}^{(i')}$ via the Euclidean distance $\left\|\mathbf{w}^{(i)} - \mathbf{w}^{(i')}\right\|_2$ between the parameters $\mathbf{w}^{(i)}$ and $\mathbf{w}^{(i')}$ of the corresponding probability distributions.

In most FL applications, the parameters of the probability distribution $p^{(i)}\big(\mathcal{D}^{(i)}; \mathbf{w}^{(i)}\big)$ underlying a local dataset are unknown.[1] However, it is often possible to estimate these parameters using established statistical techniques such as ML [23, Ch. 3]. Given the estimates $\widehat{\mathbf{w}}^{(i)}$ and $\widehat{\mathbf{w}}^{(i')}$ for the model parameters, we can then compute the discrepancy measure $d^{(i,i')} := \left\|\widehat{\mathbf{w}}^{(i)} - \widehat{\mathbf{w}}^{(i')}\right\|_2$.

[1] One exception is when the local dataset is generated by drawing i.i.d. realizations from $p^{(i)}\big(\mathcal{D}^{(i)}; \mathbf{w}^{(i)}\big)$.

Example Consider local datasets, each consisting of a single number $y^{(i)} = w^{(i)} + n^{(i)}$ with $n^{(i)} \sim \mathcal{N}(0, 1)$ and model parameter $w^{(i)}$, for $i = 1, \ldots, n$. The ML estimator for $w^{(i)}$ is then given by $\hat{w}^{(i)} = y^{(i)}$ [30, 98]. Accordingly, the resulting discrepancy measure is [99]:

$$d^{(i,i')} := \left| y^{(i)} - y^{(i')} \right|.$$

Example Consider an FL network with nodes $i \in \mathcal{V}$ that carry local datasets $\mathcal{D}^{(i)}$. Each $\mathcal{D}^{(i)}$ consists of data points with labels in the label space $\mathcal{Y}^{(i)}$. We can measure the similarity between nodes i and i' by the fraction of data points in $\mathcal{D}^{(i)} \bigcup \mathcal{D}^{(i')}$ with labels lying in $\mathcal{Y}^{(i)} \cap \mathcal{Y}^{(i')}$ [100].

Example Consider local datasets $\mathcal{D}^{(i)}$ constituted by images of handwritten digits $0, 1, \ldots, 9$. We model a local dataset using a hierarchical probabilistic model: Each node $i \in \mathcal{V}$ is assigned a deterministic but unknown distribution $\boldsymbol{\alpha}^{(i)} = \left(\alpha_0^{(i)}, \ldots, \alpha_9^{(i)}\right)$. The entry $\alpha_j^{(i)}$ is the fraction of images at node i that show digit j. We interpret the labels $y^{(i,1)}, \ldots, y^{(i,m_i)}$ as realizations of i.i.d. RVs, with values in $\{0, 1, \ldots, 9\}$ and distributed according to $\boldsymbol{\alpha}^{(i)}$. We also interpret the features as realizations of RVs having conditional probability distribution $p(\mathbf{x}|y)$, which is the same for all nodes $i \in \mathcal{V}$. We can then estimate the dissimilarity between nodes i and i' via the distance between (estimations of) the parameters $\boldsymbol{\alpha}^{(i)}$ and $\boldsymbol{\alpha}^{(i')}$.

The above examples of a discrepancy measure—based on parameter estimates of a probabilistic model—are a special case of a more general two-step approach:

- First, we assign a vector representation $\mathbf{z}^{(i)} \in \mathbb{R}^{m'}$ to each node $i \in \mathcal{V}$ [23, 101].
- Second, we define the discrepancy $d^{(i,i')}$ between nodes i and i' as the distance between the representation vectors $\mathbf{z}^{(i)}$ and $\mathbf{z}^{(i')}$, e.g.,

$$d^{(i,i')} := \left\| \mathbf{z}^{(i)} - \mathbf{z}^{(i')} \right\|.$$

We next discuss three specific implementations of the first step to obtain the representation vector for each node i.

Parametric Probabilistic Models If we use a parametric probabilistic model $p\left(\mathcal{D}^{(i)}; \mathbf{w}^{(i)}\right)$ for the local dataset $\mathcal{D}^{(i)}$, we can use an estimator $\widehat{\mathbf{w}}^{(i)}$ to obtain $\mathbf{z}^{(i)}$. One popular approach for estimating the model parameters of a probabilistic model is the ML principle [23].

Gradients We now discuss a construction for the vector representation $\mathbf{z}^{(i)} \in \mathbb{R}^{m'}$ that is inspired by the update structure of stochastic gradient descent (SGD). In particular, we define the discrepancy between two local datasets by treating them as two batches used by SGD to train a model. If these two batches consist of data points generated from similar probability distributions, their corresponding gradient approximations (5.11) are close. This suggests to use the gradient $\nabla f(\mathbf{w}')$ of the average loss (or empirical risk) $f(\mathbf{w}) := (1/|\mathcal{D}^{(i)}|) \sum_{(\mathbf{x},y) \in \mathcal{D}^{(i)}} L\left((\mathbf{x}, y), h^{(\mathbf{w})}\right)$

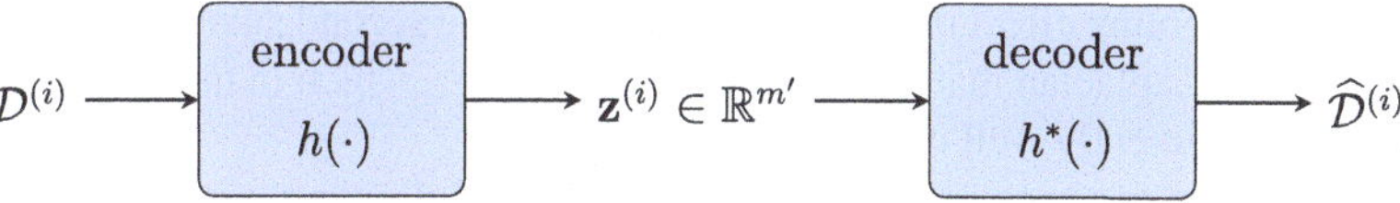

Fig. 7.2 A generic autoencoder consists of an encoder that maps the input to a latent representation and a decoder that attempts to reconstruct the original input. Both components are trained jointly by minimizing a reconstruction loss (see [23, Ch. 9]). When a local dataset is used as input, its latent representation can serve as a compact vector embedding

as a vector representation $\mathbf{z}^{(i)}$ for $\mathcal{D}^{(i)}$. We can generalize this construction, for parametric local models $\mathcal{H}^{(i)}$, by using the gradient of the local loss function:

$$\mathbf{z}^{(i)} := \nabla L_i(\mathbf{v}) . \tag{7.4}$$

Note that the construction (7.4) requires to specify the model parameters $\mathbf{v}$ at which the gradient is evaluated.

Feature learning Another approach is to use an autoencoder [101, Ch. 14] to learn an embedding of a local dataset. In particular, we feed the dataset into an encoder artificial neural network (ANN) that has been jointly trained with a decoder ANN on a suitable learning task. The encoder maps the dataset to a latent vector, or embedding, which serves as its vector representation. A generic setup is illustrated in Fig. 7.2.

7.3 Graph Learning Methods

Assume we have constructed a discrepancy measure $d^{(i,i')} \in \mathbb{R}_{+}$ that quantifies the dissimilarity between any two local datasets $\mathcal{D}^{(i)}$ and $\mathcal{D}^{(i')}$. One way to construct an FL network is by connecting each node i to its nearest neighbors, i.e., the nodes $i' \in \mathcal{V} \setminus \{i\}$ with the smallest values of $d^{(i,i')}$.

An alternative to this nearest-neighbor construction is to formulate graph learning as a constrained linear optimization problem. Let us measure the quality of a candidate edge-weight assignment $A_{i,i'} \in \mathbb{R}_{+}$ using the objective function

$$\sum_{i,i' \in \mathcal{V}} A_{i,i'} d^{(i,i')} . \tag{7.5}$$

This function penalizes large weights between nodes that are dissimilar. Without any constraints, the minimum of (7.5) is trivially achieved by setting $A_{i,i'} = 0$ for all pairs, i.e., resulting in an empty graph.

As discussed in Sect. 7.1, however, a useful FL network must contain a sufficient number of edges to ensure that GTVMin produces meaningful model parameters. In particular, the pooling effect of GTVMin depends on the second smallest eigenvalue $\lambda_2(\mathbf{L}^{(\mathcal{G})})$ of the Laplacian matrix being sufficiently large, which in turn requires that the graph is sufficiently well connected (see (7.2)).

To enforce the presence of edges, we introduce the following constraints:

$$A_{i,i} = 0, \quad \sum_{i' \neq i} A_{i,i'} = d_{\max}^{(\mathcal{G})} \quad \text{and} \quad A_{i,i'} \in [0, 1] \quad \text{for all } i, i' \in \mathcal{V}. \tag{7.6}$$

These constraints ensure that each node i has (weighted) node degree $\sum_{i' \neq i} A_{i,i'}$ equal to $d_{\max}^{(\mathcal{G})}$ and that edge weights are bounded and symmetric.

Combining the objective function (7.5) with the constraints (7.6), we arrive at the following graph learning principle:

$$\begin{aligned} \big\{\widehat{A}_{i,i'}\big\}_{i,i' \in \mathcal{V}} \in \operatorname*{argmin}_{A_{i,i'} = A_{i',i}} \quad & \sum_{i,i' \in \mathcal{V}} A_{i,i'} d^{(i,i')} \\ \text{s.t.} \quad & A_{i,i'} \in [0, 1] \quad \forall i, i' \in \mathcal{V}, \\ & A_{i,i} = 0 \quad \forall i \in \mathcal{V}, \\ & \sum_{i' \neq i} A_{i,i'} = d_{\max}^{(\mathcal{G})} \quad \forall i \in \mathcal{V}. \end{aligned} \tag{7.7}$$

This constrained minimization problem is a special case of the general quadratic program introduced in (4.14). Because the objective is linear, (7.7) is equivalent to a linear program [47, Sec. 4.3]. Approximate solutions to (7.7) can be efficiently computed using projected GD, as discussed in Sect. 4.5.

The first constraint in (7.7) bounds edge weights between 0 and 1. The second prohibits self-loops, which have no effect on the outcome of GTVMin (see (3.22)). The final constraint enforces regularity: every node has the same node degree $d^{(i)} = d_{\max}^{(\mathcal{G})}$.

While regular graphs simplify the analysis of GTVMin, they may not always be desirable in practice. In some FL applications, it may be advantageous to allow varying node degrees—such as graphs with a small number of hub nodes with high node degree [11, 99] or to minimize the total number of edges.

We can enforce an upper bound on the total number $E_{\max}$ of edges by modifying the last constraint in (7.7):

$$\begin{aligned} \widehat{A}_{i,i'} \in \operatorname*{argmin}_{A_{i,i'} = A_{i',i}} & \sum_{i,i' \in \mathcal{V}} A_{i,i'} d^{(i,i')} \\ & A_{i,i'} \in [0, 1] \text{ for all } i, i' \in \mathcal{V}, \\ & A_{i,i} = 0 \text{ for all } i \in \mathcal{V}, \\ & \sum_{i',i \in \mathcal{V}} A_{i,i'} = E_{\max}. \end{aligned} \tag{7.8}$$

The problem has a closed-form solution as explained in [99]: It is obtained by placing the edges between those pairs $i, i' \in \mathcal{V}$ that result in the smallest discrepancy $d^{(i,i')}$. However, it might still be useful to solve (7.8) via iterative optimization methods such as the gradient-based methods discussed in Chap. 4. These methods can be implemented in a fully distributed fashion as message passing over an underlying communication network [68]. This communication network might be significantly different from the learned FL network. For some FL applications, the functional connectivity of two devices i and i' reflects also a similarity between probability distributions of local datasets $\mathcal{D}^{(i)}$ and $\mathcal{D}^{(i')}$ [102].

7.4 Exercises

7.1 A Simple Ranking Approach. Consider a collection of devices $i = 1, \ldots, n = 100$, each carrying a local dataset that consists of a single vector $\mathbf{x} \in \mathbb{R}^{(m_i)}$. We interpret the vectors $\mathbf{x} \in \mathbb{R}^{m_i}$, for $i = 1, \ldots, n$, as statistically independent RVs. Moreover, the vector $\mathbf{x} \in \mathbb{R}^{m_i}$ is a realization of a multivariate normal distribution $\mathcal{N}(c_i \mathbf{1}, \mathbf{I})$ with given (fixed) quantities $c_i \in \{-1, 1\}$. We construct an FL network by determining for each node i its neighborhood $\mathcal{N}^{(i)}$ as follows:

- we randomly select a fraction $\mathcal{B}^{(i)}$ of 10% from all other nodes
- we define $\mathcal{N}^{(i)}$ as those $i' \in \mathcal{B}^{(i)}$ whose corresponding values

  $$|(1/m_i)\mathbf{1}^T \mathbf{x}^{(i)} - (1/m_{i'})\mathbf{1}^T \mathbf{x}^{(i')}|$$

  are among the three smallest.

Analyze the probability that some neighborhood $\mathcal{N}^{(i)}$ contains a node i' such that $c_i \neq c_{i'}$.

Chapter 8
Trustworthy FL

Abstract This chapter explores how federated learning (FL) systems can be designed to meet key requirements for trustworthy artificial intelligence. It draws on international policy frameworks, including those from the EU, USA, OECD, and others. The chapter begins by addressing human agency and oversight, emphasizing model simplicity, continuous validation, and ethical limitations on data use. It then analyzes the technical robustness of FL systems, including their sensitivity to data perturbations, estimation errors, and infrastructure faults. Robustness is evaluated for both the learned models and the algorithms that compute them. The chapter also highlights strategies to improve network resilience and prevent single points of failure. Privacy and data governance are discussed in light of legal constraints, with a focus on minimizing unnecessary data use and ensuring organizational accountability. The chapter closes by discussing transparency, explainability, fairness, and environmental sustainability. It proposes ways to enforce explainability through user-specific penalties in the optimization objective. Fairness is addressed through data augmentation, while societal and environmental impacts are acknowledged through system design choices. Altogether, the chapter builds a bridge between mathematical methods like GTVMin and broader concerns about AI ethics and accountability.

This chapter examines how regulatory frameworks for trustworthy AI inform the design and implementation of GTVMin-based methods. Our discussion is primarily guided by the key requirements (KRs) for trustworthy AI as formulated by the *European Union's High-Level Expert Group on AI* [103]. Comparable ethical frameworks have emerged globally, including *Australia's AI Ethics Principles* [104], the *OECD AI Principles* [105], China's governance efforts [106–108], and US developments such as the *NIST AI Risk Management Framework* [109], the *Blueprint for an AI Bill of Rights* [110], and *Executive Order 14110 on the Safe, Secure, and Trustworthy Development and Use of Artificial Intelligence* [111].

Section 8.1 examines how FL systems can support human agency and oversight, as required by the principle of respect for human autonomy within the broader framework of trustworthy AI.

A. Jung, *Federated Learning*, https://doi.org/10.1007/978-981-95-1009-2_8

Section 8.2 investigates the robustness of FL systems against different forms of perturbations. Perturbations can arise from the intrinsic variability of local datasets that are obtained from stochastic data generation processes. Another source for perturbations is imperfections of the communication links between devices. We devote Chap. 10 to perturbations that are intentional (or adversarial) during so-called cyber attacks.

Section 8.3 addresses the need for privacy protection and data governance. This includes regulatory constraints on data processing, the data minimization principle, and the organizational structures needed to enforce compliance. We devote Chap. 9 to a detailed treatment of quantitative measures for privacy leakage and techniques to mitigate it in GTVMin-based FL systems.

Section 8.4 focuses on the transparency and the explainability of GTVMin-based FL systems. We introduce quantitative metrics for subjective explainability that reflect how well personalized models align with individual users' expectations. We can incorporate these metrics into GTVMin-based methods to ensure tailored explainability for heterogeneous populations of device users.

8.1 Human Agency and Oversight

> *..AI systems should support human autonomy and decision-making, as prescribed by the principle of respect for human autonomy. This requires that AI systems should both act as enablers to a democratic, flourishing, and equitable society by supporting the user's agency and foster fundamental rights and allow for human oversight...* [103, p.15]

Human Dignity Learning personalized model parameters for recommender systems allows to boost addiction or widespread emotional manipulation resulting in harmful societal effects [112–114]. KR1 rules out certain design choices for the labels of data points. In particular, we might not use the mental and psychological characteristics of a user as the label. We should avoid loss functions that can be used to train predictors of psychological characteristics. Using personalized ML models to predict user preferences for products or susceptibility toward propaganda is also referred to as *micro-targeting* [115].

Simple Is Good Human oversight can be facilitated by relying on simple local models. Examples include linear models with few features or decision trees with a small tree depth. However, we are unaware of a widely accepted definition of when a model is simple. Loosely speaking, a simple model results in a learned hypothesis that allows humans to understand how features of a data point relate to the prediction $h(\mathbf{x})$. This notion of simplicity is closely related to the concept of explainability which we discuss in more detail in Sect. 8.4.

Continuous Monitoring In its simplest form, GTVMin-based methods involve a single training phase, i.e., learning local model parameters by solving GTVMin. However, this approach is only useful if the data can be well approximated by an i.i.d. assumption. In particular, this approach works only if the statistical properties of local datasets do not change over time. For many FL applications, this assumption is unrealistic (consider a social network which is exposed to constant change of memberships and user behavior). It is then important to continuously compute a validation error which is then used, in turn, to diagnose the overall FL system (see [23, Sec. 6.6]).

8.2 Technical Robustness and Safety

> *...Technical robustness requires that AI systems be developed with a preventative approach to risks and in a manner such that they reliably behave as intended while minimizing unintentional and unexpected harm and preventing unacceptable harm....* [103, p.16].

Practical FL systems are obtained by implementing FL algorithms in physical distributed computers [17, 18]. One example of a distributed computer is a collection of smartphones that are connected either by short-range wireless links or by a cellular network.

Distributed computers (as physical objects) typically incur imperfections, such as a temporary lack of connectivity or mobile devices that run out of battery and therefore become inactive. Moreover, the data generation processes can be subject to perturbations such as statistical anomalies or outliers. This section studies in some detail the robustness of GTVMin-based systems against different perturbations of data sources and imperfections of computational infrastructure.

Consider a GTVMin-based FL system that trains a single (global) linear model in a distributed fashion from a collection of local datasets $\mathcal{D}^{(i)}$, for $i = 1, \ldots, n$. As discussed in Sect. 6.1, this single-model FL setting uses GTVMin (3.24) over a connected FL network with a sufficiently large choice of α.

To ensure **KR2**, we need to understand the effect of perturbations on a GTVMin-based FL system. These perturbations might be intentional (or adversarial) and affect the local datasets used to evaluate the loss of local model parameters or the computational infrastructure used to implement a GTVMin-based method (see Chap. 5). We next explain how to use some of the theoretical tools from previous chapters to quantify the robustness of GTVMin-based FL systems.

8.2.1 Sensitivity Analysis

As pointed out in Chap. 3, GTVMin (3.24) can be rewritten as the minimization of a quadratic function,

$$\min_{\mathbf{w}=\mathrm{stack}\{\mathbf{w}^{(i)}\}_{i=1}^{n}} \mathbf{w}^T \mathbf{Q} \mathbf{w} + \mathbf{q}^T \mathbf{w}. \tag{8.1}$$

The matrix $\mathbf{Q}$ and vector $\mathbf{q}$ are determined by the feature matrices $\mathbf{X}^{(i)}$ and label vectors $\mathbf{y}^{(i)}$ at the nodes $i \in \mathcal{V}$ (see (2.23)). We next study the sensitivity of (the solutions of) (8.1) toward external perturbations of the label vector.[1]

Consider an additive perturbation $\widetilde{\mathbf{y}}^{(i)} := \mathbf{y}^{(i)} + \boldsymbol{\varepsilon}^{(i)}$ of the label vector $\mathbf{y}^{(i)}$. Using the perturbed label vector $\widetilde{\mathbf{y}}^{(i)}$ results also in a "perturbation" of GTVMin (8.1),

$$\min_{\mathbf{w}=\text{stack}\{\mathbf{w}^{(i)}\}} \mathbf{w}^T\mathbf{Q}\mathbf{w} + \mathbf{q}^T\mathbf{w} + \mathbf{n}^T\mathbf{w} + c. \tag{8.2}$$

An inspection of (2.23) yields that $\mathbf{n} = \left(\left(\boldsymbol{\varepsilon}^{(1)}\right)^T\mathbf{X}^{(1)}, \ldots, \left(\boldsymbol{\varepsilon}^{(n)}\right)^T\mathbf{X}^{(n)}\right)^T$. The next result provides an upper bound on the deviation between the solutions of (8.1) and (8.2).

Proposition 8.1 *Consider the GTVMin instance* (8.1) *for learning local model parameters of a linear model for each node $i \in \mathcal{V}$ of an FL network $\mathcal{G}$. We assume that the FL network is connected, i.e., $\lambda_2\big(\mathbf{L}^{(\mathcal{G})}\big) > 0$ and the local datasets are such that $\bar{\lambda}_{\min} > 0$ (see* (5.4)*). Then, the deviation between the solution $\widehat{\mathbf{w}}^{(i)}$ to* (8.1) *and the solution $\widetilde{\mathbf{w}}^{(i)}$ to the perturbed problem* (8.2) *is upper bounded as*

$$\sum_{i=1}^{n} \left\| \widehat{\mathbf{w}}^{(i)} - \widetilde{\mathbf{w}}^{(i)} \right\|_2^2 \leq \frac{\lambda_{\max}(1+\rho^2)^2}{\left[\min\{\lambda_2\big(\mathbf{L}^{(\mathcal{G})}\big)\alpha\rho^2, \bar{\lambda}_{\min}/2\}\right]^2} \sum_{i=1}^{n} \left\| \boldsymbol{\varepsilon}^{(n)} \right\|_2^2.$$

Here, we used the shorthand $\rho := \bar{\lambda}_{\min}/(4\lambda_{\max})$ (see (5.4)*).*

Proof The assumptions of Proposition 8.1 allow to apply the lower bound (5.6) on the eigenvalues of the matrix $\mathbf{Q}$ in (8.1). □

8.2.2 Estimation Error Analysis

Proposition 8.1 characterizes the sensitivity of GTVMin solutions against *external* perturbations of the local datasets. While this notion of robustness is important, it might not suffice for a comprehensive assessment of an FL system. For example, we can trivially achieve perfect robustness by delivering constant model parameters, e.g., $\widehat{\mathbf{w}}^{(i)} = \mathbf{0}$. Clearly, such an FL system is not very useful.

Another form of robustness is to ensure a small estimation error of (3.24). To study this form of robustness, we use a variant of the probabilistic model (3.32): We assume that the labels and features of data points of each local dataset $\mathcal{D}^{(i)}$, for

[1] Our study can be generalized to also take into account perturbations of the feature matrices $\mathbf{X}^{(i)}$, for $i = 1, \ldots, n$.

$i = 1, \ldots, n$, are related via

$$\mathbf{y}^{(i)} = \mathbf{X}^{(i)}\overline{\mathbf{w}} + \boldsymbol{\varepsilon}^{(i)}. \tag{8.3}$$

In contrast to Sect. 3.3.2, we assume that all components of (8.3) are deterministic. In particular, the noise term $\boldsymbol{\varepsilon}^{(i)}$ is a deterministic but unknown quantity. This term accommodates any perturbation that might arise from technical imperfections or intrinsic label noise due to random fluctuations in the labeling process.[2]

In the ideal case of no perturbation, we would have $\boldsymbol{\varepsilon}^{(i)} = \mathbf{0}$. However, we often only know some upper bound measure for the size of the perturbation, as measured, e.g., by $\left\|\boldsymbol{\varepsilon}^{(i)}\right\|_2^2$. We next present upper bounds on the estimation error $\widehat{\mathbf{w}}^{(i)} - \overline{\mathbf{w}}$ incurred by the GTVMin solutions $\widehat{\mathbf{w}}^{(i)}$.

This estimation error consists of two components, the first component being $\operatorname{avg}\{\widehat{\mathbf{w}}^{(i')}\} - \overline{\mathbf{w}}$ for each node $i \in \mathcal{V}$. Note that this error component is identical for all nodes $i \in \mathcal{V}$. The second component of the estimation error is the deviation $\widetilde{\mathbf{w}}^{(i)} := \widehat{\mathbf{w}}^{(i)} - \operatorname{avg}\{\widehat{\mathbf{w}}^{(i')}\}$ of the learned local model parameters $\widehat{\mathbf{w}}^{(i')}$, for $i' = 1, \ldots, n$, from their average $\operatorname{avg}\{\widehat{\mathbf{w}}^{(i')}\} = (1/n)\sum_{i'=1}^{n} \widehat{\mathbf{w}}^{(i')}$. As discussed in Sect. 3.3.2, these two components correspond to two orthogonal subspaces of $\mathbb{R}^{d \cdot n}$.

According to Proposition 3.1, the second error component is upper bounded as

$$\sum_{i=1}^{n} \left\|\widetilde{\mathbf{w}}^{(i)}\right\|_2^2 \leq \frac{1}{\lambda_2 \alpha} \sum_{i=1}^{n} (1/m_i) \left\|\boldsymbol{\varepsilon}^{(i)}\right\|_2^2. \tag{8.4}$$

To bound the first error component $\bar{\mathbf{c}} - \overline{\mathbf{w}}$, using the shorthand $\bar{\mathbf{c}} := \operatorname{avg}\{\widehat{\mathbf{w}}^{(i)}\}$, we first note that (see (3.24))

$$\bar{\mathbf{c}} = \operatorname*{argmin}_{\mathbf{w} \in \mathbb{R}^d} \sum_{i \in \mathcal{V}} (1/m_i) \left\|\mathbf{y}^{(i)} - \mathbf{X}^{(i)}\left(\mathbf{w} - \widetilde{\mathbf{w}}^{(i)}\right)\right\|_2^2 + \alpha \sum_{\{i,i'\} \in \mathcal{E}} A_{i,i'} \left\|\widetilde{\mathbf{w}}^{(i)} - \widetilde{\mathbf{w}}^{(i')}\right\|_2^2. \tag{8.5}$$

Using a similar argument as in the proof for Proposition 2.1, we obtain

$$\|\bar{\mathbf{c}} - \overline{\mathbf{w}}\|_2^2 \leq \left\|\sum_{i=1}^{n} (1/m_i) \left(\mathbf{X}^{(i)}\right)^T \left(\boldsymbol{\varepsilon}^{(i)} + \mathbf{X}^{(i)}\widetilde{\mathbf{w}}^{(i)}\right)\right\|_2^2 / (n\bar{\lambda}_{\min})^2. \tag{8.6}$$

[2] Consider labels obtained from physical sensing devices which are typically subject to measurement errors [116].

Here, $\bar{\lambda}_{\min}$ is the smallest eigenvalue of $(1/n)\sum_{i=1}^{n}\mathbf{Q}^{(i)}$, i.e., the average of the matrices $\mathbf{Q}^{(i)} = (1/m_i)\big(\mathbf{X}^{(i)}\big)^T\mathbf{X}^{(i)}$ over all nodes $i \in \mathcal{V}$.[3] Note that the bound (8.6) is only valid if $\bar{\lambda}_{\min} > 0$ which, in turn, implies that the solution to (8.5) is unique.

We can develop (8.6) further using

$$\begin{aligned}
&\left\| \sum_{i=1}^{n}(1/m_i)\big(\mathbf{X}^{(i)}\big)^T\big(\boldsymbol{\varepsilon}^{(i)} + \mathbf{X}^{(i)}\widetilde{\mathbf{w}}^{(i)}\big)\right\|_2 \\
&\stackrel{(a)}{\leq} \sum_{i=1}^{n}\left\|(1/m_i)\big(\mathbf{X}^{(i)}\big)^T\big(\boldsymbol{\varepsilon}^{(i)} + \mathbf{X}^{(i)}\widetilde{\mathbf{w}}^{(i)}\big)\right\|_2 \\
&\stackrel{(b)}{\leq} \sqrt{n}\sqrt{\sum_{i=1}^{n}\left\|(1/m_i)\big(\mathbf{X}^{(i)}\big)^T\big(\boldsymbol{\varepsilon}^{(i)} + \mathbf{X}^{(i)}\widetilde{\mathbf{w}}^{(i)}\big)\right\|_2^2} \\
&\stackrel{(c)}{\leq} \sqrt{n}\sqrt{\sum_{i=1}^{n}2\left\|(1/m_i)\big(\mathbf{X}^{(i)}\big)^T\boldsymbol{\varepsilon}^{(i)}\right\|_2^2 + 2\left\|(1/m_i)\big(\mathbf{X}^{(i)}\big)^T\mathbf{X}^{(i)}\widetilde{\mathbf{w}}^{(i)}\right\|_2^2} \\
&\stackrel{(d)}{\leq} \sqrt{n}\sqrt{\sum_{i=1}^{n}(2/m_i)\lambda_{\max}\left\|\boldsymbol{\varepsilon}^{(i)}\right\|_2^2 + 2\lambda_{\max}^2\left\|\widetilde{\mathbf{w}}^{(i)}\right\|_2^2}.
\end{aligned} \tag{8.7}$$

Here, step (a) uses the triangle inequality of norms, step (b) uses the Cauchy-Schwarz inequality, step (c) uses the inequality $\|\mathbf{a}+\mathbf{b}\|_2^2 \leq 2\Big(\|\mathbf{a}\|_2^2 + \|\mathbf{b}\|_2^2\Big)$, and step (d) uses the maximum eigenvalue $\lambda_{\max} := \max_{i\in\mathcal{V}}\lambda_d\big(\mathbf{Q}^{(i)}\big)$ of the matrices $\mathbf{Q}^{(i)} = (1/m_i)\big(\mathbf{X}^{(i)}\big)^T\mathbf{X}^{(i)}$ (see (5.4)).

Inserting (8.7) into (8.6) results in the upper bound

$$\begin{aligned}
\|\bar{\mathbf{c}} - \overline{\mathbf{w}}\|_2^2 &\leq 2\sum_{i=1}^{n}\left[(1/m_i)\lambda_{\max}\left\|\boldsymbol{\varepsilon}^{(i)}\right\|_2^2 + \lambda_{\max}^2\left\|\widetilde{\mathbf{w}}^{(i)}\right\|_2^2\right]/(n\bar{\lambda}_{\min}^2) \\
&\stackrel{(8.4)}{\leq} 2\big(\lambda_{\max} + (\lambda_{\max}^2/(\lambda_2\alpha))\big)\sum_{i=1}^{n}(1/m_i)\left\|\boldsymbol{\varepsilon}^{(i)}\right\|_2^2/(n\bar{\lambda}_{\min}^2).
\end{aligned} \tag{8.8}$$

The upper bound (8.8) on the estimation error of GTVMin-based methods depends on both the FL network $\mathcal{G}$ via the eigenvalue λ_2 of $\mathbf{L}^{(\mathcal{G})}$ and the feature matrices $\mathbf{X}^{(i)}$ of the local datasets (via the quantities $\lambda_{\max}$ and $\bar{\lambda}_{\min}$ as defined in (5.4)). Let

[3] We encountered the quantity $\bar{\lambda}_{\min}$ already during our discussion of gradient-based methods for solving the GTVMin instance (3.24) (see (5.4)).

us next discuss how the upper bound (8.8) might guide the choice of the FL network $\mathcal{G}$ and the features of data points in the local datasets.

According to (8.8), we should use an FL network $\mathcal{G}$ with large $\lambda_2(\mathbf{L}^{(\mathcal{G})})$ to ensure a small estimation error for GTVMin-based methods. Note that we came across the same design criterion already when discussing graph learning methods in Chap. 7. In particular, using an FL network with large $\lambda_2(\mathbf{L}^{(\mathcal{G})})$ also tends to speed up the convergence of gradient-based methods for solving GTVMin (such as Algorithm 4).

The upper bound (8.8) suggests using features that result in a small ratio $\lambda_{\max}/\bar{\lambda}_{\min}$ between the quantities $\lambda_{\max}$ and $\bar{\lambda}_{\min}$ (see (5.4)). Some feature learning methods have been proposed in order to minimize this ratio [23, 117].

8.2.3 Robustness of FL Algorithms

The previous subsections studied the robustness of GTVMin solutions against perturbations of local datasets. Ensuring trustworthy FL systems also requires robustness of FL algorithms against perturbations of their executions. It turns out that our design choices (e.g., the shape of local loss functions) for GTVMin crucially affect the robustness of the FL algorithms discussed in Sect. 5.

For ease of exposition, we will focus on FL algorithms for parametric local models that are based on the update

$$\mathbf{w}^{(i,k+1)} \in \underset{\mathbf{w}^{(i)} \in \mathbb{R}^d}{\operatorname{argmin}} \left[L_i\left(\mathbf{w}^{(i)}\right) + \sum_{i' \in \mathcal{N}^{(i)}} A_{i,i'} \phi\left(\mathbf{w}^{(i',k)} - \mathbf{w}^{(i,k)}\right) \right]. \tag{8.9}$$

Note that Algorithms 5 and 11 use (8.9) as their core computational step. We next discuss the robustness of (8.9) against perturbations of the model parameters $\mathbf{w}^{(i',k)}$ that a device receives from its neighbors $i' \in \mathcal{N}^{(i)}$. We focus on two specific choices for the GTV penalty function ϕ.

GTV Penalty $\phi(\cdot) = \|\cdot\|_2^2$ For the penalty function $\phi(\mathbf{w}^{(i)} - \mathbf{w}^{(i')}) = \left\|\mathbf{w}^{(i)} - \mathbf{w}^{(i')}\right\|_2^2$, we can rewrite (8.9) as (see Exercise 5.6)

$$\mathbf{w}^{(i,k+1)} \in \underset{\mathbf{w}^{(i)} \in \mathbb{R}^d}{\operatorname{argmin}} L_i\left(\mathbf{w}^{(i)}\right) + \alpha d^{(i)} \left\|\mathbf{w}^{(i)} - \widehat{\mathbf{w}}^{(\mathcal{N}^{(i)})}\right\|_2^2. \tag{8.10}$$

Here, we used $\widehat{\mathbf{w}}^{(\mathcal{N}^{(i)})} := (1/d^{(i)}) \sum_{i' \in \mathcal{N}^{(i)}} A_{i,i'} \mathbf{w}^{(i',k)}$ and the weighted node degree $d^{(i)} = \sum_{i' \in \mathcal{N}^{(i)}} A_{i,i'}$ (see (3.7)).

If the local loss function $L_i(\cdot)$ is convex, and under some mild technical conditions,[4] the update (8.10) is well-defined; i.e., the minimization has a unique solution [39, Ch. 6]. Moreover, the update (8.10) then coincides with an application of the proximal operator $\mathbf{prox}_{L_i(\cdot),\rho}(\cdot)$ (see (3.30)) of $L_i(\cdot)$ [40],

$$\mathbf{w}^{(i,k+1)} = \mathbf{prox}_{L_i(\cdot),\rho}(\widehat{\mathbf{w}}^{(\mathcal{N}^{(i)})}) \text{ with } \rho = 2\alpha d^{(i)}. \tag{8.11}$$

Figure 8.1 illustrates the update (8.11) as a straight line. The slope of this line indicates the robustness of (8.11) against perturbations of the received model parameters $\mathbf{w}^{(i',k)}$, for $i' \in \mathcal{N}^{(i)}$. These perturbations result in a modified input $\widetilde{\mathbf{w}}^{(i)}$ (instead of $\widehat{\mathbf{w}}^{(\mathcal{N}^{(i)})}$) for the proximal operator $\mathbf{prox}_{L_i(\cdot),\rho}(\widehat{\mathbf{w}}^{(\mathcal{N}^{(i)})})$. A natural quantitative measure for the robustness (or stability) of (8.11) is

$$\frac{\left\| \mathbf{prox}_{L_i(\cdot),\rho}(\widetilde{\mathbf{w}}^{(i)}) - \mathbf{prox}_{L_i(\cdot),\rho}(\widehat{\mathbf{w}}^{(\mathcal{N}^{(i)})}) \right\|_2}{\left\| \widetilde{\mathbf{w}}^{(i)} - \widehat{\mathbf{w}}^{(\mathcal{N}^{(i)})} \right\|_2}. \tag{8.12}$$

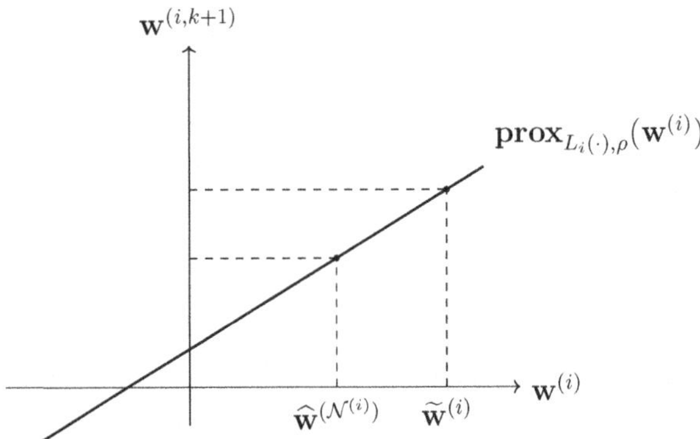

Fig. 8.1 For a convex local loss function $L_i(\cdot)$, the update (8.10) becomes the evaluation of the proximal operator $\mathbf{prox}_{L_i(\cdot),\rho}(\cdot)$ with $\rho = 2\alpha d^{(i)}$. We can measure the robustness of (8.10) by the slope of $\mathbf{prox}_{L_i(\cdot),\rho}(\cdot)$ (see (8.12))

[4] Strictly speaking, we need to require loss function $L_i(\cdot)$ to have a non-empty and closed epigraph which does not contain any non-horizontal lines [40].

It turns out that if the local loss function $L_i(\cdot)$ is strongly convex with coefficient σ, then (8.12) is upper bounded by [72, Sec. 6]

$$\frac{1}{1+(\sigma/\rho)} = \frac{1}{1+(\sigma/(2\alpha d^{(i)}))}. \tag{8.13}$$

We can interpret the quantity (8.13) as a measure for the robustness of the update (8.10). The smaller this quantity, the more robust are FL systems based on (8.10).

Note how the robustness measure (8.13) can guide the design choices for the components of GTVMin. In particular, to ensure a small value (8.13) (ensuring robustness), we should use

- a local loss function that is strongly convex with a large coefficient σ,
- an FL network with small node degrees $d^{(i)}$,
- a small value α for GTVMin parameter.

GTV Penalty $\phi(\cdot) = \|\cdot\|_2$ Let us now study (8.9) for the center node $i = 1$ of a star-shaped FL network (see Fig. 5.3). This uses the trivial local loss function $L_i(\cdot) \equiv 0$ and is connected via unit-weight edges to the peripheral nodes $i' = 2, \ldots, n$. The variation of local model parameters is measured with the penalty function $\phi(\mathbf{w}^{(i)} - \mathbf{w}^{(i')}) = \left\|\mathbf{w}^{(i)} - \mathbf{w}^{(i')}\right\|_2$. This special case of (8.9) can be written as

$$\mathbf{w}^{(1,k+1)} \in \underset{\mathbf{w}^{(i)} \in \mathbb{R}^d}{\operatorname{argmin}} \sum_{i'=2}^{n} \left\|\mathbf{w}^{(i',k)} - \mathbf{w}^{(i)}\right\|_2. \tag{8.14}$$

Note that (8.14) is nothing but the geometric median of the model parameters $\mathbf{w}^{(i',k)}$, for $i' \in \mathcal{N}^{(i)}$. The usefulness of the geometric median for robust FL has been studied recently [118].

The update (8.14) defines a non-smooth convex optimization problem. Any solution $\mathbf{w}^{(1,k+1)}$ to this problem must satisfy the subgradient optimality condition

$$\sum_{i'=2}^{n} \mathbf{g}^{(i')} = \mathbf{0} \quad \text{, with } \mathbf{g}^{(i')} = \begin{cases} \frac{\mathbf{w}^{(i',k)} - \mathbf{w}^{(1,k+1)}}{\left\|\mathbf{w}^{(i',k)} - \mathbf{w}^{(1,k+1)}\right\|_2} & \text{if } \mathbf{w}^{(i',k)} \neq \mathbf{w}^{(1,k+1)} \\ \mathbf{u} \in \mathcal{B}(1) & \text{otherwise,} \end{cases} \tag{8.15}$$

where $\mathcal{B}(1) := \{\mathbf{u} \in \mathbb{R}^d : \|\mathbf{u}\|_2 \leq 1\}$ denotes the unit Euclidean ball. Each $\mathbf{g}^{(i')}$ is a subgradient of the convex non-smooth function $f(\mathbf{w}^{(i)}) := \left\|\mathbf{w}^{(i',k)} - \mathbf{w}^{(i)}\right\|_2$.

Figure 8.2 illustrates the optimality condition (8.15) for the case where node $i = 1$ has three neighbors, two of which are trustworthy. The third neighbor is not trustworthy and may send arbitrarily corrupted model parameters. Despite such adversarial perturbations, the solution $\mathbf{w}^{(1,k+1)}$ of (8.15) cannot be arbitrarily far

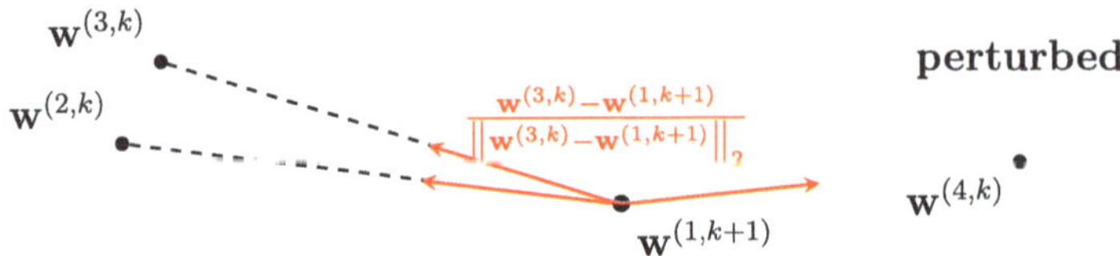

Fig. 8.2 Illustration of the (zero-subgradient) optimality condition (8.15) for the update (8.14). The arrows represent unit-norm subgradients arising from the components $\left\|\mathbf{w}^{(i',k)} - \mathbf{w}^{(i)}\right\|_2$ for $i' = 2, \ldots, n$

from the model parameters of the trustworthy neighbors, provided they form the majority.

Intuitively, if the solution were far from the honest models, then the corresponding subgradients $\mathbf{g}^{(i')}$ for the trustworthy neighbors $i' \in \mathcal{N}^{(i)}$ would point in nearly the same direction, and their sum would have a norm close to the number of honest neighbors. However, the subgradients from the non-trustworthy nodes—being unit vectors—cannot cancel this sum unless they are sufficiently many which contradicts the majority assumption. For a more detailed robustness analysis of (8.15), we refer to [119, Thm. 2.2].

8.2.4 Network Resilience

The previous sections studied the robustness of GTVMin-based methods against perturbations of local datasets (see Exercise 8.1) and in terms of ensuring a small estimation error (see (8.8)). We also need to ensure that FL systems are robust against imperfections of the computational infrastructure used to solve GTVMin. These imperfections include hardware failures, running out of battery or lack of wireless connectivity.

Chapter 5 showed how to design FL algorithms by applying gradient-based methods to solve GTVMin (3.24). We obtain practical FL systems by implementing these algorithms, such as Algorithm 4, in a particular computational infrastructure. Two important examples of such an infrastructure are mobile networks and wireless sensor networks [19, 120].

The effect of imperfections in the implementation of the GD based Algorithm 4 can be modeled as perturbed GD (4.12) from Chap. 4. We can then analyze the robustness of the resulting FL system via the convergence analysis of perturbed GD discussed in Sect. 4.4.

According to (4.13), the performance of the decentralized Algorithm 4 degrades gracefully in the presence of imperfections such as missing or faulty communication links. In contrast, the server-based implementation of FedAvg Algorithm 9 offers a single point of failure (the server).

Instead of modeling the effect of network failures as perturbed GD, we can instead interpret it as exact GD applied to a perturbed instance of GTVMin. This perturbed instance uses a pruned FL network $\widetilde{\mathcal{G}}$, consisting of edges that are still active (i.e., corresponding to active communication links).

The effectiveness of GTVMin crucially depends on the second-smallest eigenvalue λ_2 of the Laplacian matrix (3.9) associated with the FL network $\widetilde{\mathcal{G}}$ (see Sect. 3.3.2). As discussed in Sect. 7.1, λ_2 reflects how well-connected the FL network $\widetilde{\mathcal{G}}$ is. A larger λ_2 means better connectivity, which is required by GTVMin to combine the information provided by devices that work on similar learning tasks (see Sect. 6.2).

To make GTVMin robust against communication link failures, we need to design the original FL network $\mathcal{G}$ so that even if some edges are removed, the resulting $\widetilde{\mathcal{G}}$ still stays well connected—that is, λ_2 remains large enough. This idea is related to resilient network design, which studies how to build networks that stay connected even when some parts fail [121, 122].

8.3 Privacy and Data Governance

> *..privacy, a fundamental right particularly affected by AI systems. Prevention of harm to privacy also necessitates adequate data governance that covers the quality and integrity of the data used...* [103, p.17].

We have introduced GTVMin and FL networks as abstract mathematical structures for the study of FL systems. However, to obtain actual FL systems, we need to implement these mathematical concepts in a given physical hardware. These implementations incur deviations from the (idealized) GTVMin formulation (3.22) and the gradient-based methods (such as Algorithm 4) used to solve it. For example, using quantized label values results in a quantization error. Moreover, the local datasets can deviate significantly from a typical realization of i.i.d. RVs, which is referred to as statistical bias [123, Sec. 3.3.])

Data processing regulations limit the choice of the features of a data point [124–126]. In particular, the general data protection regulation (GDPR) includes a data minimization principle which requires to use only features that are relevant for predicting the label.

Data Governance Some FL applications involve local datasets that are generated by human users, i.e., personal data. Whenever personal data is used by an FL method, special care must be dedicated toward data protection regulations [126]. It is useful (or even compulsory) to designate a data protection officer and conduct impact assessments [103].

Privacy The operation of an FL system must not violate the fundamental human right to privacy [127]. One of the most important characteristics of FL, and distinguishing from distributed optimization, is the privacy-friendly exchange of information among the system components. We dedicate the entire Chap. 9 to the discussion of quantitative measures and methods for privacy protection in GTVMin-based FL systems.

8.4 Transparency

Traceability This key requirement includes the documentation of design choices (and underlying business models) for a GTVMin-based FL system. This includes the source for the local datasets, the local models, the local loss function, as well as the construction of the FL network. Moreover, the documentation should also cover the details of the implemented optimization method used to solve GTVMin. This documentation might also require the periodic storing of the model parameters along with a time stamp (*logging*).

Communication Depending on the use case, FL systems need to communicate the capabilities and limitations to their end users (e.g., of a digital health app running on a smartphone). For example, we can indicate a measure of uncertainty about the predictions delivered by the trained local models. Such an uncertainty measure can be obtained naturally from a probabilistic model for the data generation. For example, the conditional variance of the label y, given the features $\mathbf{x}$ of a random data point. Another example of an uncertainty measure is the validation error of a trained local model.

Explainability The transparency of an FL system can be facilitated by a sufficient level of explainability of the trained personalized model $\hat{h}^{(i)} \in \mathcal{H}^{(i)}$. It is important to note that the explainability of $\hat{h}^{(i)}$ is subjective: A given learned hypothesis $\hat{h}^{(i)}$ might offer a high degree of explainability to one user (a graduate student at a university) but a low degree of explainability to another user (a high-school student). We must ensure explainability or the trained models $\hat{h}^{(i)}$ for potentially different users of the devices $i = 1, \ldots, n$.

The explainability of trained ML models is closely related to their simulatability [128–130]: How well can a user anticipate (or guess) the prediction $\hat{y} = \hat{h}^{(i)}(\mathbf{x})$ delivered by $\hat{h}^{(i)}$ for a data point with features $\mathbf{x}$. We can then measure the explainability of $\hat{h}^{(i)}(\mathbf{x})$ to the user at node i by comparing the prediction $\hat{h}^{(i)}(\mathbf{x})$ with the corresponding *guess* (or *simulation*) $u^{(i)}(\mathbf{x})$.

We can enforce (subjective) explainability of FL systems by modifying the local loss functions in GTVMin. For ease of exposition, we focus on the GTVMin instance (5.1) for training local (personalized) linear models. For each node $i \in \mathcal{V}$, we construct a test-set $\mathcal{D}_t^{(i)}$ and ask user i to deliver a guess $u^{(i)}(\mathbf{x})$ for each data point in $\mathcal{D}_t^{(i)}$.[5]

We measure the (subjective) explainability of a linear hypothesis with model parameters $\mathbf{w}^{(i)}$ by

$$(1/|\mathcal{D}_t^{(i)}|) \sum_{\mathbf{x} \in \mathcal{D}_t^{(i)}} \left(u^{(i)}(\mathbf{x}) - \mathbf{x}^T \mathbf{w}^{(i)} \right)^2. \tag{8.16}$$

It seems natural to add this measure as a penalty term to the local loss function in (5.1), resulting in the new loss function

$$L_i\left(\mathbf{w}^{(i)}\right) := \underbrace{(1/m_i) \left\| \mathbf{y}^{(i)} - \mathbf{X}^{(i)} \mathbf{w}^{(i)} \right\|_2^2}_{\text{training error}} + \underbrace{\rho\, (1/|\mathcal{D}_t^{(i)}|) \sum_{\mathbf{x} \in \mathcal{D}_t^{(i)}} (u^{(i)}(\mathbf{x}) - \mathbf{x}^T \mathbf{w}^{(i)})^2}_{\text{subjective explainability}}. \tag{8.17}$$

The regularization parameter ρ controls the preference for a high subjective explainability of the hypothesis $h^{(i)}(\mathbf{x}) = \left(\mathbf{w}^{(i)}\right)^T \mathbf{x}$ over a small training error [130]. It can be shown that (8.17) is the average weighted squared error loss of $h^{(i)}(\mathbf{x})$ on an augmented version of $\mathcal{D}^{(i)}$. This augmented version includes the data point $\left(\mathbf{x}, u^{(i)}(\mathbf{x})\right)$ for each data point $\mathbf{x}$ in the test-set $\mathcal{D}_t^{(i)}$.

So far, we have focused on the problem of explaining (the predictions of) a trained personalized model to some user. The general idea is to provide partial information, in the form of some explanation, about the learned hypothesis map $\hat{h}$. Explanations should help the user to anticipate the prediction $\hat{h}(\mathbf{x})$ for any given data point. Instead of explaining a given trained model $\hat{h}$, it might be more useful to explain an entire FL algorithm.

Mathematically, we can interpret an FL algorithm as a map $\mathcal{A}$ that reads in local datasets and delivers learned hypothesis maps $\hat{h}^{(i)}$. We can explain an FL algorithm by providing partial information about this map $\mathcal{A}$. Thus, mathematically speaking, the problem of explaining a learned hypothesis is essentially the same as the problem of explaining an entire FL algorithm: Provide partial information about a map such that the user can anticipate the results of applying the map to arbitrary arguments. However, a description of the map $\mathcal{A}$ is typically more complex, in a quantitative sense, than a learned hypothesis map.

[5] We only use the features of the data points in $\mathcal{D}_t^{(i)}$; i.e., this dataset can be constructed from unlabeled data.

```python
from sklearn.datasets import load_iris
from sklearn.model_selection import train_test_split
from sklearn.tree import DecisionTreeClassifier
from sklearn.metrics import accuracy_score

# Load the Iris dataset
data = load_iris()
X = data.data
y = data.target

# Split the dataset into training and test sets
X_train, X_test, y_train, y_test = train_test_split(X, y,
    test_size=0.3, random_state=42)

# Create a Decision Tree classifier
clf = DecisionTreeClassifier(random_state=42)

# Train the classifier
clf.fit(X_train, y_train)

# Make predictions on the test data
y_pred = clf.predict(X_test)

# Calculate accuracy
accuracy = accuracy_score(y_test, y_pred)
accuracy
```

Fig. 8.3 Python code for an ML method that trains a decision tree on the *Iris* dataset

The different complexity levels of maps to be explained require different forms of explanation. For example, we could explain an FL algorithm using a pseudo-code such as Algorithm 4. Figure 8.3 illustrates another form of explanation, i.e., a code fragment written in the programming language Python.

8.5 Diversity, Nondiscrimination, and Fairness

> *…we must enable inclusion and diversity throughout the entire AI system's life cycle…this also entails ensuring equal access through inclusive design processes as well as equal treatment.*" [103, p.18].

The local datasets used for the training of local models should be carefully selected to not enforce existing discrimination. In a healthcare application, there might be significantly more training data for patients of a specific gender, resulting in models that perform best for that specific gender at the cost of worse performance for the minority [123, Sec. 3.3.].

Fairness is also important for ML methods used to determine credit score and, in turn, if a loan should be granted or not [131]. Here, we must ensure that ML methods do not discriminate against customers based on ethnicity or race. To this end, we could augment data points by modifying any features that mainly reflect the ethnicity or race of a customer (see Fig. 8.4).

8.6 Societal and Environmental Well-Being

> *…Sustainability and ecological responsibility of AI systems should be encouraged, and research should be fostered into AI solutions addressing areas of global concern, such as, for instance, the Sustainable Development Goals.*"[103, p.19].

Society FL systems might be used to deliver personalized recommendations to users within a social media application (social network). These recommendations might be (fake) news used to boost polarization and, in the extreme case, social unrest [132].

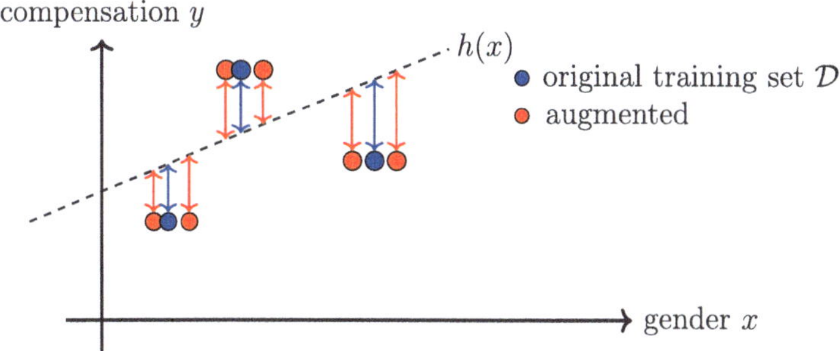

Fig. 8.4 We can improve the fairness of an ML method by augmenting the training set using perturbations of an irrelevant feature such as the gender of a person for which we want to predict the adequate compensation as the label

Environment Chapter 5 discussed FL algorithms that were obtained by applying gradient-based methods to solve GTVMin. These methods require computational resources to compute local updates for model parameters and to share them across the edges of the FL network. Computation and communication require energy which should be generated in an environmental-friendly fashion [133].

8.7 Exercises

8.1 Robustness of GTVMin. Discuss the robustness of GTVMin (3.24) for training local linear models. In particular, which attack is more effective (detrimental): perturbing the labels, the features of data points in the local datasets or perturbing the FL network, e.g., by removing (or adding) edges.

8.2 Subjectively Explainable FL. Consider GTVMin (3.24) to train local linear models with model parameters $\mathbf{w}^{(i)}$. The local datasets are modeled as (3.32). Each local model has a user that is characterized by the user signal $u(\mathbf{x}) := \mathbf{x}^T\mathbf{u}^{(i)}$. To ensure subjective explainability of local model with model parameters $\mathbf{w}^{(i)}$, we require the deviation $(1/m_i)\left\|\widetilde{\mathbf{X}}^{(i)}\left(\mathbf{w}^{(i)} - \mathbf{u}^{(i)}\right)\right\|_2^2$ to be sufficiently small. Here, we used the feature matrix $\widetilde{\mathbf{X}}^{(i)}$ obtained from the realization of m_i i.i.d. RVs with common probability distribution $\mathcal{N}(\mathbf{0}, \mathbf{I})$. We then add this deviation to the local loss functions resulting in using the augmented loss function (8.17) used in (3.24). Study, either analytically or by numerical experiments, the effect of varying levels of explainability (via the parameter ρ in (8.17)) on the estimation error $\widehat{\mathbf{w}}^{(i)} - \overline{\mathbf{w}}^{(i)}$.

Chapter 9
Privacy Protection in FL

Abstract This chapter examines how federated learning (FL) systems can be designed to protect user privacy. It begins by discussing the risks of privacy leakage when model parameters are shared across devices, even without transmitting raw data. The chapter then introduces formal measures to quantify privacy, including sensitivity analysis and differential privacy (DP). It presents two main approaches to enforcing DP in FL: adding random noise to model updates (postprocessing) and modifying data before training (preprocessing). The chapter also introduces Rényi differential privacy as a more flexible alternative to standard DP. In addition, it explores private feature learning, which involves transforming features to reduce the risk of exposing sensitive attributes while preserving task performance. The chapter concludes with practical strategies, including the privacy funnel method and optimal linear transformations, to balance utility and privacy. These tools provide a foundation for building FL systems that comply with data protection regulations while maintaining high model quality.

The core idea of FL is to share information contained in collections of local datasets to improve the training of personalized ML models. Chapter 5 discussed FL algorithms that share information in the form of model parameters that are computed from the local loss function. Each node $i \in \mathcal{V}$ receives the current model parameters of other nodes and, after executing a local update, shares its new model parameters with other nodes.

Depending on the design choices for GTVMin-based methods, sharing model parameters allows to reconstruct local loss functions and, in turn, to estimate private information about individual data points which represent human patients [134]. Thus, the bad news is that FL systems will almost inevitably incur some leakage of private information. The good news is, however, that the extent of privacy leakage can be controlled by (i) careful design choices for GTVMin and (ii) applying modifications to basic FL algorithms from Chap. 5.

A. Jung, *Federated Learning*, https://doi.org/10.1007/978-981-95-1009-2_9

This chapter revolves around two main questions:

- How can we measure privacy leakage in an FL system?
- How can we control (minimize) privacy leakage of an FL system?

Section 9.1 addresses the first question, while Sects. 9.2 and 9.3 address the second question.

9.1 Measuring Privacy Leakage

Consider an FL system designed to train personalized models for users indexed by $i = 1, \ldots, n$, each equipped with a heart rate sensor. Every user i generates a local dataset $\mathcal{D}^{(i)}$, consisting of time-stamped heart rate measurements. A single data point corresponds to one physical activity, such as a 50-minute run. The features of such a data point includes a time series of GPS coordinates, while the label may be the average heart rate recorded during the activity. We assume that this average heart rate is private and should not be disclosed to third parties.[1]

To enhance learning, the FL system incorporates expert-provided information in the form of pairwise similarity scores $A_{i,i'}$ between users i and i', based on characteristics such as body weight and height. These similarity scores are used to regularize the learning process.

Using an FL algorithm—such as Algorithm 4—we aim to learn, for each user i, personalized model parameters $\mathbf{w}^{(i)}$ for an AI-powered healthcare assistant [135]. This algorithm can be represented as a map $\mathcal{A}(\cdot)$ that takes as input the collection of local datasets

$$\mathcal{D} := \left\{\mathcal{D}^{(i)}\right\}_{i=1}^{n}$$

and delivers learned model parameters

$$\mathcal{A}(\mathcal{D}) := \left(\widehat{\mathbf{w}}^{(1)}, \ldots, \widehat{\mathbf{w}}^{(n)}\right).$$

Figure 9.1 illustrates the mapping from local datasets to learned model parameters that is implemented by an FL algorithm.

A privacy-preserving FL system should not allow to infer, solely from the learned model parameters, the average heart rate $y^{(i,r)}$ during a specific single activity r of a specific user i. Mathematically, we must ensure that the map $\mathcal{A}$ is not invertible: The learned model parameters (or hypothesis) should not change if we were to apply the FL algorithm to a perturbed dataset that includes a different value for the average heart rate $y^{(i,r)}$.

[1] For instance, individuals may not wish to share heart rate profiles with potential employers.

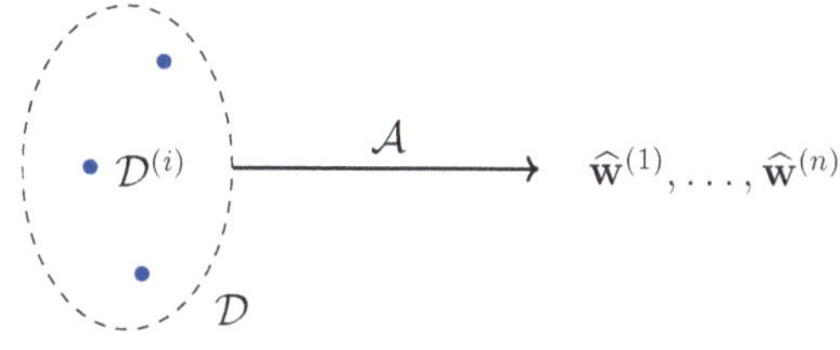

Fig. 9.1 An FL algorithm maps the local datasets $\mathcal{D}^{(i)}$ to the learned model parameters $\widehat{\mathbf{w}}^{(i)}$, for $i = 1, \ldots, n$

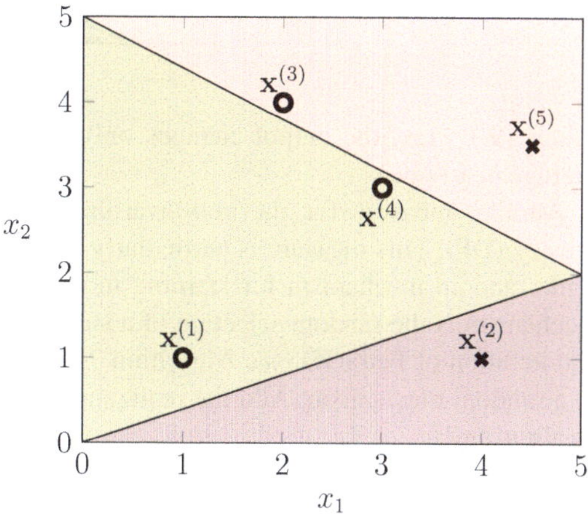

Fig. 9.2 Scatterplot of a dataset used to train a decision tree. We indicate the decision regions along with the labels of data points (via their markers)

Figure 9.2 depicts the decision regions of a decision tree. This decision tree has been trained by (approximately) solving ERM with a training set that consists of four data points. Each data point is characterized by a feature vector $\mathbf{x}^{(r)} = \left(x_1^{(r)}, x_2^{(r)}\right)^T$ and a binary label $y^{(r)} \in \{\circ, \times\}$, for $r = 1, \ldots, 5$. If an attacker would know the label values of $\mathbf{x}^{(1)}, \mathbf{x}^{(4)}$, it could infer the label of $\mathbf{x}^{(2)}$ based on the decision regions.

The sole requirement for an FL algorithm $\mathcal{A}$ to be not invertible is not sufficient in general. Indeed, we can easily make any algorithm $\mathcal{A}$ by simple pre- or post-processing techniques whose effect is limited to irrelevant regions of the input space. Note that the input space is the space of all possible datasets. The level of privacy protection offered by $\mathcal{A}$ can be characterized by a measure of its non-invertibility (or non-injectivity).

A simple measure of non-invertibility is the sensitivity of the output $\mathcal{A}(\mathcal{D})$ against varying the heart rate value $y^{(i,r)}$,

$$\frac{\left\|\mathcal{A}(\mathcal{D}) - \mathcal{A}(\mathcal{D}')\right\|_2}{\varepsilon}. \tag{9.1}$$

Here, $\mathcal{D}$ denotes some given collection of local datasets and $\mathcal{D}'$ is a modified dataset. In particular, $\mathcal{D}'$ is obtained by replacing the actual average heart rate $y^{(i,r)}$ with the modified value $y^{(i,r)} + \varepsilon$. The privacy protection offered by $\mathcal{A}$ is higher for smaller

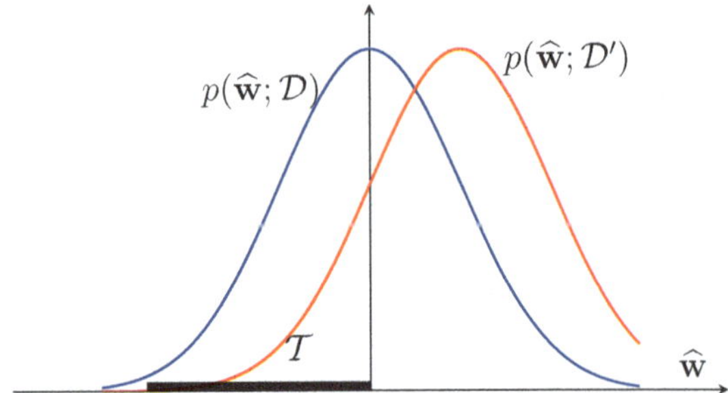

Fig. 9.3 Probability distributions of the learned model parameters $\widehat{\mathbf{w}} = \left(\widehat{\mathbf{w}}^{(1)}, \ldots, \widehat{\mathbf{w}}^{(n)}\right)$ delivered by some FL algorithm for two different input datasets, denoted by $\mathcal{D}'$ and $\mathcal{D}$

values (9.1); i.e., the output changes only a little when varying the value of the average heart rate.

Another measure for the non-invertibility of $\mathcal{A}$ is referred to as differential privacy (DP). This measure is particularly useful for stochastic algorithms that use some random mechanism for learning model parameters. One example of such a mechanism is the random selection of a subset data points that form a batch within one iteration of FedSGD (see Algorithm 7). Section 9.2 discusses another example of a random mechanism: Add the realization of a RV to the intermediate results of an algorithm.

A stochastic algorithm $\mathcal{A}$ can be described by a probability distribution $p(\widehat{\mathbf{w}}; \mathcal{D})$ over the possible values of the learned model parameters $\widehat{\mathbf{w}}$. Figure 9.3 illustrates a stochastic algorithm along with the associated probability distribution $p(\widehat{\mathbf{w}}; \mathcal{D})$.[2] This probability distribution is parametrized by the dataset $\mathcal{D}$ that is fed as input to the algorithm $\mathcal{A}$. Figure 9.3 depicts the probability distributions of an algorithm for two different choices $\mathcal{D}, \mathcal{D}'$ of the input dataset.

DP measures the non-invertibility of a stochastic algorithm $\mathcal{A}$ via the similarity of the probability distributions obtained for two datasets $\mathcal{D}, \mathcal{D}'$ that are considered as adjacent (or neighboring) [123, 138]. Typically, we consider $\mathcal{D}'$ as adjacent to $\mathcal{D}$ if it is obtained by modifying the features or label of a single data point in $\mathcal{D}$.

As a case in point, consider data points representing physical activities which are characterized by a binary feature $x_j \in \{0, 1\}$ that indicates an excessively high average heart rate during the activity. We could then define neighboring datasets by changing the feature x_j of a single data point. In general, the notion of neighboring datasets is a design choice used in the definition of quantitative measures for privacy protection. An FL algorithm ensures privacy protection if there is no statistical test that allows to reliably distinguish between neighboring input datasets. Figure 9.3 illustrates the acceptance region $\mathcal{T}$ that defines a statistical test.

[2] For more details about the concept of a measurable space, we refer to the literature [28, 136, 137].

The de facto standard for quantifying privacy leakage in ML and FL systems is the following definition:

Definition 9.1 (From [138]) A stochastic algorithm $\mathcal{A}$ is (ε, δ)-DP if, for any two neighbouring datasets $\mathcal{D}, \mathcal{D}'$,

$$\text{Prob}\{\mathcal{A}(\mathcal{D}) \in \mathcal{S}\} \leq \exp(\varepsilon)\text{Prob}\{\mathcal{A}(\mathcal{D}') \in \mathcal{S}\} + \delta$$

holds for every measurable set $\mathcal{S}$.

Definition 9.1 formalizes the notion that the presence or absence of an individual data point (representing, e.g., human individual) in a dataset $\mathcal{D}$ should not significantly affect the probability distribution of the output $\mathcal{A}(\mathcal{D})$. The notion of (ε, δ)-DP is widely adopted in FL applications [138–141]. The US Census Bureau adopted (ε, δ)-DP for the 2020 census [141]. The National Institute of Standards and Technology (NIST) has published some guidance for evaluating and implementing DP mechanisms in government and industry settings [142].

Besides (ε, δ)-DP, there have also been proposed other measures for privacy leakage. These measures differ by how they quantify precisely the similarity between probability distributions $p(\widehat{\mathbf{w}}; \mathcal{D})$ and $p(\widehat{\mathbf{w}}; \mathcal{D}')$ induced by neighboring datasets [143]. One such alternative measure is the Rényi divergence of order $\alpha > 1$,

$$D_{\alpha}\Big(p(\widehat{\mathbf{w}}; \mathcal{D}) \big\| p(\widehat{\mathbf{w}}; \mathcal{D}')\Big) := \frac{1}{\alpha - 1}\mathbb{E}_{p(\widehat{\mathbf{w}}; \mathcal{D}')}\left[\left(\frac{dp(\widehat{\mathbf{w}}; \mathcal{D})}{dp(\widehat{\mathbf{w}}; \mathcal{D}')}\right)^{\alpha}\right].$$

The Rényi divergence allows to define the following variant of DP [143, 144].

Definition 9.2 (From [138]) An algorithm $\mathcal{A}$ is (α, γ)-RDP if, for any two neighboring datasets $\mathcal{D}$ and $\mathcal{D}'$,

$$D_{\alpha}\Big(p(\widehat{\mathbf{w}}; \mathcal{D}) \big\| p(\widehat{\mathbf{w}}; \mathcal{D}')\Big) \leq \gamma.$$

A recent use case of (α, γ)-RDP is the analysis of DP guarantees offered by variants of SGD [143]. This analysis uses the fact that (α, γ)-RDP implies (ε, δ)-DP for suitable choices of ε, δ [143].

One important property of the DP notions in Definitions 9.1 and 9.2 is that they are preserved by post-processing:

Proposition 9.1 *Consider an FL system $\mathcal{A}$ that is applied to some dataset $\mathcal{D}$ and some (possibly stochastic) map $\mathcal{B}$ that does not depend on $\mathcal{D}$. If $\mathcal{A}$ is (ε, δ)-DP (or (α, γ)-RDP), then so is also the composition $\mathcal{B} \circ \mathcal{A}$.*

Proof See, e.g., [138, Sec. 2.3]. □

According to Proposition (9.1), the level of DP offered by an algorithm $\mathcal{A}$ does not deteriorate by any post-processing of its output. It seems almost natural to make this immunity against post-processing a defining property of any useful notion of DP [144]. However, due to Proposition (9.1), this property is already "built-in" into the Definitions 9.1 and 9.2.

Operational Meaning of DP The mathematically precise formulation of DP in Definition 9.1 is somewhat abstract. It is instructive to interpret (ε, δ)-DP from the perspective of hypothesis testing [142]: We use the output $\widehat{\mathbf{w}} \in \mathbb{R}^d$ of algorithm $\mathcal{A}$ to test if the underlying dataset fed into $\mathcal{A}$ was $\mathcal{D}$ or if it was a neighboring dataset $\mathcal{D}'$ [145]. Such a statistical test uses a region $\mathcal{T} \subseteq \mathbb{R}^d$ and to declare

- "dataset $\mathcal{D}$ seems to be used" if $\widehat{\mathbf{w}} \in \mathcal{T}$, or
- "dataset $\mathcal{D}'$ seems to be used" if $\widehat{\mathbf{w}} \notin \mathcal{T}$.

The performance of a test $\mathcal{T}$ is characterized by two error probabilities:

- The probability of declaring $\mathcal{D}'$ but actually $\mathcal{D}$ was fed into $\mathcal{A}$, which is $P_{\mathcal{D}\rightarrow\mathcal{D}'} := 1 - \int_{\mathcal{T}} p(\widehat{\mathbf{w}}; \mathcal{D})$.
- The probability of declaring $\mathcal{D}$ but actually $\mathcal{D}'$ was fed into $\mathcal{A}$, which is $P_{\mathcal{D}'\rightarrow\mathcal{D}} := \int_{\mathcal{T}} p(\widehat{\mathbf{w}}; \mathcal{D}')$.

For a privacy-preserving algorithm $\mathcal{A}$, there should be no test $\mathcal{T}$ for which both $P_{\mathcal{D}\rightarrow\mathcal{D}'}$ and $P_{\mathcal{D}'\rightarrow\mathcal{D}}$ are simultaneously small (close to 0). This intuition can be made precise as follows (see [146, Thm. 2.1.], [142] or [147]): If an algorithm $\mathcal{A}$ is (ε, δ)-DP, then

$$\exp(\varepsilon) P_{\mathcal{D}\rightarrow\mathcal{D}'} + P_{\mathcal{D}'\rightarrow\mathcal{D}} \geq 1 - \delta. \tag{9.2}$$

Thus, if $\mathcal{A}$ is (ε, δ)-DP with a small ε, δ (close to 0), then (9.2) implies $P_{\mathcal{D}\rightarrow\mathcal{D}'} + P_{\mathcal{D}'\rightarrow\mathcal{D}} \approx 1$.

9.2 Ensuring Differential Privacy

Depending on the underlying local datasets, local models, and optimization method, a GTVMin-based method $\mathcal{A}$ might already ensure DP by design. A basic means of ensuring DP is through careful feature selection (or learning) for the local datasets (see Fig. 9.5). Random sampling used by SGD-based algorithms can also provide a certain level of DP [148, 149].

According to Proposition 9.1, DP can also be actively ensured by applying pre- and post-processing techniques to the input and output of an FL algorithm $\mathcal{A}$. Mathematically, the map $\mathcal{A}$ is concatenated with the maps $\mathcal{I}$ and $\mathcal{O}$. These maps represent the pre- and post-processing and are typically stochastic, i.e., defined by a conditional probability distribution. The concatenation results in a new algorithm

$\mathcal{A}' := \mathcal{O} \circ \mathcal{A} \circ \mathcal{I}$. In summary, for a given dataset $\mathcal{D}$, the new (privacy-enhanced) algorithm $\mathcal{A}'$ produces learned model parameters by

- apply the preprocessing $\mathcal{I}(\mathcal{D})$,
- compute $\mathcal{A}(\mathcal{I}(\mathcal{D}))$ using the original algorithm,
- and finally apply the post-processing $\mathcal{O}(\mathcal{A}(\mathcal{I}(\mathcal{D})))$, yielding $\mathcal{A}'(\mathcal{D})$.

Post-processing Maybe the most widely used post-processing technique for DP is to add *some noise* [138],

$$\mathcal{O}(\mathcal{A}) := \mathcal{A} + \mathbf{n}, \text{ with noise } \mathbf{n} = \left(n_1, \ldots, n_{nd}\right)^T, n_1, \ldots, n_{nd} \overset{i.i.d.}{\sim} p(n). \quad (9.3)$$

Note that the post-processing (9.3) is parametrized by the choice of the probability distribution $p(n)$ of the noise entries. Two important choices are the Laplacian distribution $p(n) := \frac{1}{2b} \exp\left(-\frac{|n|}{b}\right)$ and the normal distribution $p(n) := \frac{1}{\sqrt{2\pi\sigma^2}} \exp\left(-\frac{n^2}{2\sigma^2}\right)$ (i.e., using Gaussian noise $n \sim \mathcal{N}(0, \sigma^2)$).

When using Gaussian noise $n \sim \mathcal{N}(0, \sigma^2)$ in (9.3), the variance σ^2 can be chosen based on the sensitivity

$$\Delta_2(\mathcal{A}) := \max_{\mathcal{D}, \mathcal{D}'} \left\| \mathcal{A}(\mathcal{D}) - \mathcal{A}(\mathcal{D}') \right\|_2. \quad (9.4)$$

Here, the maximum is over all pairs of neighboring datasets $\mathcal{D}, \mathcal{D}'$. Adding Gaussian noise with variance $\sigma^2 > \sqrt{2 \ln(1.25/\delta)} \cdot \Delta_2(\mathcal{A})/\varepsilon$ ensures that $\mathcal{A}$ is (ε, δ)-DP [138, Thm. 3.22]. It might be difficult to evaluate the sensitivity (9.4) for a given FL algorithm $\mathcal{A}$ [150]. For a GTVMin-based method, i.e., $\mathcal{A}(\mathcal{D})$ is a solution to (3.22), we can upper bound $\Delta_2(\mathcal{A})$ via a perturbation analysis similar in spirit to the proof of Proposition 8.1.

Preprocessing Instead of ensuring DP via post-processing the output of an FL algorithm $\mathcal{A}$, we can ensure DP by applying a preprocessing map $\mathcal{I}(\mathcal{D})$ to the dataset $\mathcal{D}$. The result of the preprocessing is a new dataset $\widehat{\mathcal{D}} = \mathcal{I}(\mathcal{D})$ which can be made available (publicly!) to any algorithm $\mathcal{A}$ that has no direct access to $\mathcal{D}$. According to Proposition 9.1, as long as the preprocessing map $\mathcal{I}$ is (ε, δ)-DP (see Definition 9.1), so will be the composition $\mathcal{A} \circ \mathcal{I}$.

As for post-processing, one important approach to preprocessing is to "add" or "inject" noise. This results in a stochastic preprocessing map $\widehat{\mathcal{D}} = \mathcal{I}(\mathcal{D})$ that is characterized by a probability distribution. The noise mechanisms used for preprocessing might be different from just adding the realization of a RV (see (9.3)):[3]

[3] Can you think of a simple preprocessing map that is deterministic and guarantees maximum DP?

- For a classification method with a discrete label space $\mathcal{Y} = \{1, \ldots, K\}$, we can inject noise by replacing the true label of a data point with a randomly selected element of $\mathcal{Y}$ [151, Mechanism 1]. The noise injection might also include the replacement of the features of a data point by a realization of a RV whose probability distribution is somehow matched to the dataset $\mathcal{D}$ [151, Mechanism 2].
- Another form of noise injection is to construct $\mathcal{I}(\mathcal{D})$ by randomly selecting data points from the original (private) dataset $\mathcal{D}$ [152]. Note that such noise injection is naturally provided by SGD methods (see, e.g., step 4 of Algorithm 6).

How To Be Sure? Consider some algorithm $\mathcal{A}$, possibly obtained by pre- and post-processing techniques, that is claimed to be (ε, δ)-DP. In practice, we might not know the detailed implementation of the algorithm. For example, we might not have access to the noise generation mechanism used in the pre- or post-processing steps. How can we verify a claim about DP of algorithm $\mathcal{A}$ without having access to the detailed implementation of $\mathcal{A}$? One approach could be to apply the algorithm to synthetic datasets $\mathcal{D}^{(1)}_{\rm syn}, \ldots, \mathcal{D}^{(L)}_{\rm syn}$ that differ only in some private attribute of a single data point. We can then try to predict the private attribute $s^{(r)}$ of the dataset $\mathcal{D}^{(r)}_{\rm syn}$ by applying a learned hypothesis $\widehat{h}$ to the output $\mathcal{A}\big(\mathcal{D}^{(r)}_{\rm syn}\big)$ delivered by the *algorithm under test* $\mathcal{A}$. The hypothesis $\widehat{h}$ might be learned by an ERM-based method (see Algorithm 1) using a training set consisting of pairs $\left(\mathcal{A}\big(\mathcal{D}^{(r)}_{\rm syn}\big), s^{(r)}\right)$ for some $r \in \{1, \ldots, L\}$.

9.3 Private Feature Learning

Section 9.2 discussed preprocessing techniques that ensure DP of an FL algorithm. We next discuss preprocessing techniques that are not directly motivated from a DP perspective. Instead, we cast privacy-friendly preprocessing of a dataset as a feature learning problem [23, Ch. 9].

Consider a data point characterized by a feature vector $\mathbf{x} \in \mathbb{R}^d$ and a label $y \in \mathbb{R}$. Moreover, each data point is characterized by a private attribute s. We want to learn a (potentially stochastic) feature map $\mathbf{\Phi} : \mathbb{R}^d \rightarrow \mathbb{R}^{d'}$ such that the new features $\mathbf{z} = \mathbf{\Phi}(\mathbf{x}) \in \mathbb{R}^{d'}$ do not allow to accurately predict the private attribute s. Trivially, we can make the accurate prediction of s from $\mathbf{\Phi}(\mathbf{x})$ impossible by using a constant map, e.g., $\mathbf{\Phi}(\mathbf{x}) = 0$. However, we still want the new features $\mathbf{z} = \mathbf{\Phi}(\mathbf{x})$ to allow for a sufficiently accurate prediction (using a suitable hypothesis) of the label y.

Privacy Funnel To quantify the predictability of the private attribute s solely from the transformed features $\mathbf{z} = \phi(\mathbf{x})$, we can use the i.i.d. assumption as a simple but useful probabilistic model. Indeed, we can then use the MI $I\left(s; \mathbf{\Phi}(\mathbf{x})\right)$ as a measure for the predictability of s from $\mathbf{\Phi}(\mathbf{x})$. A small value of $I\left(s; \mathbf{\Phi}(\mathbf{x})\right)$ indicates that it is difficult to predict the private attribute s solely from $\mathbf{\Phi}(\mathbf{x})$, i.e., a high level of privacy

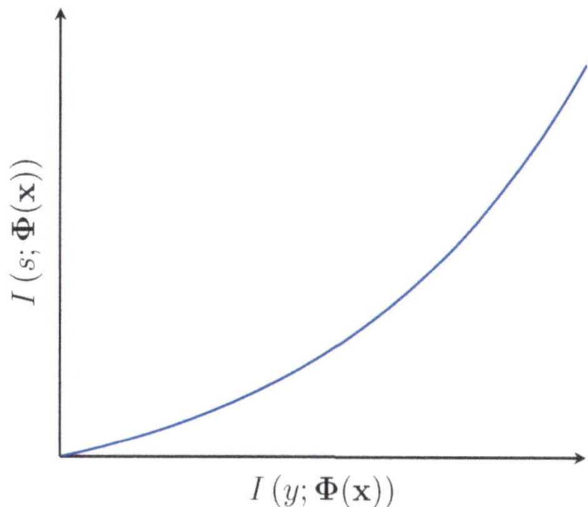

Fig. 9.4 The solutions of the privacy funnel (9.5) trace out (for varying constraint R) a curve in the plane spanned by the values of $I(s; \mathbf{\Phi}(\mathbf{x}))$ (measuring the privacy leakage) and $I(y; \mathbf{\Phi}(\mathbf{x}))$ (measuring the usefulness of the transformed features for predicting the label)

protection.[4] Similarly, we can use the MI $I(y; \mathbf{\Phi}(\mathbf{x}))$ to measure the predictability of the label y from $\mathbf{\Phi}(\mathbf{x})$. A large value $I(y; \mathbf{\Phi}(\mathbf{x}))$ indicates that $\mathbf{\Phi}(\mathbf{x})$ allows to accurately predict y (which is of course preferable).

It seems natural to use a feature map $\mathbf{\Phi}(\mathbf{x})$ that optimally balances a small $I(s; \mathbf{\Phi}(\mathbf{x}))$, i.e., a sufficiently large privacy protection, with a sufficiently large $I(y; \mathbf{\Phi}(\mathbf{x}))$ to allow for an accurate prediction of y. The mathematically precise formulation of this plan is known as the privacy funnel [154, Eq. (2)],

$$\min_{\mathbf{\Phi}(\cdot)} I(s; \mathbf{\Phi}(\mathbf{x})) \text{ such that } I(y; \mathbf{\Phi}(\mathbf{x})) \geq R. \tag{9.5}$$

Figure 9.4 illustrates the solution of (9.5) for varying R, i.e., the minimum value of $I(y; \mathbf{\Phi}(\mathbf{x}))$.

Optimal Private Linear Transformation The privacy funnel (9.5) uses the MI $I(s; \mathbf{\Phi}(\mathbf{x}))$ to quantify the privacy leakage of a feature map $\mathbf{\Phi}(\mathbf{x})$. An alternative measure for the privacy leakage is the minimum reconstruction error $s - \hat{s}$. The reconstruction $\hat{s}$ is obtained by applying a reconstruction map $r(\cdot)$ to the transformed features $\mathbf{\Phi}(\mathbf{x})$. If the joint probability distribution $p(s, \mathbf{x})$ is a multivariate normal distribution and the $\mathbf{\Phi}(\cdot)$ is a linear map (of the form $\mathbf{\Phi}(\mathbf{x}) := \mathbf{F}\mathbf{x}$ with some matrix $\mathbf{F}$), then the optimal reconstruction map is again linear [30].

We would like to find the linear feature map $\mathbf{\Phi}(\mathbf{x}) := \mathbf{F}\mathbf{x}$ such that for any linear reconstruction map $\mathbf{r}$ (resulting in $\hat{s} := \mathbf{r}^T\mathbf{F}\mathbf{x}$) the expected squared error $\mathbb{E}\{(s-\hat{s})^2\}$ is large. The smallest possible expected squared error loss

$$\varepsilon(\mathbf{F}) := \min_{\mathbf{r}\in\mathbb{R}^{d'}} \mathbb{E}\{(s - \mathbf{r}^T\mathbf{F}\mathbf{x})^2\}$$

[4] The relation between MI-based privacy measures and DP has been studied in some detail recently [153].

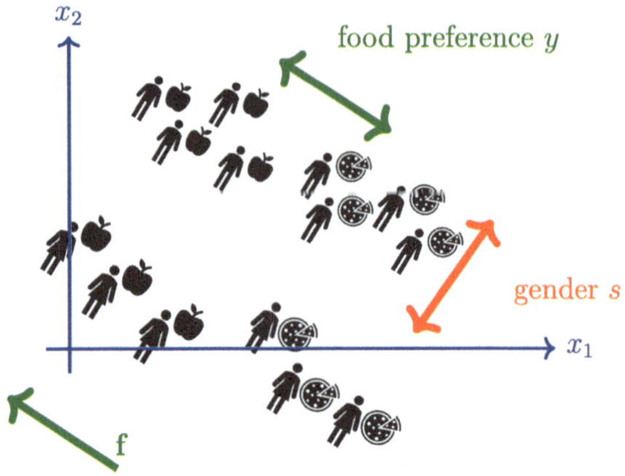

Fig. 9.5 A toy dataset $\mathcal{D}$ whose data points represent customers, each characterized by features $\mathbf{x} = (x_1, x_2)^T$. These raw features carry information about a sensitive attribute s (gender) and the label y (food preference) of a person. The scatterplot that we can find a linear feature transformation $\mathbf{F} := \mathbf{f}^T \in \mathbb{R}^{1 \times 2}$ resulting in a new feature $z := \mathbf{F}\mathbf{x}$ that does not allow to predict s while still allowing to predict y

measures the level of privacy protection offered by the new features $\mathbf{z} = \mathbf{F}\mathbf{x}$. The larger the value $\varepsilon(\mathbf{F})$, the more privacy protection is offered. It can be shown that $\varepsilon(\mathbf{F})$ is maximized by any $\mathbf{F}$ that is orthogonal to the cross-covariance vector $\mathbf{c}_{\mathbf{x},s} := \mathbb{E}\{\mathbf{x}s\}$, i.e., whenever $\mathbf{F}\mathbf{c}_{\mathbf{x},s} = \mathbf{0}$. One specific choice for $\mathbf{F}$ that satisfies this orthogonality condition is

$$\mathbf{F} = \mathbf{I} - (1/\left\|\mathbf{c}_{\mathbf{x},s}\right\|_2^2)\mathbf{c}_{\mathbf{x},s}\mathbf{c}_{\mathbf{x},s}^T. \tag{9.6}$$

Figure 9.5 illustrates a dataset for which we want to find a linear feature map $\mathbf{F}$ such that the new features $\mathbf{z} = \mathbf{F}\mathbf{x}$ do not allow to accurately predict a sensitive attribute.

9.4 Exercises

9.1 Where Is *Alice*? Consider a device, named *Alice*, that implements an asynchronous variant of Algorithm 5 (see (5.28) and (5.29)). The local dataset of the device consists of temperature measurements obtained from some FMI weather station. Assuming that no other device interacts with *Alice* except for your device, named *Bob*. Develop a software for *Bob* that interacts with *Alice*, according to (5.28), in order to determine at which FMI station we can find *Alice*.

9.2 Linear Discriminant Analysis with privacy protection. Consider a binary classification problem with data points characterized by a feature vector $\mathbf{x} \in \mathbb{R}^d$ and a binary label $y \in \{-1, 1\}$. Each data point has a sensitive attribute $s = \mathbf{F}\mathbf{x}$, obtained by applying a fixed matrix $\mathbf{F}$ to the feature vector $\mathbf{x}$. We use a probabilistic model—interpreting data points $(\mathbf{x}, y)$ as i.i.d. realizations of a RV—with the feature vector having multivariate normal distribution $\mathcal{N}\left(\boldsymbol{\mu}^{(y)}, \mathbf{C}^{(y)}\right)$ conditioned on y. The label is uniformly distributed over the label space $\{-1, 1\}$. Try to find a vector $\mathbf{a}$ such that the transformed feature vector $z' := \mathbf{a}^T\mathbf{x}$ optimally balances the privacy leakage (information carried by z' about s) with the information carried by z' about the label y.

9.3 Where Are You? Consider a social media post of a friend that is traveling across Finland. This post includes a snapshot of a temperature measurement and a clock. Can you guess the latitude and longitude of the location where your friend took this snapshot? We can use ERM to do this: Use Algorithm 1 to learn a vector-valued hypothesis $\widehat{h}$ for predicting latitude and longitude from the time and value of a temperature measurement. Use the weather recordings at FMI stations to construct a training set and a validation set.

9.4 Ensuring Privacy with Preprocessing. Repeat the privacy attack described in Exercise 9.3 but this time using a preprocessed version of the raw data. The preprocessing can be implemented either via randomly selecting a subset of data points in the raw dataset or by adding noise to their features and labels. How well can one predict the latitude and longitude from the time and value of a temperature measurement using a hypothesis $\widehat{\mathbf{h}}$ learned from the perturbed data?

9.5 Private Feature Learning. Download hourly weather observations during April 2023 at FMI station *Kustavi Isokari*. You can access these observations here https://en.ilmatieteenlaitos.fi/download-observations. Each time period of 1 hour corresponds to a data point that is characterized by the following features:

- x_1 = Average temperature [°C]
- x_2 = Maximum temperature [°C]
- x_3 = Minimum temperature [°C]
- x_4 = Average relative humidity [%],
- x_5 = Wind speed [m/s],
- x_6 = Maximum wind speed [m/s],
- x_7 = Average wind direction [°],
- x_8 = Maximum gust speed [m/s],
- x_9 = Precipitation [mm],
- x_{10} = Average air pressure [hPa]
- x_{11} = hour of the day $(1, \ldots, 24)$.

The goal of this exercise is to learn a linear feature transformation $\mathbf{z} = \mathbf{F}\mathbf{x}$ such that the new features do not allow to recover the hour of the day x_{11} (which is considered a private attribute s of the data point). However, the new features should still allow to reconstruct the average temperature x_1.

We construct the matrix $\mathbf{F}$ according to (9.6) by replacing the exact cross-covariance vector $\mathbf{c}_{\mathbf{x},s}$ with an estimate (or approximation) $\hat{\mathbf{c}}_{\mathbf{x},s}$. This estimate is computed as follows:

1. read all data points and construct a feature matrix $\mathbf{X} \in \mathbb{R}^{m \times 11}$ with m being the total number of data points
2. remove the sample means from each feature, resulting in the centered feature matrix

$$\widehat{\mathbf{X}} := \mathbf{X} - (1/m)\mathbf{1}\mathbf{1}^T\mathbf{X}\,, \mathbf{1} := \big(1, \ldots, 1\big)^T \in \mathbb{R}^m.$$

3. extract the sensitive attribute or each data point and store it in the vector

$$\mathbf{s} := \big(\hat{x}_1^{(1)}, \hat{x}_1^{(2)}, \ldots, \hat{x}_1^{(m)}\big)^T.$$

4. compute the empirical cross-covariance vector

$$\hat{\mathbf{c}}_{\mathbf{x},s} := (1/m)\big(\widehat{\mathbf{X}}\big)^T \mathbf{s}$$

The matrix $\mathbf{F}$ obtained from (9.6) by replacing $\mathbf{c}_{\mathbf{x},s}$ with $\hat{\mathbf{c}}_{\mathbf{x},s}$ is then used to compute the privacy-preserving features $\mathbf{z}^{(r)} = \mathbf{F}\mathbf{x}^{(r)}$ for $r = 1, \ldots, m$. To verify if these new features are indeed privacy-preserving, we use linear regression (as implemented by the `LinearRegression` class of the Python package `scikit-learn`) to learn the model parameters of a linear model to predict the sensitive attribute $s^{(r)} = x_1^{(r)}$ (the hour of the day during which the measurement has been taken) from the features $\mathbf{z}^{(r)}$.

Chapter 10
Cybersecurity in FL: Attacks and Defenses

Abstract This chapter examines how federated learning (FL) systems can be compromised by adversaries and how to defend against such attacks. It begins by introducing two basic attack vectors: model poisoning, where attackers tamper with model updates, and data poisoning, where they manipulate local datasets. The chapter then categorizes attacks based on their goals, including denial-of-service, backdoor, and model inversion attacks. Each attack type targets a different vulnerability in the FL workflow. Next, the chapter describes how to design GTVMin-based FL methods that are more robust. Defense strategies include clipping or trimming suspicious updates, using robust loss functions, and adding noise to maintain differential privacy. The chapter concludes by highlighting that well-designed FL systems must balance robustness, privacy protection, and resilience to ensure secure and reliable learning across distributed devices.

FL, like ML more broadly, fundamentally relies on externally provided data. In most ML applications, the computational device that trains a model rarely has direct access to the raw data points of the training set. Instead, training often proceeds on preprocessed data supplied by external sources or curated databases.

As a case in point, consider an ML application for animal health care based on monitoring livestock in remote regions. Direct access to raw data points, such as those depicted in Fig. 10.1, would require physically visiting distant pastures with specialized measurement equipment such as stomach sensors. Instead, developers typically rely on external databases assembled by researchers or veterinarians who collected the data on-site.

This reliance on external data is even more pronounced in FL systems. One of the primary purposes of FL is to leverage the information contained in the local datasets of many interconnected devices, which form an FL network. However, this raises a critical question: *How can we be confident that every device behaves as intended and faithfully follows the agreed-upon FL algorithm?*

A. Jung, *Federated Learning*, https://doi.org/10.1007/978-981-95-1009-2_10

Fig. 10.1 In many ML applications, such as monitoring livestock in remote regions, direct access to raw data points is impractical. ML methods often rely on external databases curated by third parties, introducing potential vulnerabilities

Except in the rare case where we have full control over every device in the FL network, it is essential to design FL systems that are robust against potential attacks. Here, an attack refers to the intentional perturbation (or manipulation) of FL system parts.

This chapter is structured as follows: Sect. 10.1 discusses how such attacks can be carried out by perturbing different components of an FL system. Section 10.2 distinguishes different attack types according to their objectives. Section 10.3 provides some guidance on the design choices for GTVMin-based methods to ensure robustness against attacks.

10.1 A Simple Attack Model

Consider an FL system that implements one of the FL algorithms discussed in Chap. 5. As discussed in Sect. 5.7, these algorithms share a common form. Many widely used FL algorithms for parametric local models (with model parameters belonging to $\mathbb{R}^d$) compute and share the results (across the edges of the FL network) of local updates

$$\mathbf{w}^{(i,k+1)} = \underset{\mathbf{w}^{(i)} \in \mathbb{R}^d}{\operatorname{argmin}} \left[L_i\left(\mathbf{w}^{(i)}\right) + \sum_{i' \in \mathcal{N}^{(i)}} A_{i,i'} \phi\left(\mathbf{w}^{(i',k)} - \mathbf{w}^{(i,k)}\right) \right]. \tag{10.1}$$

During time-instant k, device i solves (10.1) in order to obtain new model parameters $\mathbf{w}^{(i,k+1)}$. Carefully note that update (10.1) involves the model parameters $\mathbf{w}^{(i',k)}$ at neighbors $i' \in \mathcal{N}^{(i')}$. In practice, these model parameters need to be communicated over some physical channel (e.g., a short-range wireless link) between device i and device i'. Figure 10.2 illustrates the information flow during the local update (10.1).

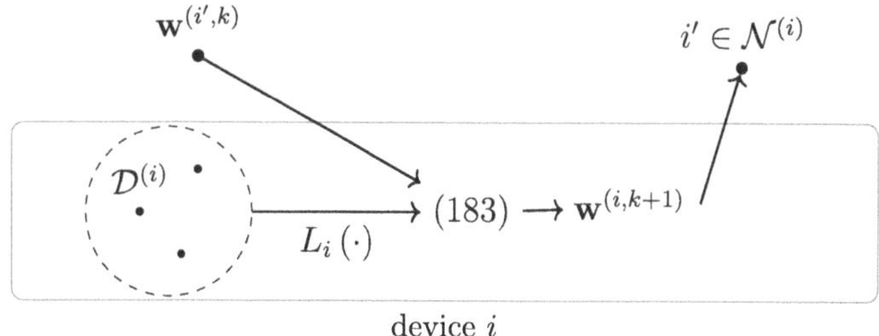

Fig. 10.2 A GTVMin-based FL system from the perspective of a specific device i

From the viewpoint of a specific device i, control is typically limited to the local loss function $L_i\left(\mathbf{w}^{(i)}\right)$, which is often computed as the average loss over the local dataset.[1] In contrast, the model parameters $\mathbf{w}^{(i',k)}$ received from neighboring devices may be unreliable: They can be intentionally perturbed (or poisoned). In what follows, we describe two major classes of attacks that exploit different parts of the FL system to manipulate the shared model parameters $\mathbf{w}^{(i',k)}$ and thereby influence the local update step (10.1).

10.1.1 *Model Poisoning*

If an attacker has control over some of the communication links within an FL system, it can directly manipulate the model parameters shared between nodes. A model poisoning attack on the update (10.1) replaces the vector $\mathbf{w}^{(i',k)}$, for some $i' \in \mathcal{N}^{(i)}$ with a perturbed vector $\widetilde{\mathbf{w}}^{(i',k)}$. We have already discussed the robustness of the update (10.1), for specific choices of ϕ, against perturbations in Sect. 8.2.3.

10.1.2 *Data Poisoning*

Consider an attacker with access to the local datasets of a subset of devices $\mathcal{W} \subset \mathcal{V}$ in the FL network. By poisoning the local datasets at these compromised nodes, the attacker can manipulate the corresponding local updates (10.1). Protecting a given device i from such poisoning is nontrivial, especially when the attacker can exploit software vulnerabilities, such as those in smartphone operating systems [155].

The impact of the poisoned updates propagates from nodes $i' \in \mathcal{W}$ through the edges of the FL network during successive update steps. As a result, even nodes

[1] This assumption may not always hold in practice—for instance, the FL application might not be granted full access to the operating system of device i (e.g., a smartphone).

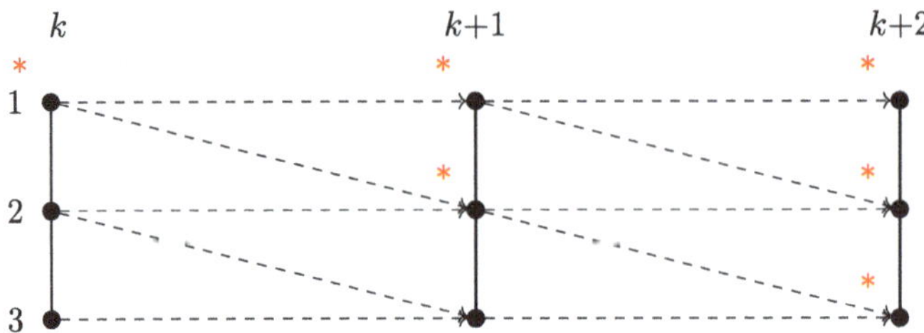

Fig. 10.3 Propagation of the effect of a data poisoning attack that perturbs the update (10.1) of $i = 1$ during time k

whose local datasets remain clean can eventually be affected—provided they are connected to $\mathcal{W}$. In fact, if the FL network $\mathcal{G}$ is connected, the influence of poisoned updates can reach all nodes within a number of steps proportional to the graph's diameter.

Figure 10.3 illustrates this phenomenon in a chain-structured FL network with three nodes $i = 1, 2, 3$ connected by unit-weight edges $\mathcal{E} = \{1, 2\}, \{2, 3\}$. The attacker poisons the local dataset $\mathcal{D}^{(1)}$ at node $i = 1$ at time $k - 1$, resulting in a perturbed update at time k. This perturbation influences node $i = 2$ at time $k + 1$ and subsequently node $i = 3$ at time $k + 2$. The affected updates are marked by a red star ($*$) in Fig. 10.3.

Data poisoning can consist of adding the realization of RVs to the features and label of a data point: We poison a data point by replacing its features $\mathbf{x}$ and label y with $\widetilde{\mathbf{x}} := \mathbf{x} + \Delta\mathbf{x}$ and $\tilde{y} = y + \Delta y$.

For FL applications with local models being used for predicting the label of a data point, we further distinguish between the following data poisoning strategies [156]:

- **Label Poisoning.** The attacker manipulates the labels of data points in the training set.
- **Clean-Label attack.** The attacker leaves the labels untouched and only manipulates the features of data points in the training set.

The effect of data poisoning is that the original local loss functions $L_i(\cdot)$ in GTVMin (3.24) are replaced by perturbed local loss functions $\tilde{L}_i(\cdot)$. The degree of perturbation depends on the fraction of poisoned data points as well as the choice of the loss function used to measure the prediction error.

Different loss functions provide varying levels of robustness against data poisoning. For example, using the absolute error loss yields increased robustness against perturbations of the label values of a few data points, compared to the squared error loss (see Exercise 10.4). Another class of robust loss functions is obtained by including a penalty term (as in regularization).

10.2 Attack Types

Based on their objective, we distinguish the following attacks on FL systems: denial-of-service attacks, backdoor attacks, and privacy (or model inversion) attacks [157].

- **Denial-of-service attack.** A denial-of-service attack manipulates $\mathbf{w}^{(i',k)}$ in (10.1) to nudge the updates $\mathbf{w}^{(i,k+1)}$ toward model parameters $\overline{\mathbf{w}}^{(i)}$ with a large local loss. In other words, the resulting hypothesis $\bar{h}^{(i)}$ delivers poor predictions for the data points in the local dataset $\mathcal{D}^{(i)}$ (see Fig. 10.4) [158].
- **Backdoor attack.** This attack manipulates $\mathbf{w}^{(i',k)}$ in (10.1) to nudge the updates $\mathbf{w}^{(i,k+1)}$ toward model parameters $\widetilde{\mathbf{w}}^{(i)}$ with a small loss on the local dataset but highly irregular predictions for specific feature vectors. In other words, the hypothesis $\tilde{h}^{(i)}$ "behaves well" on $\mathcal{D}^{(i)}$ but delivers prespecified predictions on a subset $\mathcal{K} \subseteq \mathcal{X}$ of the feature space. We can interpret the subset $\mathcal{K}$ as a backdoor which is opened by any data point with a feature vector $\mathbf{x} \in \mathcal{K}$ (see Fig. 10.4) [159].
- **Privacy (or model inversion) attack.** This attack manipulates $\mathbf{w}^{(i',k)}$ in (10.1) such that the updates $\mathbf{w}^{(i,k+1)}$ maximally leak information about sensitive attributes of data points stored at device i. One approach is to force another device i' to learn a copy of model parameters $\mathbf{w}^{(i)}$ by designing trivial local loss functions and manipulating the structure of the FL network (see Exercise 9.1). Once obtained, the copied model parameters can be probed to reveal private information. A notable class of privacy attacks is model inversion, where an attacker tries to reconstruct feature vectors of data points [160].

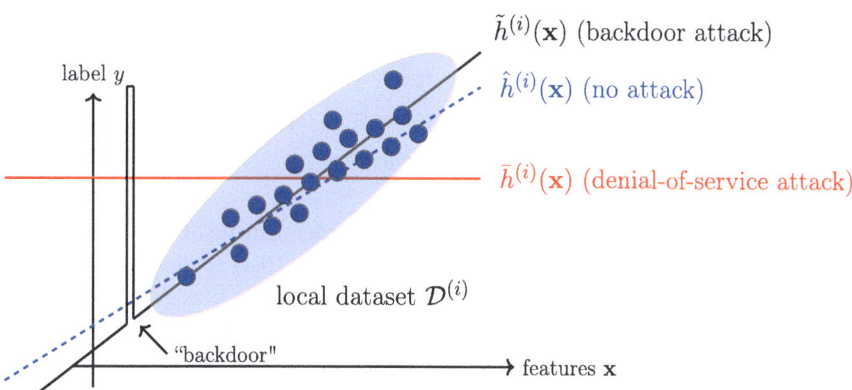

Fig. 10.4 A local dataset $\mathcal{D}^{(i)}$ along with three hypothesis maps learned via iterating (10.1) under three attack scenarios

10.3 Making FL Robust Against Attacks

We next discuss how to make the update (10.1) more robust against the attacks discussed in Sect. 10.2. Our focus will be on GTVMin-based methods using the GTV penalty $\phi(\cdot) = \|\cdot\|_2^2$. For this choice, (10.1) can be written as (see (8.10))

$$\mathbf{w}^{(i,k+1)} \in \underset{\mathbf{w}^{(i)} \in \mathbb{R}^d}{\operatorname{argmin}} L_i\left(\mathbf{w}^{(i)}\right) + \alpha d^{(i)} \left\|\mathbf{w}^{(i)} - \widehat{\mathbf{w}}^{(\mathcal{N}^{(i)})}\right\|_2^2,$$
$$\text{with } \widehat{\mathbf{w}}^{(\mathcal{N}^{(i)})} := (1/d^{(i)}) \sum_{i' \in \mathcal{N}^{(i)}} A_{i,i'} \mathbf{w}^{(i',k)}. \tag{10.2}$$

Here, we used the weighted node degree $d^{(i)} = \sum_{i' \in \mathcal{N}^{(i)}} A_{i,i'}$ (see (3.7)).

The update (10.2) can be attacked via manipulating the model parameters $\mathbf{w}^{(i',k)}$ and, in turn, their average $\widehat{\mathbf{w}}^{(\mathcal{N}^{(i)})}$. Consider an attack that perturbs up to $\eta \cdot |\mathcal{N}^{(i)}|$ of these model parameters (see Fig. 10.5). It turns out that an effective defense against these perturbations is to replace the average by [161]

$$(1/d^{(i)}) \sum_{i' \in \mathcal{N}^{(i)}} A_{i,i'} \tau(\mathbf{w}^{(i',k)}), \tag{10.3}$$

with some generalized threshold (or clipping) function τ. The literature on robust FL has studied different constructions of τ [161, 162]. Intuitively, the threshold function τ should not change clean model parameters $\mathbf{w}^{(i',k)}$ but also limit the impact of perturbed $\mathbf{w}^{(i',k)}$.

For the special case of model dimension, i.e., each local model is parametrized by a single number $w \in \mathbb{R}$, one useful choice for τ in (10.3) is

$$\tau(w) = \begin{cases} \tau_u & \text{for } w \geq \tau_u \\ w & \text{for } w \in [\tau_l, \tau_u] \\ a & \text{for } w \leq \tau_l. \end{cases} \tag{10.4}$$

A natural choice for the thresholds τ_l, τ_u is to use order statistic of the values $\mathbf{w}^{(i',k)}$ for $i' \in \mathcal{N}^{(i)}$. In particular, the upper threshold τ_u in (10.4) is chosen such that it is exceeded by $\mathbf{w}^{(i',k)}$ only for a small number of neighbors $i' \in \mathcal{N}^{(i)}$. The lower threshold τ_l in (10.4) is chosen analogously (see Fig. 10.5). The robustness of using (10.4) in the averaging step (10.3) has been studied in [161].

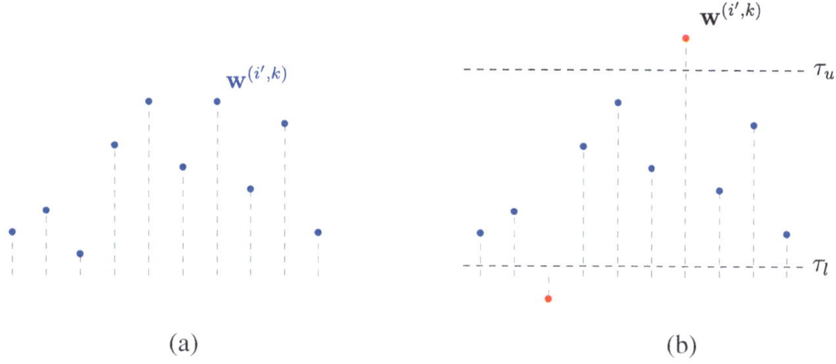

Fig. 10.5 An attack on (10.1) perturbs (adversarially) a fraction η of the received model parameters $\mathbf{w}^{(i',k)}$. (**a**) Original ("clean") model parameters. (**b**) Poisoned model parameters

Another important choice for the threshold function τ in (10.3) is

$$\tau(w) = c \begin{cases} w & \text{if } w \in \mathcal{T} \\ 0 & \text{otherwise,} \end{cases}$$

$$\text{with } c = \frac{|\mathcal{N}^{(i)}|}{|\{i' \in \mathcal{N}^{(i)} : \mathbf{w}^{(i',k)} \in \mathcal{T}\}|}. \tag{10.5}$$

Inserting (10.5) into (10.3) yields the trimmed mean [163]. Indeed, the effect of (10.5) is that the average (10.3) is computed over a subset (or trimmed version) $\mathcal{T}$ of $\mathbf{w}^{(i',k)}$ for $i' \in \mathcal{N}^{(i)}$. Different constructions for the subset $\mathcal{T}$ in (10.5) have been studied in the literature on robust FL [164–166]. One such construction is based on the order statistic of $\mathbf{w}^{(i',k)}$, for $i' \in \mathcal{N}^{(i)}$, by excluding the most extreme values [167].

Note that (10.5) is defined for scalar model parameters $\mathbf{w}^{(i',k)} \in \mathbb{R}$ (i.e., for local models with dimension $d = 1$). We can generalize (10.5) to higher dimensions $d > 1$ by applying it separately to each entry $w_1^{(i,k)}, \ldots, w_d^{(i,k)}$ of the model parameters $\mathbf{w}^{(i,k)}$. The robustness of GTVMin-based methods using the averaging step (10.3) has been studied in [167].

So far, our discussion focused on protecting the update (10.2) (which is the core step of GTVMin-based methods that use the GTV penalty $\phi(\cdot) = \|\cdot\|_2^2$) against denial-of-service attacks and backdoor attacks. We now discuss how to protect (10.2) against privacy attacks.

For a fixed time k, we can ensure a prescribed level of DP by replacing the update (10.2) with a noisy version

$$\mathbf{w}^{(i,k+1)} + \sigma \cdot \mathbf{n}^{(k)}, \text{ with scaled noise } \sigma \cdot \mathbf{n}^{(k)}. \tag{10.6}$$

This noisy update (10.6) is then shared with the neighbors $i' \in \mathcal{N}^{(i)}$. Any (or even each) of these neighbors could be involved in a privacy attack that aims to learn a sensitive attribute of the local dataset $\mathcal{D}^{(i)}$.

The noise term $\mathbf{n}^{(k)}$ in (10.6) is drawn independently for each time k from a prescribed probability distribution, such as the Laplace distribution or the normal distribution [138]. A key challenge for implementing (10.6) is to find a useful choice for the noise strength σ. Increasing σ results in stronger privacy protection but typically degrades the accuracy of the trained local models [142]. However, choosing σ too small can result in insufficient privacy protection.

The minimum value σ required to ensure (ε, δ)-DP (see Definition 9.1) with prescribed values $\varepsilon, \delta \geq 0$ depends on

- how the shape of the local loss function $L_i(\cdot)$ changes when data points are added to (or removed from) the local dataset $\mathcal{D}^{(i)}$ (see [168]),
- the value of the GTVMin parameter α,
- the number of time instants k during which the update (10.2) is executed and the noisy result (10.6) shared with the neighbors [146, 169].

10.4 Exercises

10.1 Model inversion for linear regression. Consider an ERM-based method for training a linear model using plain GD. Assume that the model parameters are initialized to zero, $\mathbf{w}^{(0)} = \mathbf{0} \in \mathbb{R}^d$, and that the training error $\widehat{L}(\mathbf{w})$ is the average squared error loss on a training set,

$$\mathcal{D} = \left\{ \left(\mathbf{x}^{(1)}, y^{(1)}\right), \ldots, \left(\mathbf{x}^{(m)}, y^{(m)}\right) \right\}.$$

Suppose an attacker can observe the sequence of gradients, $\nabla \widehat{L}\left(\mathbf{w}^{(k)}\right)$ computed during the first few iterations $k = 0, 1, \ldots$. To what extent is it possible, based solely on the observed gradients and the knowledge of zero initialization, to reconstruct the training set?

10.2 Denial-of-service attack. Construct an FL network of FMI stations, and store it as a `networkx.Graph()` object. Implement Algorithm 4 to learn, for each node $i = 1, \ldots, n$, the model parameters of a linear model. Launch a denial-of-service attack by poisoning the local datasets at increasingly many nodes $i' \neq 1$. The goal of the attack is to increase the validation error of the learned model parameters $\mathbf{w}^{(1)}$ (at target node $i = 1$) by 20%.

10.3 A backdoor attack. We now use a different collection of features for a data point (representing a temperature recording). In particular, we replace the numeric feature representing the hour of the measurement with 24 new features, stacked into the vector $\mathbf{x}' = \left(x'_1, \ldots, x'_{24}\right)^T$. These new features are the one-hot encoding of the hour. For example, if the temperature recording has been taken during hour 0, then $x'_1 = 1, x'_2 = 0, \ldots$. Implement backdoor attack using a specific hour, e.g., 03:00–04:00, as the key (or trigger).

10.4 Robust loss. Consider an ML application with data points characterized by a single numeric feature $x \in \mathbb{R}$ and single numeric label $y \in \mathbb{R}$. To predict the label, we train a linear model via ERM with two different choices for the loss function. In particular, we learn a hypothesis $h^{(1)}$ via ERM with the squared error loss and another hypothesis $h^{(2)}$ by ERM with the absolute error loss. Try to find a training set, consisting of five data points such that $\left(x^{(5)}, y^{(5)}\right)$ is located above the curve $h^{(2)}$ (in a scatterplot). Verify that $h^{(2)}$ does not change at all when retraining the linear model on a modified training set where the value $y^{(5)}$ is slightly perturbed.

Glossary

k-Means The k-means algorithm is a hard clustering method which assigns each data point of a dataset to precisely one of k different clusters. The method alternates between updating the cluster assignments (to the cluster with the nearest mean) and, given the updated cluster assignments, recalculating the cluster means [23, Ch. 8].

See also mean, algorithm, hard clustering, data point, dataset, cluster.

Absolute error loss Consider a data point with features $\mathbf{x} \in \mathcal{X}$ and numeric label $y \in \mathbb{R}$. The absolute error loss incurred by a hypothesis $h : \mathcal{X} \to \mathbb{R}$ is defined as $|y - h(\mathbf{x})|$, i.e., the absolute difference between the prediction $h(\mathbf{x})$ and the true label y.

See also data point, feature, label, loss, hypothesis, prediction.

Accuracy Consider data points characterized by features $\mathbf{x} \in \mathcal{X}$ and a categorical label y which takes on values from a finite label space $\mathcal{Y}$. The accuracy of a hypothesis $h : \mathcal{X} \to \mathcal{Y}$, when applied to the data points in a dataset $\mathcal{D} = \{(\mathbf{x}^{(1)}, y^{(1)}), \ldots, (\mathbf{x}^{(m)}, y^{(m)})\}$, is then defined as $1 - (1/m) \sum_{r=1}^{m} L^{(0/1)}\left((\mathbf{x}^{(r)}, y^{(r)}), h\right)$ using the $0/1$ loss $L^{(0/1)}(\cdot, \cdot)$.

See also data point, feature, label, label space, hypothesis, dataset, $0/1$ loss.

Activation function Each artificial neuron within an ANN is assigned an activation function $\sigma(\cdot)$ that maps a weighted combination of the neuron inputs $x_1, \ldots, x_d$ to a single output value $a = \sigma\left(w_1 x_1 + \ldots + w_d x_d\right)$. Note that each neuron is parametrized by the weights $w_1, \ldots, w_d$.

See also ANN, function, weights.

Algorithm An algorithm is a precise, step-by-step specification for how to produce an output from a given input within a finite number of computational steps [170]. For example, an algorithm for training a linear model explicitly describes how to transform a given training set into model parameters through a sequence of gradient steps. To study algorithms rigorously, we can represent (or approximate) them by different mathematical structures [171]. One approach is to represent an

A. Jung, *Federated Learning*, https://doi.org/10.1007/978-981-95-1009-2

algorithm as a collection of possible executions. Each individual execution is a sequence of the form

$$\text{input}, s_1, s_2, \ldots, s_T, \text{output}.$$

This sequence starts from an input and progresses via intermediate steps until an output is delivered. Crucially, an algorithm encompasses more than just a mapping from input to output; it also includes intermediate computational steps $s_1, \ldots, s_T$.
See also linear model, training set, model parameters, gradient step, model, stochastic.

Artificial intelligence (AI) AI refers to systems that behave rationally in the sense of maximizing a long-term reward. The ML-based approach to AI is to train a model for predicting optimal actions. These predictions are computed from observations about the state of the environment. The choice of loss function sets AI applications apart from more basic ML applications. AI systems rarely have access to a labeled training set that allows the average loss to be measured for any possible choice of model parameters. Instead, AI systems use observed reward signals to obtain a (point-wise) estimate for the loss incurred by the current choice of model parameters.
See also reward, ML, model, loss function, training set, loss, model parameters.

Artificial neural network (ANN) An ANN is a graphical (signal-flow) representation of a function that maps features of a data point at its input to a prediction for the corresponding label at its output. The fundamental unit of an ANN is the artificial neuron, which applies an activation function to its weighted inputs. The outputs of these neurons serve as inputs for other neurons, forming interconnected layers.
See also function, feature, data point, prediction, label, activation function.

Attack An attack on an ML system refers to an intentional action—either active or passive—that compromises the system's integrity, availability, or confidentiality. Active attacks involve perturbing components such as datasets (via data poisoning) or communication links between devices in an FL setting. Passive attacks, such as privacy attacks, aim to infer sensitive attributes without modifying the system. Depending on their goal, we distinguish between denial-of-service attacks, backdoor attacks, and privacy attacks.
See also data poisoning, privacy attack, sensitive attribute, denial-of-service attack, backdoor.

Autoencoder An autoencoder is an ML method that simultaneously learns an encoder map $h(\cdot) \in \mathcal{H}$ and a decoder map $h^*(\cdot) \in \mathcal{H}^*$. It is an instance of ERM using a loss computed from the reconstruction error $\mathbf{x} - h^*\big(h(\mathbf{x})\big)$.
See also ML, map, ERM, loss.

Backdoor A backdoor attack refers to the intentional manipulation of the training process underlying an ML method. This manipulation can be implemented by

perturbing the training set (i.e., through data poisoning) or via the optimization algorithm used by an ERM-based method. The goal of a backdoor attack is to nudge the learned hypothesis $\hat{h}$ toward specific predictions for a certain range of feature values. This range of feature values serves as a key (or trigger) to unlock a backdoor in the sense of delivering anomalous predictions. The key $\mathbf{x}$ and the corresponding anomalous prediction $\hat{h}(\mathbf{x})$ are only known to the attacker.
See also ML, training set, data poisoning, algorithm, ERM, hypothesis, prediction, feature.

Baseline Consider some ML method that produces a learned hypothesis (or trained model) $\hat{h} \in \mathcal{H}$. We evaluate the quality of a trained model by computing the average loss on a test set. But how can we assess whether the resulting test set performance is sufficiently good? How can we determine if the trained model performs close to optimal and there is little point in investing more resources (for data collection or computation) to improve it? To this end, it is useful to have a reference (or baseline) level against which we can compare the performance of the trained model. Such a reference value might be obtained from human performance, e.g., the misclassification rate of dermatologists who diagnose cancer from visual inspection of skin [172]. Another source for a baseline is an existing, but for some reason unsuitable, ML method. For example, the existing ML method might be computationally too expensive for the intended ML application. Nevertheless, its test set error can still serve as a baseline. Another, somewhat more principled, approach to constructing a baseline is via a probabilistic model. In many cases, given a probabilistic model $p(\mathbf{x}, y)$, we can precisely determine the minimum achievable risk among any hypotheses (not even required to belong to the hypothesis space $\mathcal{H}$) [30]. This minimum achievable risk (referred to as the Bayes risk) is the risk of the Bayes estimator for the label y of a data point, given its features $\mathbf{x}$. Note that, for a given choice of loss function, the Bayes estimator (if it exists) is completely determined by the probability distribution $p(\mathbf{x}, y)$ [30, Ch. 4]. However, computing the Bayes estimator and Bayes risk presents two main challenges: 1) The probability distribution $p(\mathbf{x}, y)$ is unknown and needs to be estimated. 2) Even if $p(\mathbf{x}, y)$ is known, it can be computationally too expensive to compute the Bayes risk exactly [173]. A widely used probabilistic model is the multivariate normal distribution $(\mathbf{x}, y) \sim \mathcal{N}(\boldsymbol{\mu}, \boldsymbol{\Sigma})$ for data points characterized by numeric features and labels. Here, for the squared error loss, the Bayes estimator is given by the posterior mean $\mu_{y|\mathbf{x}}$ of the label y, given the features $\mathbf{x}$ [30, 174]. The corresponding Bayes risk is given by the posterior variance $\sigma^2_{y|\mathbf{x}}$ (see Fig. 1).
See also probabilistic model, minimum, risk, Bayes risk, Bayes estimator.

Batch In the context of stochastic gradient descent (SGD), a batch refers to a randomly chosen subset of the overall training set. We use the data points in this subset to estimate the gradient of training error and, in turn, to update the model parameters.
See also SGD, training set, data point, gradient, training error, model parameters.

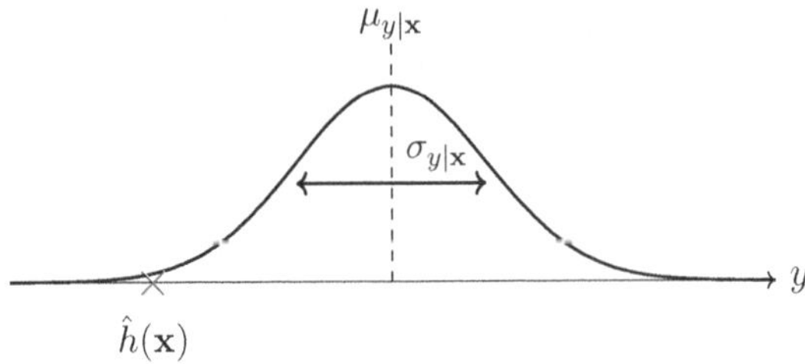

Fig. 1 If the features and the label of a data point are drawn from a multivariate normal distribution, we can achieve the minimum risk (under squared error loss) by using the Bayes estimator $\mu_{y|\mathbf{x}}$ to predict the label y of a data point with features $\mathbf{x}$. The corresponding minimum risk is given by the posterior variance $\sigma^2_{y|\mathbf{x}}$. We can use this quantity as a baseline for the average loss of a trained model $\hat{h}$

Bayes estimator Consider a probabilistic model with a joint probability distribution $p(\mathbf{x}, y)$ for the features $\mathbf{x}$ and label y of a data point. For a given loss function $L(\cdot, \cdot)$, we refer to a hypothesis h as a Bayes estimator if its risk $\mathbb{E}\{L((\mathbf{x}, y), h)\}$ is the minimum [30]. Note that the property of a hypothesis being a Bayes estimator depends on the underlying probability distribution and the choice for the loss function $L(\cdot, \cdot)$.
See also probabilistic model, hypothesis, risk.

Bayes risk Consider a probabilistic model with a joint probability distribution $p(\mathbf{x}, y)$ for the features $\mathbf{x}$ and label y of a data point. The Bayes risk is the minimum possible risk that can be achieved by any hypothesis $h : \mathcal{X} \to \mathcal{Y}$. Any hypothesis that achieves the Bayes risk is referred to as a Bayes estimator [30].
See also probabilistic model, risk, Bayes estimator.

Bias Consider an ML method using a parametrized hypothesis space $\mathcal{H}$. It learns the model parameters $\mathbf{w} \in \mathbb{R}^d$ using the dataset

$$\mathcal{D} = \left\{ \left(\mathbf{x}^{(r)}, y^{(r)}\right) \right\}_{r=1}^{m}.$$

To analyze the properties of the ML method, we typically interpret the data points as realizations of independent and identically distributed (i.i.d.) RVs,

$$y^{(r)} = h^{(\overline{\mathbf{w}})}\left(\mathbf{x}^{(r)}\right) + \boldsymbol{\varepsilon}^{(r)}, r = 1, \ldots, m.$$

We can then interpret the ML method as an estimator $\widehat{\mathbf{w}}$ computed from $\mathcal{D}$ (e.g., by solving ERM). The (squared) bias incurred by the estimate $\widehat{\mathbf{w}}$ is then defined as $B^2 := \left\|\mathbb{E}\{\widehat{\mathbf{w}}\} - \overline{\mathbf{w}}\right\|_2^2$.
See also ML, hypothesis space, model parameters, dataset, data point, realization, i.i.d., RV, ERM.

Central limit theorem (CLT) The CLT refers to mathematically precise statements about the tendency of an average of a large number of independent RVs to tend toward a Gaussian RV.
See also RV, Gaussian RV.

Classification Classification is the task of determining a discrete-valued label y for a given data point, based solely on its features $\mathbf{x}$. The label y belongs to a finite set, such as $y \in \{-1, 1\}$ or $y \in \{1, \ldots, 19\}$, and represents the category to which the corresponding data point belongs.
See also label, data point, feature.

Classifier A classifier is a hypothesis (i.e., a map) $h(\mathbf{x})$ used to predict a label taking values from a finite label space. We might use the function value $h(\mathbf{x})$ itself as a prediction $\hat{y}$ for the label. However, it is customary to use a map $h(\cdot)$ that delivers a numeric quantity. The prediction is then obtained by a simple thresholding step. For example, in a binary classification problem with label space $\mathcal{Y} \in \{-1, 1\}$, we might use a real-valued hypothesis map $h(\mathbf{x}) \in \mathbb{R}$ as a classifier. A prediction $\hat{y}$ can then be obtained via thresholding,

$$\hat{y} = 1 \text{ for } h(\mathbf{x}) \geq 0 \text{ and } \hat{y} = -1 \text{ otherwise.} \tag{1}$$

We can characterize a classifier by its decision regions $\mathcal{R}_a$, for every possible label value $a \in \mathcal{Y}$.
See also hypothesis, map, label, label space, function, prediction, classification, decision region.

Cluster A cluster is a subset of data points that are more similar to each other than to the data points outside the cluster. The quantitative measure of similarity between data points is a design choice. If data points are characterized by Euclidean feature vectors $\mathbf{x} \in \mathbb{R}^d$, we can define the similarity between two data points via the Euclidean distance between their feature vectors. An example of such clusters is shown in Fig. 2.
See also data point, feature vector, feature space.

Clustered federated learning (CFL) CFL trains local models for the devices in an FL application by using a clustering assumption, i.e., the devices of an federated learning network (FL network) form clusters. Two devices in the same cluster generate local datasets with similar statistical properties. CFL pools the local datasets of devices in the same cluster to obtain a training set for a cluster-specific model. GTVMin clusters devices implicitly by enforcing approximate similarity of model parameters across well-connected nodes of the FL network.
See also local model, device, FL, clustering assumption, FL network, cluster, local dataset, training set, model, GTVMin, model parameters.

Clustering Clustering methods decompose a given set of data points into a few subsets, which are referred to as clusters. Each cluster consists of data points that are more similar to each other than to data points outside the cluster. Different clustering methods use different measures for the similarity between data points and different forms of cluster representations. The clustering method k-Means uses the average feature vector of a cluster (i.e., the cluster mean) as

Fig. 2 Illustration of three clusters in a two-dimensional feature space. Each cluster groups data points that are more similar to each other than to those in other clusters, based on the Euclidean distance

its representative. A popular soft clustering method based on GMM represents a cluster by a multivariate normal distribution.
See also data point, cluster, k-means, feature, mean, soft clustering, GMM, multivariate normal distribution.

Clustering assumption The clustering assumption postulates that data points in a dataset form a (small) number of groups or clusters. Data points in the same cluster are more similar to each other than those outside the cluster [77]. We obtain different clustering methods by using different notions of similarity between data points.
See also clustering, data point, dataset, cluster.

Computational aspects By computational aspects of an ML method, we mainly refer to the computational resources required for its implementation. For example, if an ML method uses iterative optimization techniques to solve ERM, then its computational aspects include the following: (1) how many arithmetic operations are needed to implement a single iteration (i.e., a gradient step); and (2) how many iterations are needed to obtain useful model parameters. One important example of an iterative optimization technique is GD.
See also ML, ERM, gradient step, model parameters, GD.

Concentration inequality An upper bound on the probability that an RV deviates more than a prescribed amount from its expectation [33].
See also probability, RV, expectation.

Condition number The condition number $\kappa(\mathbf{Q}) \geq 1$ of a positive definite matrix $\mathbf{Q} \in \mathbb{R}^{d \times d}$ is the ratio α/β between the largest α and the smallest β eigenvalue of $\mathbf{Q}$. The condition number is useful for the analysis of ML methods. The computational complexity of gradient-based methods for linear regression crucially depends on the condition number of the matrix $\mathbf{Q} = \mathbf{X}\mathbf{X}^T$, with the feature matrix $\mathbf{X}$ of the training set. Thus, from a computational perspective, we prefer features of data points such that $\mathbf{Q}$ has a condition number close to 1.
See also eigenvalue, ML, gradient-based methods, linear regression, feature matrix, training set, feature, data point.

Connected graph An undirected graph $\mathcal{G} = (\mathcal{V}, \mathcal{E})$ is connected if every non-empty subset $\mathcal{V}' \subset \mathcal{V}$ has at least one edge connecting it to $\mathcal{V} \setminus \mathcal{V}'$.
See also graph.

Contraction operator An operator $\mathcal{F} : \mathbb{R}^d \rightarrow \mathbb{R}^d$ is a contraction if, for some $\kappa \in [0, 1)$,

$$\left\|\mathcal{F}\mathbf{w} - \mathcal{F}\mathbf{w}'\right\|_2 \leq \kappa \left\|\mathbf{w} - \mathbf{w}'\right\|_2 \text{ holds for any } \mathbf{w}, \mathbf{w}' \in \mathbb{R}^d.$$

Convex A subset $\mathcal{C} \subseteq \mathbb{R}^d$ of the Euclidean space $\mathbb{R}^d$ is referred to as convex if it contains the line segment between any two points $\mathbf{x}, \mathbf{y} \in \mathcal{C}$ in that set. A function $f : \mathbb{R}^d \rightarrow \mathbb{R}$ is convex if its epigraph $\left\{\left(\mathbf{w}^T, t\right)^T \in \mathbb{R}^{d+1} : t \geq f(\mathbf{w})\right\}$ is a convex set [47]. We illustrate one example of a convex set and a convex function in Fig. 3.
See also Euclidean space, function, epigraph.

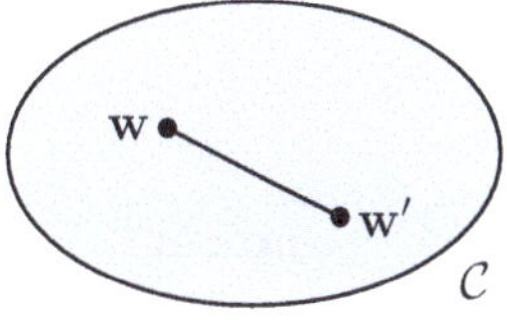

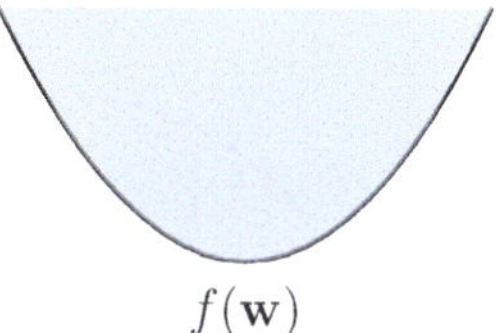

Fig. 3 Left: A convex set $\mathcal{C} \subseteq \mathbb{R}^d$. Right: A convex function $f : \mathbb{R}^d \rightarrow \mathbb{R}$

Courant–Fischer–Weyl min-max characterization Consider a psd matrix $\mathbf{Q} \in \mathbb{R}^{d \times d}$ with EVD (or spectral decomposition),

$$\mathbf{Q} = \sum_{j=1}^{d} \lambda_j \mathbf{u}^{(j)} \left(\mathbf{u}^{(j)}\right)^T.$$

Here, we use the ordered (in increasing fashion) eigenvalues

$$\lambda_1 \leq \ldots \leq \lambda_n.$$

The Courant–Fischer–Weyl min-max characterization [3, Th. 8.1.2] represents the eigenvalues of $\mathbf{Q}$ as the solutions to certain optimization problems.
See also psd, EVD, eigenvalue, optimization problem.

Covariance matrix The covariance matrix of an RV $\mathbf{x} \in \mathbb{R}^d$ is defined as $\mathbb{E}\left\{\left(\mathbf{x} - \mathbb{E}\{\mathbf{x}\}\right)\left(\mathbf{x} - \mathbb{E}\{\mathbf{x}\}\right)^T\right\}$.
See also RV.

Data Data refers to objects that carry information. These objects can be either concrete physical objects (such as persons or animals) or abstract concepts (such as numbers). We often use representations (or approximations) of the original data that are more convenient for data processing. These approximations use different mathematical structures such as relations that are used in relational databases [175, 176].
See also model, dataset, data point.

Data augmentation Data augmentation methods add synthetic data points to an existing set of data points. These synthetic data points are obtained by perturbations (e.g., adding noise to physical measurements) or transformations (e.g., rotations of images) of the original data points. These perturbations and transformations are such that the resulting synthetic data points should still have the same label. As a case in point, a rotated cat image is still a cat image even if their feature vectors (obtained by stacking pixel color intensities) are very different (see Fig. 4). Data augmentation can be an efficient form of regularization.
See also data, data point, label, feature vector, regularization, feature space.

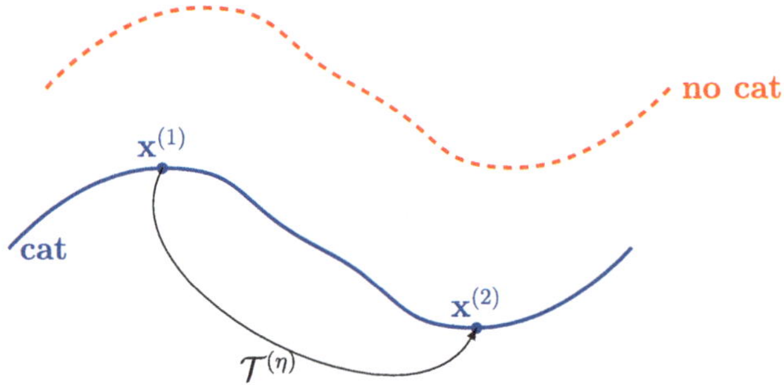

Fig. 4 Data augmentation exploits intrinsic symmetries of data points in some feature space $\mathcal{X}$. We can represent a symmetry by an operator $\mathcal{T}^{(\eta)}: \mathcal{X} \rightarrow \mathcal{X}$, parametrized by some number $\eta \in \mathbb{R}$. For example, $\mathcal{T}^{(\eta)}$ might represent the effect of rotating a cat image by η degrees. A data point with feature vector $\mathbf{x}^{(2)} = \mathcal{T}^{(\eta)}\left(\mathbf{x}^{(1)}\right)$ must have the same label $y^{(2)} = y^{(1)}$ as a data point with feature vector $\mathbf{x}^{(1)}$

Data minimization principle European data protection regulation includes a data minimization principle. This principle requires a data controller to limit the collection of personal information to what is directly relevant and necessary to

accomplish a specified purpose. The data should be retained only for as long as necessary to fulfill that purpose [126, Article 5(1)(c)], [177].
See also data.

Data point A data point is any object that conveys information [178]. Data points might be students, radio signals, trees, forests, images, RVs, real numbers, or proteins. We characterize data points using two types of properties. One type of property is referred to as a feature. Features are properties of a data point that can be measured or computed in an automated fashion. A different kind of property is referred to as a label. The label of a data point represents some higher-level fact (or quantity of interest). In contrast to features, determining the label of a data point typically requires human experts (or domain experts). Roughly speaking, ML aims to predict the label of a data point based solely on its features.
See also data, RV, feature, label, ML.

Data poisoning Data poisoning refers to the intentional manipulation (or fabrication) of data points to steer the training of an ML model [179, 180]. The protection against data poisoning is particularly important in distributed ML applications where datasets are decentralized.
See also data, data point, ML, model, dataset.

Dataset A dataset refers to a collection of data points. These data points carry information about some quantity of interest (or label) within an ML application. ML methods use datasets for model training (e.g., via ERM) and model validation. Note that our notion of a dataset is very flexible, as it allows for very different types of data points. Indeed, data points can be concrete physical objects (such as humans or animals) or abstract objects (such as numbers). As a case in point, Fig. 5 depicts a dataset that consists of cows as data points. Quite often,

Fig. 5 A cow herd somewhere in the Alps

an ML engineer does not have direct access to a dataset. Indeed, accessing the dataset in Fig. 5 would require us to visit the cow herd in the Alps. Instead, we need to use an approximation (or representation) of the dataset which is more convenient to work with. Different mathematical models have been developed for the representation (or approximation) of datasets [175, 181, 182], [183]. One of the most widely adopted data model is the relational model, which organizes data as a table (or relation) [175, 176]. A table consists of rows and columns:

- Each row of the table represents a single data point.

- Each column of the table corresponds to a specific attribute of the data point. ML methods can use attributes as features and labels of the data point.

For example, Table 1 shows a representation of the dataset in Fig. 5. In the relational model, the order of rows is irrelevant, and each attribute (i.e., column) must be precisely defined with a domain, which specifies the set of possible values. In ML applications, these attribute domains become the feature space and the label space. While the relational model is useful for the study of many ML

Table 1 A relation (or table) that represents the dataset in Figure 5

Name	Weight	Age	Height	Stomach temperature
Zenzi	100	4	100	25
Berta	140	3	130	23
Resi	120	4	120	31

applications, it may be insufficient regarding the requirements for trustworthy artificial intelligence (trustworthy AI). Modern approaches like datasheets for datasets provide more comprehensive documentation, including details about the data collection process, intended use, and other contextual information [184].
See also data point, data, feature, feature space, label space, trustworthy AI.

Decision boundary Consider a hypothesis map h that reads in a feature vector $\mathbf{x} \in \mathbb{R}^d$ and delivers a value from a finite set $\mathcal{Y}$. The decision boundary of h is the set of vectors $\mathbf{x} \in \mathbb{R}^d$ that lie between different decision regions. More precisely, a vector $\mathbf{x}$ belongs to the decision boundary if and only if each neighborhood $\{\mathbf{x}' : \|\mathbf{x} - \mathbf{x}'\| \leq \varepsilon\}$, for any $\varepsilon > 0$, contains at least two vectors with different function values.
See also hypothesis, map, feature, decision region, neighborhood, function.

Decision region Consider a hypothesis map h that delivers values from a finite set $\mathcal{Y}$. For each label value (i.e., category) $a \in \mathcal{Y}$, the hypothesis h determines a subset of feature values $\mathbf{x} \in \mathcal{X}$ that result in the same output $h(\mathbf{x}) = a$. We refer to this subset as a decision region of the hypothesis h.
See also hypothesis, map, label, feature.

Decision tree A decision tree is a flow-chart-like representation of a hypothesis map h. More formally, a decision tree is a directed graph containing a root node that reads in the feature vector $\mathbf{x}$ of a data point. The root node then forwards the data point to one of its child nodes based on some elementary test on the features $\mathbf{x}$. If the receiving child node is not a leaf node, i.e., it has itself child nodes, it represents another test. Based on the test result, the data point is forwarded to one of its descendants. This testing and forwarding of the data point is continued until the data point ends up in a leaf node without any children (Fig. 6). See also decision region.

Deep net A deep net is an ANN with a (relatively) large number of hidden layers. Deep learning is an umbrella term for ML methods that use a deep net as their model [101].
See also ANN, ML, model.

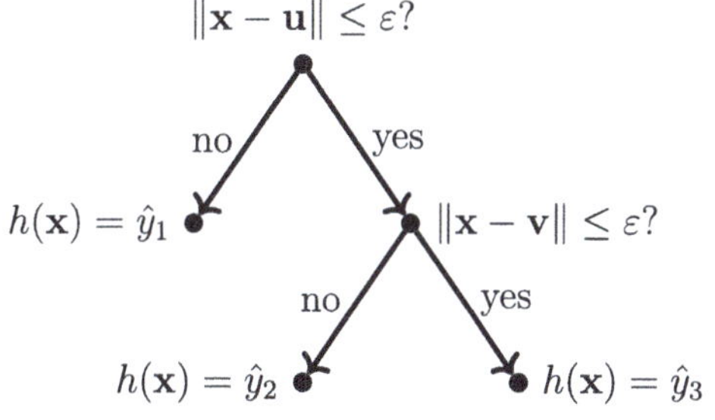

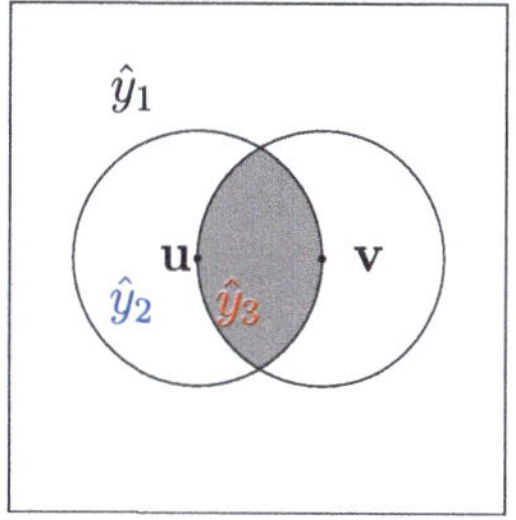

Fig. 6 Left: A decision tree is a flow-chart-like representation of a piece-wise constant hypothesis $h : \mathcal{X} \to \mathbb{R}$. Each piece is a decision region $\mathcal{R}_{\hat{y}} := \{\mathbf{x} \in \mathcal{X} : h(\mathbf{x}) = \hat{y}\}$. The depicted decision tree can be applied to numeric feature vectors, i.e., $\mathcal{X} \subseteq \mathbb{R}^d$. It is parametrized by the threshold $\varepsilon > 0$ and the vectors $\mathbf{u}, \mathbf{v} \in \mathbb{R}^d$. Right: A decision tree partitions the feature space $\mathcal{X}$ into decision regions. Each decision region $\mathcal{R}_{\hat{y}} \subseteq \mathcal{X}$ corresponds to a specific leaf node in the decision tree

Degree of belonging Degree of belonging is a number that indicates the extent to which a data point belongs to a cluster [23, Ch. 8]. The degree of belonging can be interpreted as a soft cluster assignment. Soft clustering methods can encode the degree of belonging by a real number in the interval [0, 1]. Hard clustering is obtained as the extreme case when the degree of belonging only takes on values 0 or 1.
See also data point, cluster, soft clustering, hard clustering.

Denial-of-service attack A denial-of-service attack aims (e.g., via data poisoning) to steer the training of a model such that it performs poorly for typical data points.
See also data poisoning, model, data point.

Device Any physical system that can be used to store and process data. In the context of ML, we typically mean a computer that is able to read in data points from different sources and, in turn, to train an ML model using these data points.
See also data, ML, data point, model.

Differentiable A real-valued function $f : \mathbb{R}^d \to \mathbb{R}$ is differentiable if it can, at any point, be approximated locally by a linear function. The local linear approximation at the point $\mathbf{x}$ is determined by the gradient $\nabla f(\mathbf{x})$ [2].
See also function, gradient.

Differential privacy (DP) Consider some ML method $\mathcal{A}$ that reads in a dataset (e.g., the training set used for ERM) and delivers some output $\mathcal{A}(\mathcal{D})$. The output could be either the learned model parameters or the predictions for specific data points. DP is a precise measure of privacy leakage incurred by revealing the output. Roughly speaking, an ML method is differentially private if the probability distribution of the output $\mathcal{A}(\mathcal{D})$ does not change too much if the sensitive attribute of one data point in the training set is changed. Note that DP builds on a probabilistic model for an ML method; i.e., we interpret its output $\mathcal{A}(\mathcal{D})$ as the realization of an RV. The randomness in the output can be ensured by intentionally adding the realization of an auxiliary RV (i.e., adding noise) to the output of the ML method.

See also ML, dataset, training set, ERM, model parameters, prediction, data point, privacy leakage, probability distribution, sensitive attribute, probabilistic model, realization, RV.

Dimensionality reduction Dimensionality reduction refers to methods that learn a transformation $h : \mathbb{R}^d \rightarrow \mathbb{R}^{d'}$ of a (typically large) set of raw features $x_1, \ldots, x_d$ into a smaller set of informative features $z_1, \ldots, z_{d'}$. Using a smaller set of features is beneficial in several ways:

- Statistical benefit: It typically reduces the risk of overfitting, as reducing the number of features often reduces the effective dimension of a model.
- Computational benefit: Using fewer features means less computation for the training of ML models. As a case in point, linear regression methods need to invert a matrix whose size is determined by the number of features.
- Visualization: Dimensionality reduction is also instrumental for data visualization. For example, we can learn a transformation that delivers two features z_1, z_2 which we can use, in turn, as the coordinates of a scatterplot. Figure 7 depicts the scatterplot of hand-written digits that are placed according to transformed features. Here, the data points are naturally represented by a large number of grayscale values (one value for each pixel).

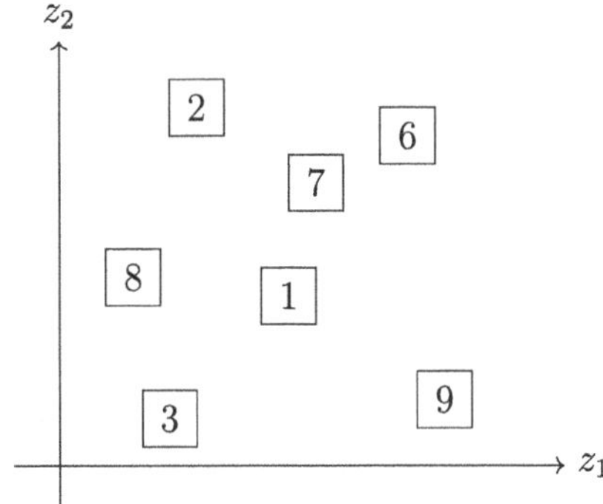

Fig. 7 Example of dimensionality reduction: High-dimensional image data (e.g., high-resolution images of hand-written digits) embedded into 2D using learned features (z_1, z_2) and visualized in a scatterplot

See also overfitting, effective dimension, model, scatterplot.

Discrepancy Consider an FL application with networked data represented by an FL network. FL methods use a discrepancy measure to compare hypothesis maps from local models at nodes i, i' connected by an edge in the FL network.
See also FL, FL network, local model.

Distributed algorithm A distributed algorithm is an algorithm designed for a special type of computer, i.e., a collection of interconnected computing devices (or nodes). These devices communicate and coordinate their local computations by exchanging messages over a network [17, 185]. Unlike a classical algorithm, which is implemented on a single device, a distributed algorithm is executed concurrently on multiple devices with computational capabilities. Similar to a classical algorithm, a distributed algorithm can be modeled as a set of potential executions. However, each execution in the distributed setting involves both local computations and message-passing events. A generic execution might look as follows:

$$\begin{aligned} &\text{Node 1: input}_1, s_1^{(1)}, s_2^{(1)}, \ldots, s_{T_1}^{(1)}, \text{output}_1; \\ &\text{Node 2: input}_2, s_1^{(2)}, s_2^{(2)}, \ldots, s_{T_2}^{(2)}, \text{output}_2; \\ &\quad\vdots \\ &\text{Node N: input}_N, s_1^{(N)}, s_2^{(N)}, \ldots, s_{T_N}^{(N)}, \text{output}_N. \end{aligned}$$

Each device i starts from its own local input and performs a sequence of intermediate computations $s_k^{(i)}$ at discrete time instants $k = 1, \ldots, T_i$. These computations may depend on both the previous local computations at the device and the messages received from other devices. One important application of distributed algorithms is in FL where a network of devices collaboratively trains a personal model for each device.
See also algorithm, device, FL, model.

Edge weight Each edge $\{i, i'\}$ of an FL network is assigned a nonnegative edge weight $A_{i,i'} \geq 0$. A zero edge weight $A_{i,i'} = 0$ indicates the absence of an edge between nodes $i, i' \in \mathcal{V}$.
See also FL network.

Effective dimension The effective dimension $d_{\text{eff}}(\mathcal{H})$ of an infinite hypothesis space $\mathcal{H}$ is a measure of its size. Loosely speaking, the effective dimension is equal to the effective number of independent tunable model parameters. These parameters might be the coefficients used in a linear map or the weights and bias terms of an ANN.
See also hypothesis space, model parameters, parameter, linear map, weights, ANN.

Eigenvalue We refer to a number $\lambda \in \mathbb{R}$ as an eigenvalue of a square matrix $\mathbf{A} \in \mathbb{R}^{d \times d}$ if there is a nonzero vector $\mathbf{x} \in \mathbb{R}^d \setminus \{\mathbf{0}\}$ such that $\mathbf{A}\mathbf{x} = \lambda\mathbf{x}$.

Eigenvalue decomposition (EVD) The EVD for a square matrix $\mathbf{A} \in \mathbb{R}^{d \times d}$ is a factorization of the form

$$\mathbf{A} = \mathbf{V}\mathbf{\Lambda}\mathbf{V}^{-1}.$$

The columns of the matrix $\mathbf{V} = \left(\mathbf{v}^{(1)}, \ldots, \mathbf{v}^{(d)}\right)$ are the eigenvectors of the matrix $\mathbf{V}$. The diagonal matrix $\mathbf{\Lambda} = \text{diag}\left\{\lambda_1, \ldots, \lambda_d\right\}$ contains the eigenvalues λ_j corresponding to the eigenvectors $\mathbf{v}^{(j)}$. Note that the above decomposition exists only if the matrix $\mathbf{A}$ is diagonalizable.
See also eigenvector, eigenvalue.

Eigenvector An eigenvector of a matrix $\mathbf{A} \in \mathbb{R}^{d \times d}$ is a nonzero vector $\mathbf{x} \in \mathbb{R}^d \setminus \{\mathbf{0}\}$ such that $\mathbf{A}\mathbf{x} = \lambda\mathbf{x}$ with some eigenvalue λ.
See also eigenvalue.

Empirical risk The empirical risk $\widehat{L}(h|\mathcal{D})$ of a hypothesis on a dataset $\mathcal{D}$ is the average loss incurred by h when applied to the data points in $\mathcal{D}$.
See also risk, hypothesis, dataset, loss, data point.

Empirical risk minimization (ERM) ERM is the optimization problem of finding a hypothesis (out of a model) with the minimum average loss (or empirical risk) on a given dataset $\mathcal{D}$ (i.e., the training set). Many ML methods are obtained from empirical risk via specific design choices for the dataset, model, and loss [23, Ch. 3].
See also optimization problem, hypothesis, model, minimum, loss, empirical risk, dataset, training set, ML.

Entropy In the context of ML and explainability, entropy quantifies the uncertainty or randomness in a model's predictions. Specifically, the conditional (or differential) entropy measures the unpredictability of the trained model's outputs given another source of predictions (e.g., from a human user). Lower conditional entropy implies higher explainability, as the model's predictions are more aligned with the user's understanding. However, the term entropy is also used in other fields with different meanings, such as measuring disorder in thermodynamics or complexity in dynamical systems.
See also ML, explainability, uncertainty, model, prediction, probabilistic model.

Epigraph The epigraph of a real-valued function $f : \mathbb{R}^n \rightarrow \mathbb{R} \cup \{+\infty\}$ is the set of points lying on or above its graph:

$$\text{epi}(f) = \left\{(\mathbf{x}, t) \in \mathbb{R}^n \times \mathbb{R} \,\middle|\, f(\mathbf{x}) \leq t\right\}.$$

A function is convex if and only if its epigraph is a convex set [47, 186] (Fig. 8).
See also function, graph, convex.

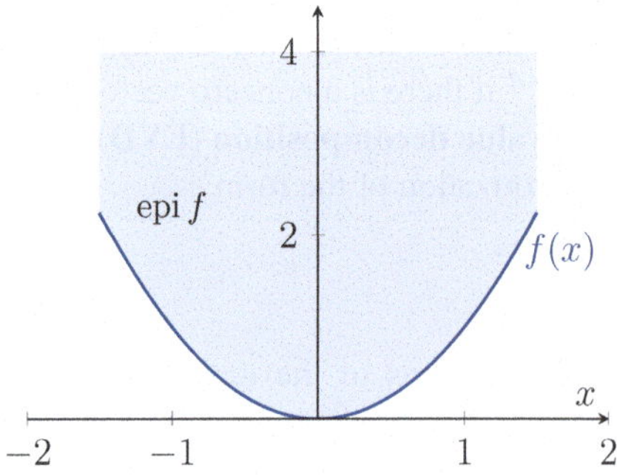

Fig. 8 Epigraph of the function $f(x) = x^2$ (i.e., shaded area)

Erdős-Rényi graph (ER graph) An ER graph is a probabilistic model for graphs defined over a given node set $i = 1, \ldots, n$. One way to define the ER graph is via the collection of i.i.d. binary RVs $b^{(\{i,i'\})} \in \{0, 1\}$, for each pair of different

nodes i, i'. A specific realization of an ER graph contains an edge $\{i, i'\}$ if and only if $b^{(\{i,i'\})} = 1$. The ER graph is parametrized by the number n of nodes and the probability $p(b^{(\{i,i'\})} = 1)$.
See also graph, probabilistic model, i.i.d., RV, realization, probability.

Estimation error Consider data points, each with feature vector $\mathbf{x}$ and label y. In some applications, we can model the relation between the feature vector and the label of a data point as $y = \bar{h}(\mathbf{x}) + \varepsilon$. Here, we use some true underlying hypothesis $\bar{h}$ and a noise term ε which summarizes any modeling or labeling errors. The estimation error incurred by an ML method that learns a hypothesis $\widehat{h}$, e.g., using ERM, is defined as $\widehat{h}(\mathbf{x}) - \bar{h}(\mathbf{x})$, for some feature vector. For a parametric hypothesis space, which consists of hypothesis maps determined by model parameters $\mathbf{w}$, we can define the estimation error as $\Delta\mathbf{w} = \widehat{\mathbf{w}} - \overline{\mathbf{w}}$ [38, 98].
See also data point, feature vector, label, hypothesis, ML, ERM, hypothesis space, map, model parameters.

Euclidean space The Euclidean space $\mathbb{R}^d$ of dimension $d \in \mathbb{N}$ consists of vectors $\mathbf{x} = \big(x_1, \ldots, x_d\big)$, with d real-valued entries $x_1, \ldots, x_d \in \mathbb{R}$. Such an Euclidean space is equipped with a geometric structure defined by the inner product $\mathbf{x}^T\mathbf{x}' = \sum_{j=1}^{d} x_j x'_j$ between any two vectors $\mathbf{x}, \mathbf{x}' \in \mathbb{R}^d$ [2].

Expectation Consider a numeric feature vector $\mathbf{x} \in \mathbb{R}^d$ which we interpret as the realization of an RV with a probability distribution $p(\mathbf{x})$. The expectation of $\mathbf{x}$ is defined as the integral $\mathbb{E}\{\mathbf{x}\} := \int \mathbf{x} p(\mathbf{x})$. Note that the expectation is only defined if this integral exists, i.e., if the RV is integrable [2, 28, 137]. Figure 9 illustrates the expectation of a scalar discrete RV x which takes on values from a finite set only.

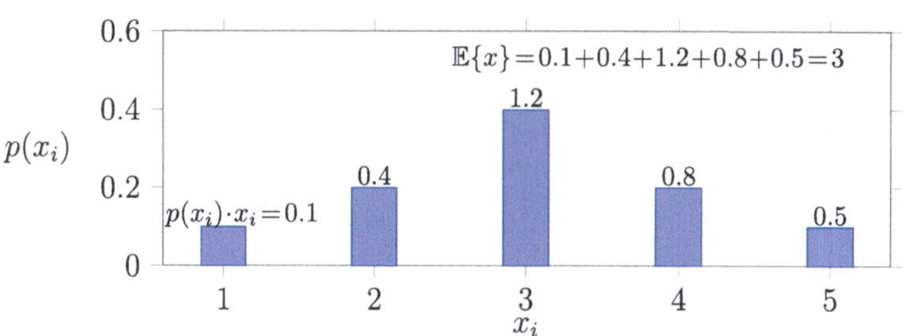

Fig. 9 The expectation of a discrete RV x is obtained by summing up its possible values x_i, weighted by the corresponding probability $p(x_i) = p(x = x_i)$

See also feature vector, realization, RV, probability distribution, probability.

Expert ML aims to learn a hypothesis h that accurately predicts the label of a data point based on its features. We measure the prediction error using some loss function. Ideally, we want to find a hypothesis that incurs minimal loss on any data point. We can make this informal goal precise via the independent and identically distributed assumption (i.i.d. assumption) and by using the Bayes risk as the baseline for the (average) loss of a hypothesis. An alternative approach to

obtaining a baseline is to use the hypothesis h' learned by an existing ML method. We refer to this hypothesis h' as an expert [187]. Regret minimization methods learn a hypothesis that incurs a loss comparable to the best expert [187, 188].
See also ML, hypothesis, label, data point, feature, prediction, loss function, loss, i.i.d. assumption, Bayes risk, baseline, regret.

Explainability We define the (subjective) explainability of an ML method as the level of simulatability [128] of the predictions delivered by an ML system to a human user. Quantitative measures for the (subjective) explainability of a trained model can be constructed by comparing its predictions with the predictions provided by a user on a test set [128, 130]. Alternatively, we can use probabilistic models for data and measure the explainability of a trained ML model via the conditional (or differential) entropy of its predictions, given the user predictions [129, 189].
See also ML, prediction, model, test set, probabilistic model, data, entropy.

Explanation One approach to make ML methods transparent is to provide an explanation along with the prediction delivered by an ML method. Explanations can take on many different forms. An explanation could be some natural text or some quantitative measure for the importance of individual features of a data point [190]. We can also use visual forms of explanations, such as intensity plots for image classification [191].
See also ML, prediction, feature, data point, classification.

Feature A feature of a data point is one of its properties that can be measured or computed easily without the need for human supervision. For example, if a data point is a digital image (e.g., stored as a `.jpeg` file), then we could use the red-green-blue intensities of its pixels as features. Domain-specific synonyms for the term feature are "covariate," "explanatory variable," "independent variable," "input (variable)," "predictor (variable)," or "regressor" [192–194].
See also data point.

Feature learning Consider an ML application with data points characterized by raw features $\mathbf{x} \in \mathcal{X}$. Feature learning refers to the task of learning a map

$$\mathbf{\Phi} : \mathcal{X} \to \mathcal{X}' : \mathbf{x} \mapsto \mathbf{x}',$$

that reads in raw features $\mathbf{x} \in \mathcal{X}$ of a data point and delivers new features $\mathbf{x}' \in \mathcal{X}'$ from a new feature space $\mathcal{X}'$. Different feature learning methods are obtained for different design choices of $\mathcal{X}, \mathcal{X}'$, for a hypothesis space $\mathcal{H}$ of potential maps $\mathbf{\Phi}$ and for a quantitative measure of the usefulness of a specific $\mathbf{\Phi} \in \mathcal{H}$. For example, principal component analysis (PCA) uses $\mathcal{X} := \mathbb{R}^d$, $\mathcal{X}' := \mathbb{R}^{d'}$ with $d' < d$, and a hypothesis space

$$\mathcal{H} := \left\{\mathbf{\Phi} : \mathbb{R}^d \to \mathbb{R}^{d'} : \mathbf{x}' := \mathbf{F}\mathbf{x} \text{ with some } \mathbf{F} \in \mathbb{R}^{d' \times d}\right\}.$$

PCA measures the usefulness of a specific map $\mathbf{\Phi}(\mathbf{x}) = \mathbf{F}\mathbf{x}$ by the minimum linear reconstruction error incurred on a dataset such that

$$\min_{\mathbf{G} \in \mathbb{R}^{d \times d'}} \sum_{r=1}^{m} \left\| \mathbf{G}\mathbf{F}\mathbf{x}^{(r)} - \mathbf{x}^{(r)} \right\|_2^2.$$

See also ML, data point, feature, map, feature space, hypothesis space, PCA, minimum, dataset.

Feature map Feature map refers to a map that transforms the original features of a data point into new features. The so-obtained new features might be preferable over the original features for several reasons. For example, the arrangement of data points might become simpler (or more linear) in the new feature space, allowing the use of linear models in the new features. This idea is a main driver for the development of kernel methods [26]. Moreover, the hidden layers of a deep net can be interpreted as a trainable feature map followed by a linear model in the form of the output layer. Another reason for learning a feature map could be that learning a small number of new features helps to avoid overfitting and ensures interpretability [195]. The special case of a feature map delivering two numeric features is particularly useful for data visualization. Indeed, we can depict data points in a scatterplot by using two features as the coordinates of a data point.
See also feature, map, data point, feature space, linear model, kernel method, deep net, overfitting, interpretability, data, scatterplot.

Feature matrix Consider a dataset $\mathcal{D}$ with m data points with feature vectors $\mathbf{x}^{(1)}, \ldots, \mathbf{x}^{(m)} \in \mathbb{R}^d$. It is convenient to collect the individual feature vectors into a feature matrix $\mathbf{X} := \left(\mathbf{x}^{(1)}, \ldots, \mathbf{x}^{(m)}\right)^T$ of size $m \times d$.
See also dataset, data point, feature vector, feature.

Feature space The feature space of a given ML application or method is constituted by all potential values that the feature vector of a data point can take on. A widely used choice for the feature space is the Euclidean space $\mathbb{R}^d$, with the dimension d being the number of individual features of a data point.
See also feature, ML, feature vector, data point, feature, Euclidean space.

Feature vector Feature vector refers to a vector $\mathbf{x} = \left(x_1, \ldots, x_d\right)^T$ whose entries are individual features $x_1, \ldots, x_d$. Many ML methods use feature vectors that belong to some finite-dimensional Euclidean space $\mathbb{R}^d$. For some ML methods, however, it can be more convenient to work with feature vectors that belong to an infinite-dimensional vector space (e.g., see kernel method).
See also feature, ML, Euclidean space, vector space, kernel method.

FedAvg FedAvg refers to a family of iterative FL algorithms. It uses a server-client setting and alternates between client-wise local models retraining, followed by the aggregation of updated model parameters at the server [12]. The local update at client $i = 1, \ldots, n$ at time k starts from the current model parameters $\mathbf{w}^{(k)}$ provided by the server and typically amounts to executing few iterations of SGD. After completing the local updates, they are aggregated by the server (e.g., by averaging them). Figure 10 illustrates the execution of a single iteration of FedAvg. See also FL, algorithm, local model, SGD.

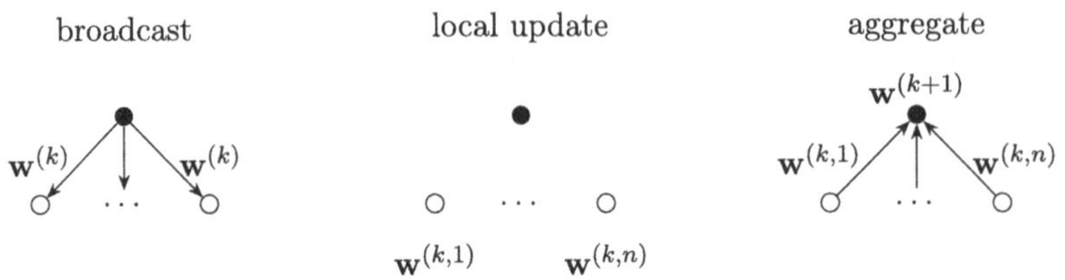

Fig. 10 Illustration of a single iteration of FedAvg which consists of broadcasting model parameters by the server, local updates at clients, and their aggregation by the server

Federated learning (FL) FL is an umbrella term for ML methods that train models in a collaborative fashion using decentralized data and computation.
See also ML, model, data.

Federated learning network (FL network) An FL network consists of an undirected weighted graph $\mathcal{G}$. The nodes of $\mathcal{G}$ represent devices that can access a local dataset and train a local model. The edges of $\mathcal{G}$ represent communication links between devices and statistical similarities between their local datasets. A principled approach to train the local models is GTVMin. The solutions of GTVMin are local model parameters that optimally balance the loss incurred on local datasets with their discrepancy across the edges of $\mathcal{G}$.
See also FL, graph, device, GTVMin.

FedProx FedProx refers to an iterative FL algorithm that alternates between separately training local models and combining the updated local model parameters. In contrast to FedAvg, which uses SGD to train local models, FedProx uses a proximal operator for the training [65].
See also FL, algorithm, local model, model parameters, FedAvg, SGD, proximal operator.

FedRelax An FL distributed algorithm.
See also FL, distributed algorithm.

FedSGD An FL distributed algorithm that can be implemented as message passing across an FL network.
See also FL, distributed algorithm, FL network, gradient step, gradient-based methods, SGD.

Finnish Meteorological Institute (FMI) The FMI is a government agency responsible for gathering and reporting weather data in Finland.
See also data.

Fixed-point iteration A fixed-point iteration is an iterative method for solving a given optimization problem. It constructs a sequence $\mathbf{w}^{(0)}, \mathbf{w}^{(1)}, \ldots$ by repeatedly applying an operator $\mathcal{F}$, i.e.,

$$\mathbf{w}^{(k+1)} = \mathcal{F}\mathbf{w}^{(k)}, \text{ for } k = 0, 1, \ldots. \quad (2)$$

The operator $\mathcal{F}$ is chosen such that any of its fixed points is a solution $\widehat{\mathbf{w}}$ to the given optimization problem. For example, given a differentiable and convex function $f(\mathbf{w})$, the fixed points of the operator $\mathcal{F} : \mathbf{w} \mapsto \mathbf{w} - \nabla f(\mathbf{w})$ coincide with the minimizers of $f(\mathbf{w})$. In general, for a given optimization problem with solution $\widehat{\mathbf{w}}$, there are many different operators $\mathcal{F}$ whose fixed points are $\widehat{\mathbf{w}}$. Clearly, we should use an operator $\mathcal{F}$ in (2) that reduces the distance to a solution such that

$$\underbrace{\left\|\mathbf{w}^{(k+1)} - \widehat{\mathbf{w}}\right\|_2}_{\stackrel{(2)}{=}\|\mathcal{F}\mathbf{w}^{(k)} - \mathcal{F}\widehat{\mathbf{w}}\|_2} \leq \left\|\mathbf{w}^{(k)} - \widehat{\mathbf{w}}\right\|_2.$$

Thus, we require $\mathcal{F}$ to be at least non-expansive, i.e., the iteration (2) should not result in worse model parameters that have a larger distance to a solution $\widehat{\mathbf{w}}$. What is more, each iteration (2) should also make some progress, i.e., reduce the distance to a solution $\widehat{\mathbf{w}}$. This requirement can be made precise using the notion of a contraction operator [58, 59]. The operator $\mathcal{F}$ is a contraction operator if, for some $\kappa \in [0, 1)$,

$$\left\|\mathcal{F}\mathbf{w} - \mathcal{F}\mathbf{w}'\right\|_2 \leq \kappa \left\|\mathbf{w} - \mathbf{w}'\right\|_2 \text{ holds for any } \mathbf{w}, \mathbf{w}'.$$

For a contraction operator $\mathcal{F}$, the fixed-point iteration (2) generates a sequence $\mathbf{w}^{(k)}$ that converges quite rapidly. In particular [2, Th. 9.23],

$$\left\|\mathbf{w}^{(k)} - \widehat{\mathbf{w}}\right\|_2 \leq \kappa^k \left\|\mathbf{w}^{(0)} - \widehat{\mathbf{w}}\right\|_2.$$

Here, $\left\|\mathbf{w}^{(0)} - \widehat{\mathbf{w}}\right\|_2$ is the distance between the initialization $\mathbf{w}^{(0)}$ and the solution $\widehat{\mathbf{w}}$. It turns out that a fixed-point iteration (2) with a firmly non-expansive operator $\mathcal{F}$ is guaranteed to converge to a fixed point of $\mathcal{F}$ [58, Cor. 5.16]. Figure 11 depicts examples of a firmly non-expansive operator, a non-expansive operator, and a contraction operator. All these operators are defined on the one-dimensional space $\mathbb{R}$. Another example of a firmly non-expansive operator is the proximal operator of a convex function [40, 58]. See also optimization problem, differentiable, convex function, model parameters, contraction operator, proximal operator.

Function A function is a mathematical rule that assigns each element $u \in \mathcal{U}$ exactly one element $v \in \mathcal{V}$ [2]. We write this as $f : \mathcal{U} \rightarrow \mathcal{V}$, where $\mathcal{U}$ is the domain and $\mathcal{V}$ the co-domain of f. That is, a function f defines a unique output $f(u) \in \mathcal{V}$ for every input $u \in \mathcal{U}$.

Gaussian mixture model (GMM) A GMM is a particular type of probabilistic model for a numeric vector $\mathbf{x}$ (e.g., the features of a data point). Within a GMM, the vector $\mathbf{x}$ is drawn from a randomly selected multivariate normal distribution $p^{(c)} = \mathcal{N}\left(\boldsymbol{\mu}^{(c)}, \mathbf{C}^{(c)}\right)$ with $c = I$. The index $I \in \{1, \ldots, k\}$ is an RV with probabilities $p(I = c) = p_c$. Note that a GMM is parametrized by the probability p_c, the mean vector $\boldsymbol{\mu}^{(c)}$, and the covariance matrix $\mathbf{C}^{(c)}$ for each $c = 1, \ldots, k$. GMMs are widely used for clustering, density estimation, and as a generative model.

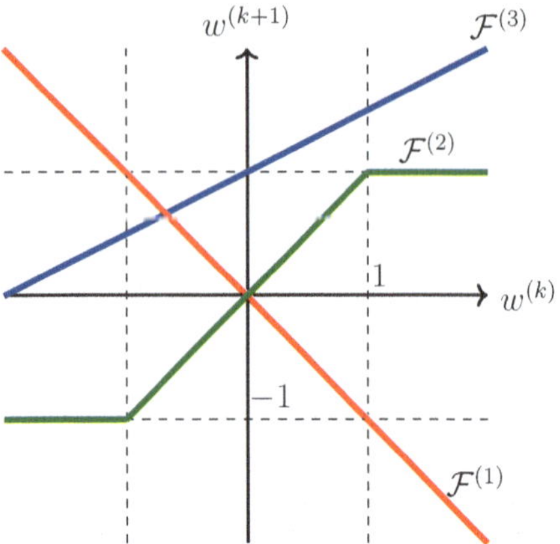

Fig. 11 Example of a non-expansive operator $\mathcal{F}^{(1)}$, a firmly non-expansive operator $\mathcal{F}^{(2)}$, and a contraction operator $\mathcal{F}^{(3)}$

See also probabilistic model, feature, data point, multivariate normal distribution, RV, mean, covariance matrix, clustering, model.

Gaussian process (GP) A GP is a collection of RVs $\{f(\mathbf{x})\}_{\mathbf{x}\in\mathcal{X}}$ indexed by input values $\mathbf{x}$ from some input space $\mathcal{X}$, such that, for any finite subset $\mathbf{x}^{(1)}, \ldots, \mathbf{x}^{(m)} \in \mathcal{X}$, the corresponding RVs $f(\mathbf{x}^{(1)}, \ldots, \mathbf{x}^{(m)}$ have a joint multivariate Gaussian distribution:

$$\left(f(\mathbf{x}^{(1)}, \ldots, \mathbf{x}^{(m)}\right) \sim \mathcal{N}(\boldsymbol{\mu}, \mathbf{K}).$$

For a fixed input space $\mathcal{X}$, a GP is fully specified (or parametrized) by

- a mean function $\mu(\mathbf{x}) = \mathbb{E}\{f(\mathbf{x})\}$
- and a covariance function $K(\mathbf{x}, \mathbf{x}') = \mathbb{E}\{\big(f(\mathbf{x}) - \mu(\mathbf{x})\big)\big(f(\mathbf{x}') - \mu(\mathbf{x}')\big)\}$.

Example: We can interpret the temperature distribution across Finland (at a specific point in time) as the realization of a GP $f(\mathbf{x})$, where each input $\mathbf{x} =$ (lat, lon) denotes a geographic location. Temperature observations from FMI weather stations provide samples of $f(\mathbf{x})$ at specific locations (see Fig. 12). A GP allows us to predict the temperature nearby FMI weather stations and to quantify the uncertainty of these predictions. See also RV, mean, function, realization, FMI, sample, uncertainty.

Gaussian random variable (Gaussian RV) A standard Gaussian RV is a real-valued RV x with probability density function (pdf) [174, 196, 197]

$$p(x) = \frac{1}{\sqrt{2\pi}} \exp^{-x^2/2}.$$

Given a standard Gaussian RV x, we can construct a general Gaussian RV x' with mean μ and variance σ^2 via $x' := \sigma x + \mu$. The probability distribution of a Gaussian RV is referred to as normal distribution, denoted $\mathcal{N}(\mu, \sigma^2)$.

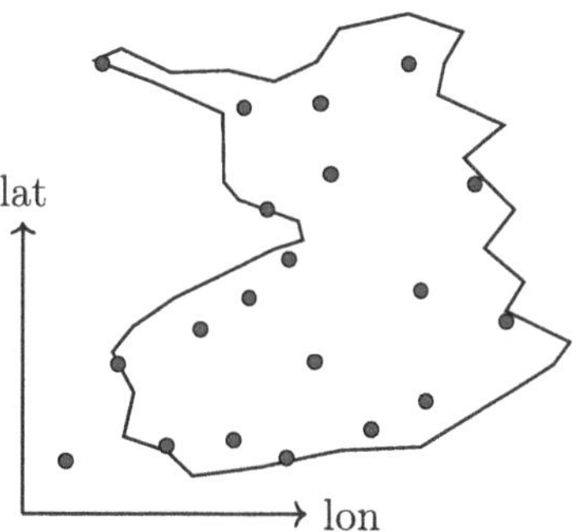

Fig. 12 We can interpret the temperature distribution over Finland as a realization of a GP indexed by geographic coordinates and sampled at FMI weather stations (indicated by blue dots)

A Gaussian random vector $\mathbf{x} \in \mathbb{R}^d$ with covariance matrix $\mathbf{C}$ and mean $\boldsymbol{\mu}$ can be constructed as [174, 197, 198]

$$\mathbf{x} := \mathbf{A}\mathbf{z} + \boldsymbol{\mu},$$

where $\mathbf{z} := \left(z_1, \ldots, z_d\right)^T$ is a vector of i.i.d. standard Gaussian RVs and $\mathbf{A} \in \mathbb{R}^{d \times d}$ is any matrix satisfying $\mathbf{A}\mathbf{A}^T = \mathbf{C}$. The probability distribution of a Gaussian random vector is referred to as the multivariate normal distribution, denoted $\mathcal{N}(\boldsymbol{\mu}, \mathbf{C})$.

Gaussian random vectors arise as finite-dimensional marginals of GPs, which define consistent joint Gaussian distributions over arbitrary (potentially infinite) index sets [199].

Gaussian RVs are widely used probabilistic models in the statistical analysis of ML methods. Their significance arises partly from the central limit theorem (CLT), which is a mathematically precise formulation of the following rule of thumb: The average of a large number of independent RVs (not necessarily Gaussian themselves) tends toward a Gaussian RV [200].

Compared to other probability distributions, the multivariate normal distribution is also distinct in that—in a mathematically precise sense—represents maximum uncertainty. Among all continuous random vectors with a given covariance matrix $\mathbf{C}$, the Gaussian random vector $\mathbf{x} \sim \mathcal{N}(\boldsymbol{\mu}, \mathbf{C})$ maximizes differential entropy [178, Th. 8.6.5]. This makes Gaussian distributions a natural choice for capturing uncertainty (or lack of knowledge) in the absence of additional structural information.

See also RV, pdf, mean, variance, probability distribution, covariance matrix, i.i.d., multivariate normal distribution, GP, probabilistic model, ML, CLT, entropy.

General Data Protection Regulation (GDPR) The GDPR was enacted by the European Union (EU), effective from May 25, 2018 [126]. It safeguards the privacy and data rights of individuals in the EU. The GDPR has significant implications for how data is collected, stored, and used in ML applications. Key provisions include the following:

- Data minimization principle: ML systems should only use the necessary amount of personal data for their purpose.
- Transparency and explainability: ML systems should enable their users to understand how the systems make decisions that impact the users.
- Data subject rights: Users should get an opportunity to access, rectify, and delete their personal data as well as to object to automated decision-making and profiling.
- Accountability: Organizations must ensure robust data security and demonstrate compliance through documentation and regular audits.

See also data, ML, data minimization principle, transparency, explainability.

Generalization Generalization refers to the ability of a model trained on a training set to make accurate predictions on new, unseen data points. This is a central goal of ML and AI, i.e., to learn patterns that extend beyond the training set. Most ML systems use ERM to learn a hypothesis $\hat{h} \in \mathcal{H}$ by minimizing the average loss over a training set of data points $\mathbf{z}^{(1)}, \ldots, \mathbf{z}^{(m)}$, which is denoted $\mathcal{D}^{(\text{train})}$. However, success on the training set does not guarantee success on unseen data—this discrepancy is the challenge of generalization.

To study generalization mathematically, we need to formalize the notion of "unseen" data. A widely used approach is to assume a probabilistic model for data generation, such as the i.i.d. assumption. Here, we interpret data points as independent RVs with an identical probability distribution $p(\mathbf{z})$. This probability distribution, which is assumed fixed but unknown, allows us to define the risk of a trained model $\hat{h}$ as the expected loss

$$\bar{L}(\hat{h}) = \mathbb{E}_{\mathbf{z} \sim p(\mathbf{z})} \{ L(\hat{h}, \mathbf{z}) \}.$$

The difference between risk $\bar{L}(\hat{h})$ and empirical risk $\widehat{L}(\hat{h}|\mathcal{D}^{(\text{train})})$ is known as the generalization gap. Tools from probability theory, such as concentration inequalities and uniform convergence, allow us to bound this gap under certain conditions [201].

Generalization without probability: Probability theory is one way to study how well a model generalizes beyond the training set, but it is not the only way. Another option is to use simple, deterministic changes to the data points in the training set. The basic idea is that a good model $\hat{h}$ should be robust, i.e., its prediction $\hat{h}(\mathbf{x})$ should not change much if we slightly change the features $\mathbf{x}$ of a data point $\mathbf{z}$.

For example, an object detector trained on smartphone photos should still detect the object if a few random pixels are masked [202]. Similarly, it should deliver the same result if we rotate the object in the image [101] (Fig. 13). See also model, training set, prediction, data point, ML, AI, ERM, hypothesis, loss, data, probabilistic model, i.i.d. assumption, RV, probability distribution, risk, empirical risk, generalization gap, probability, concentration inequality, feature.

Generalization gap The difference between the performance of a trained model on the training set and its performance on other data points (such as those in a validation set).

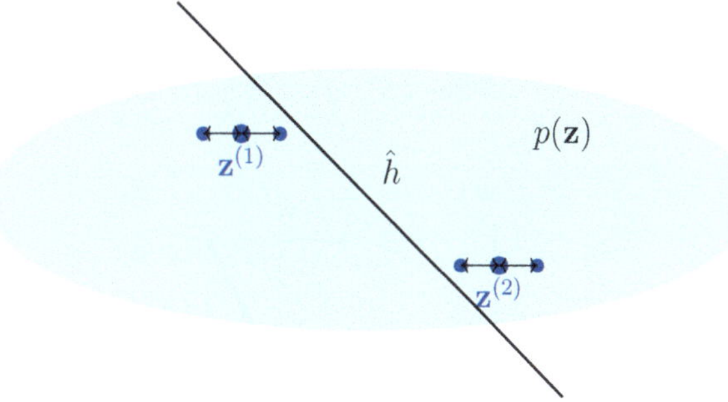

Fig. 13 Two data points $\mathbf{z}^{(1)}, \mathbf{z}^{(2)}$ that are used as a training set to learn a hypothesis $\hat{h}$ via ERM. We can evaluate $\hat{h}$ outside $\mathcal{D}^{(\text{train})}$ either by an i.i.d. assumption with some underlying probability distribution $p(\mathbf{z})$ or by perturbing the data points

See also model, training set, data point, validation set, hypothesis, decision tree, generalization, gradient-based methods, ERM, smooth, loss function, GD, model parameters, empirical risk, gradient, loss, gradient step.

Generalized total variation (GTV) GTV is a measure of the variation of trained local models $h^{(i)}$ (or their model parameters $\mathbf{w}^{(i)}$) assigned to the nodes $i = 1, \ldots, n$ of an undirected weighted graph $\mathcal{G}$ with edges $\mathcal{E}$. Given a measure $d^{(h,h')}$ for the discrepancy between hypothesis maps h, h', the GTV is

$$\sum_{\{i,i'\} \in \mathcal{E}} A_{i,i'} d^{(h^{(i)}, h^{(i')})}.$$

Here, $A_{i,i'} > 0$ denotes the weight of the undirected edge $\{i, i'\} \in \mathcal{E}$.
See also local model, model parameters, graph, discrepancy, hypothesis, map.

Generalized total variation minimization (GTVMin) GTVMin is an instance of RERM using the GTV of local model parameters as a regularizer [34].
See also RERM, GTV, model parameters, regularizer.

Gradient For a real-valued function $f : \mathbb{R}^d \to \mathbb{R} : \mathbf{w} \mapsto f(\mathbf{w})$, if a vector $\mathbf{g}$ exists such that $\lim_{\mathbf{w} \to \mathbf{w}'} \frac{f(\mathbf{w}) - \left(f(\mathbf{w}') + \mathbf{g}^T(\mathbf{w} - \mathbf{w}')\right)}{\|\mathbf{w} - \mathbf{w}'\|} = 0$, it is referred to as the gradient of f at $\mathbf{w}'$. If it exists, the gradient is unique and denoted $\nabla f(\mathbf{w}')$ or $\nabla f(\mathbf{w})\big|_{\mathbf{w}'}$ [2].
See also function.

Gradient descent (GD) GD is an iterative method for finding the minimum of a differentiable function $f(\mathbf{w})$ of a vector-valued argument $\mathbf{w} \in \mathbb{R}^d$. Consider a current guess or approximation $\mathbf{w}^{(k)}$ for the minimum of the function $f(\mathbf{w})$. We would like to find a new (better) vector $\mathbf{w}^{(k+1)}$ that has a smaller objective value $f(\mathbf{w}^{(k+1)}) < f\left(\mathbf{w}^{(k)}\right)$ than the current guess $\mathbf{w}^{(k)}$. We can achieve this typically by using a gradient step

$$\mathbf{w}^{(k+1)} = \mathbf{w}^{(k)} - \eta \nabla f(\mathbf{w}^{(k)}) \tag{3}$$

with a sufficiently small step size $\eta > 0$. Figure 14 illustrates the effect of a single GD step (3). See also minimum, differentiable, gradient step, step size, gradient.

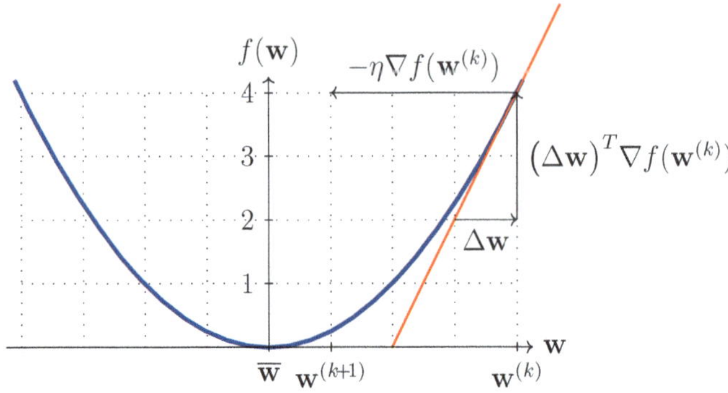

Fig. 14 A single gradient step (3) toward the minimizer $\overline{\mathbf{w}}$ of $f(\mathbf{w})$

Gradient step Given a differentiable real-valued function $f(\cdot) : \mathbb{R}^d \rightarrow \mathbb{R}$ and a vector $\mathbf{w} \in \mathbb{R}^d$, the gradient step updates $\mathbf{w}$ by adding the scaled negative gradient $\nabla f(\mathbf{w})$ to obtain the new vector (see Fig. 15)

$$\widehat{\mathbf{w}} := \mathbf{w} - \eta \nabla f(\mathbf{w}). \quad (4)$$

Mathematically, the gradient step is an operator $\mathcal{T}^{(f,\eta)}$ that is parametrized by the function f and the step size η. Note that the gradient step (4) optimizes

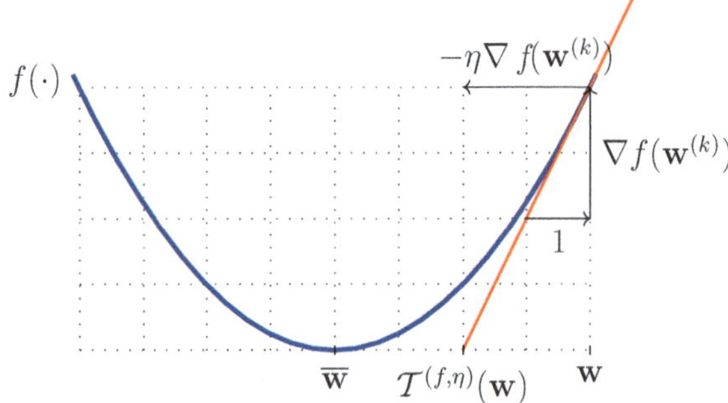

Fig. 15 The basic gradient step (4) maps a given vector $\mathbf{w}$ to the updated vector $\mathbf{w}'$. It defines an operator $\mathcal{T}^{(f,\eta)}(\cdot) : \mathbb{R}^d \rightarrow \mathbb{R}^d : \mathbf{w} \mapsto \widehat{\mathbf{w}}$

locally— in a neighborhood whose size is determined by the step size η—a linear approximation to the function $f(\cdot)$. A natural generalization of (4) is to locally optimize the function itself—instead of its linear approximation—such that

$$\widehat{\mathbf{w}} = \underset{\mathbf{w}' \in \mathbb{R}^d}{\operatorname{argmin}} f(\mathbf{w}') + (1/\eta) \left\| \mathbf{w} - \mathbf{w}' \right\|_2^2. \quad (5)$$

We intentionally use the same symbol η for the parameter in (5) as we used for the step size in (4). The larger the η, we choose in (5), the more progress the update will make toward reducing the function value $f(\widehat{\mathbf{w}})$. Note that, much like the gradient step (4), also the update (5) defines an operator that is parametrized by the function $f(\cdot)$ and the learning rate η. For a convex function $f(\cdot)$, this operator is known as the proximal operator of $f(\cdot)$ [40].

See also differentiable, function, gradient, step size, neighborhood, generalization, learning rate, convex, proximal operator.

Gradient-based methods Gradient-based methods are iterative techniques for finding the minimum (or maximum) of a differentiable objective function of the model parameters. These methods construct a sequence of approximations to an optimal choice for model parameters that results in a minimum (or maximum) value of the objective function. As their name indicates, gradient-based methods use the gradients of the objective function evaluated during previous iterations to construct new, (hopefully) improved model parameters. One important example of a gradient-based method is GD.

See also gradient, minimum, maximum, differentiable, objective function, model parameters, GD.

Graph A graph $\mathcal{G} = (\mathcal{V}, \mathcal{E})$ is a pair that consists of a node set $\mathcal{V}$ and an edge set $\mathcal{E}$. In its most general form, a graph is specified by a map that assigns each edge $e \in \mathcal{E}$ a pair of nodes [46]. One important family of graphs is simple undirected graphs. A simple undirected graph is obtained by identifying each edge $e \in \mathcal{E}$ with two different nodes $\{i, i'\}$. Weighted graphs also specify numeric weights A_e for each edge $e \in \mathcal{E}$.

See also map, weights.

Hard clustering Hard clustering refers to the task of partitioning a given set of data points into (a few) nonoverlapping clusters. The most widely used hard clustering method is k-means.

See also clustering, data point, cluster, k-means.

Hilbert space A Hilbert space is a complete inner product space [1]. That is, it is a vector space equipped with an inner product between pairs of vectors, and it satisfies the additional requirement of completeness; i.e., every Cauchy sequence of vectors converges to a limit within the space. A canonical example of a Hilbert space is the Euclidean space $\mathbb{R}^d$, for some dimension d, consisting of vectors $\mathbf{u} = \left(u_1, \ldots, u_d\right)^T$ and the standard inner product $\mathbf{u}^T\mathbf{v}$.

See also vector space, Euclidean space.

Horizontal federated learning (HFL) HFL uses local datasets constituted by different data points but uses the same features to characterize them [76]. For example, weather forecasting uses a network of spatially distributed weather (observation) stations. Each weather station measures the same quantities, such as daily temperature, air pressure, and precipitation. However, different weather stations measure the characteristics or features of different spatiotemporal regions. Each spatiotemporal region represents an individual data point, each

characterized by the same features (e.g., daily temperature or air pressure).
See also semi-supervised learning (SSL), FL, vertical federated learning (VFL).

Hypothesis A hypothesis refers to a map (or function) $h : \mathcal{X} \rightarrow \mathcal{Y}$ from the feature space $\mathcal{X}$ to the label space $\mathcal{Y}$. Given a data point with features $\mathbf{x}$, we use a hypothesis map h to estimate (or approximate) the label y using the prediction $\hat{y} = h(\mathbf{x})$. ML is all about learning (or finding) a hypothesis map h such that $y \approx h(\mathbf{x})$ for any data point (having features $\mathbf{x}$ and label y).
See also map, function, feature space, label space, data point, feature, label, prediction, ML.

Hypothesis space Every practical ML method uses a hypothesis space (or model) $\mathcal{H}$. The hypothesis space of an ML method is a subset of all possible maps from the feature space to the label space. The design choice of the hypothesis space should take into account available computational resources and statistical aspects. If the computational infrastructure allows for efficient matrix operations, and there is an (approximately) linear relation between a set of features and a label, a useful choice for the hypothesis space might be the linear model.
See also ML, hypothesis, model, map, feature space, label space, statistical aspects, feature, label, linear model.

Independent and identically distributed (i.i.d.) It can be useful to interpret data points $\mathbf{z}^{(1)}, \ldots, \mathbf{z}^{(m)}$ as realizations of i.i.d. RVs with a common probability distribution. If these RVs are continuous-valued, their joint pdf is $p\left(\mathbf{z}^{(1)}, \ldots, \mathbf{z}^{(m)}\right) = \prod_{r=1}^{m} p\left(\mathbf{z}^{(r)}\right)$, with $p(\mathbf{z})$ being the common marginal pdf of the underlying RVs.
See also data point, realization, RV, probability distribution, pdf.

Independent and identically distributed assumption (i.i.d. assumption) The i.i.d. assumption interprets data points of a dataset as the realizations of i.i.d. RVs.
See also i.i.d., data point, dataset, realization, RV.

Interpretability An ML method is interpretable for a specific user if they can well anticipate the predictions delivered by the method. The notion of interpretability can be made precise using quantitative measures of the uncertainty about the predictions [129].
See also ML, prediction, uncertainty.

Jacobi method The Jacobi method is an algorithm for solving systems of linear equations (i.e., a linear system) of the form $\mathbf{A}\mathbf{x} = \mathbf{b}$. Here, $\mathbf{A} \in \mathbb{R}^{d \times d}$ is a square matrix with nonzero main diagonal entries. The method constructs a sequence $\mathbf{x}^{(0)}, \mathbf{x}^{(1)}, \ldots$ by updating each entry of $\mathbf{x}^{(k)}$ according to

$$x_i^{(k+1)} = \frac{1}{a_{ii}} \left(b_i - \sum_{j \neq i} a_{ij} x_j^{(k)} \right).$$

Carefully note that all entries $x_1^{(k)}, \ldots, x_d^{(k)}$ are updated simultaneously. The above iteration converges to a solution, i.e., $\lim_{k \to \infty} \mathbf{x}^{(k)} = \mathbf{x}$, under certain conditions on the matrix $\mathbf{A}$, e.g., being strictly diagonally dominant or symmetric positive definite [3, 21, 203]. Jacobi-type methods are appealing for large linear systems due to their parallelizable structure [17]. We can interpret the Jacobi method as a fixed-point iteration. Indeed, using the decomposition $\mathbf{A} = \mathbf{D} + \mathbf{R}$, with $\mathbf{D}$ being the diagonal of $\mathbf{A}$, allows us to rewrite the linear equation $\mathbf{A}\mathbf{x} = \mathbf{b}$ as a fixed-point equation

$$\mathbf{x} = \underbrace{\mathbf{D}^{-1}(\mathbf{b} - \mathbf{R}\mathbf{x})}_{\mathcal{F}\mathbf{x}},$$

which leads to the iteration $\mathbf{x}^{(k+1)} = \mathbf{D}^{-1}(\mathbf{b} - \mathbf{R}\mathbf{x}^{(k)})$.
For example, for the linear equation

$$\mathbf{A}\mathbf{x} = \mathbf{b}, \quad \text{where} \quad \mathbf{A} = \begin{bmatrix} a_{11} & a_{12} & a_{13} \\ a_{21} & a_{22} & a_{23} \\ a_{31} & a_{32} & a_{33} \end{bmatrix}, \quad \mathbf{b} = \begin{bmatrix} b_1 \\ b_2 \\ b_3 \end{bmatrix},$$

the Jacobi method updates each component of $\mathbf{x}$ as follows:

$$\begin{aligned} x_1^{(k+1)} &= \frac{1}{a_{11}} \left(b_1 - a_{12}x_2^{(k)} - a_{13}x_3^{(k)} \right), \\ x_2^{(k+1)} &= \frac{1}{a_{22}} \left(b_2 - a_{21}x_1^{(k)} - a_{23}x_3^{(k)} \right), \\ x_3^{(k+1)} &= \frac{1}{a_{33}} \left(b_3 - a_{31}x_1^{(k)} - a_{32}x_2^{(k)} \right). \end{aligned}$$

See also algorithm, fixed-point iteration, optimization method.

Kernel Consider data points characterized by a feature vector $\mathbf{x} \in \mathcal{X}$ with a generic feature space $\mathcal{X}$. A (real-valued) kernel $K : \mathcal{X} \times \mathcal{X} \to \mathbb{R}$ assigns each pair of feature vectors $\mathbf{x}, \mathbf{x}' \in \mathcal{X}$ a real number $K(\mathbf{x}, \mathbf{x}')$. The value $K(\mathbf{x}, \mathbf{x}')$ is often interpreted as a measure for the similarity between $\mathbf{x}$ and $\mathbf{x}'$. Kernel methods use a kernel to transform the feature vector $\mathbf{x}$ to a new feature vector $\mathbf{z} = K(\mathbf{x}, \cdot)$. This new feature vector belongs to a linear feature space $\mathcal{X}'$ which is (in general) different from the original feature space $\mathcal{X}$. The feature space $\mathcal{X}'$ has a specific mathematical structure; i.e., it is a reproducing kernel Hilbert space [26, 204]. See also feature vector, feature space, kernel method, Hilbert space.

Kernel method A kernel method is an ML method that uses a kernel K to map the original (i.e., raw) feature vector $\mathbf{x}$ of a data point to a new (transformed) feature vector $\mathbf{z} = K(\mathbf{x}, \cdot)$ [26, 204]. The motivation for transforming the feature vectors is that, by using a suitable kernel, the data points have a "more pleasant" geometry in the transformed feature space. For example, in a binary classification problem, using transformed feature vectors $\mathbf{z}$ might allow us to use linear models, even if the data points are not linearly separable in the original feature space (see Fig. 16). See also kernel, feature vector, feature space, linear classifier.

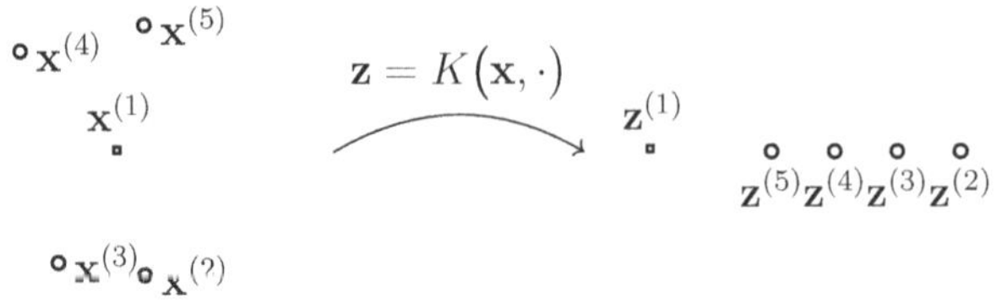

Fig. 16 Five data points characterized by feature vectors $\mathbf{x}^{(r)}$ and labels $y^{(r)} \in \{\circ, \square\}$, for $r = 1, \ldots, 5$. With these feature vectors, there is no way to separate the two classes by a straight line (representing the decision boundary of a linear classifier). In contrast, the transformed feature vectors $\mathbf{z}^{(r)} = K\big(\mathbf{x}^{(r)}, \cdot\big)$ allow us to separate the data points using a linear classifier

Label A higher-level fact or quantity of interest associated with a data point. For example, if the data point is an image, the label could indicate whether the image contains a cat or not. Synonyms for label, commonly used in specific domains, include "response variable," "output variable," and "target" [192–194].
See also data point.

Label space Consider an ML application that involves data points characterized by features and labels. The label space is constituted by all potential values that the label of a data point can take on. Regression methods, aiming at predicting numeric labels, often use the label space $\mathcal{Y} = \mathbb{R}$. Binary classification methods use a label space that consists of two different elements, e.g., $\mathcal{Y} = \{-1, 1\}$, $\mathcal{Y} = \{0, 1\}$, or $\mathcal{Y} = \{\text{"cat image"}, \text{"no cat image"}\}$.
See also ML, data point, feature, label, regression, classification.

Labeled datapoint A data point whose label is known or has been determined by some means which might require human labor.
See also data point, label.

Laplacian matrix The structure of a graph $\mathcal{G}$, with nodes $i = 1, \ldots, n$, can be analyzed using the properties of special matrices that are associated with $\mathcal{G}$. One such matrix is the graph Laplacian matrix $\mathbf{L}^{(\mathcal{G})} \in \mathbb{R}^{n \times n}$, which is defined for an undirected and weighted graph [205, 206]. It is defined element-wise as (see Fig. 17)

$$L^{(\mathcal{G})}_{i,i'} := \begin{cases} -A_{i,i'} & \text{for } i \neq i', \{i, i'\} \in \mathcal{E}, \\ \sum_{i'' \neq i} A_{i,i''} & \text{for } i = i', \\ 0 & \text{else.} \end{cases} \tag{6}$$

Here, $A_{i,i'}$ denotes the edge weight of an edge $\{i, i'\} \in \mathcal{E}$. See also graph, edge weight.

Law of large numbers The law of large numbers refers to the convergence of the average of an increasing (large) number of i.i.d. RVs to the mean of their common probability distribution. Different instances of the law of large numbers are obtained by using different notions of convergence [197].
See also i.i.d., RV, mean, probability distribution.

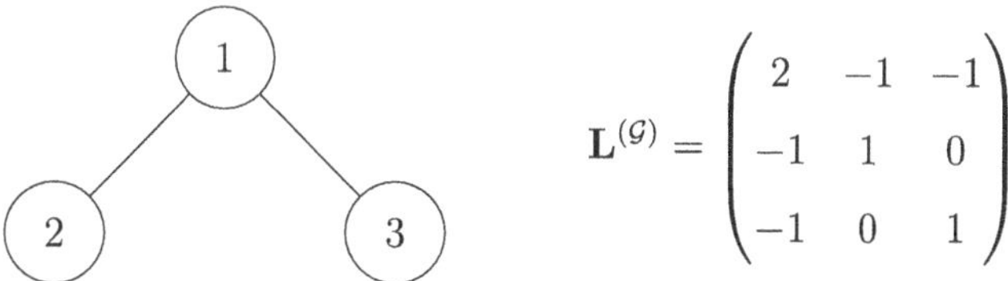

Fig. 17 Left: Some undirected graph $\mathcal{G}$ with three nodes $i = 1, 2, 3$. Right: The Laplacian matrix $\mathbf{L}^{(\mathcal{G})} \in \mathbb{R}^{3 \times 3}$ of $\mathcal{G}$

Learning rate Consider an iterative ML method for finding or learning a useful hypothesis $h \in \mathcal{H}$. Such an iterative method repeats similar computational (update) steps that adjust or modify the current hypothesis to obtain an improved hypothesis. One well-known example of such an iterative learning method is GD and its variants, SGD and projected GD. A key parameter of an iterative method is the learning rate. The learning rate controls the extent to which the current hypothesis can be modified during a single iteration. A well-known example of such a parameter is the step size used in GD [23, Ch. 5].
See also ML, hypothesis, GD, SGD, projected GD, parameter, step size.

Learning task Consider a dataset $\mathcal{D}$ constituted by several data points, each of them characterized by features $\mathbf{x}$. For example, the dataset $\mathcal{D}$ might be constituted by the images of a particular database. Sometimes it might be useful to represent a dataset $\mathcal{D}$, along with the choice of features, by a probability distribution $p(\mathbf{x})$. A learning task associated with $\mathcal{D}$ consists of a specific choice for the label of a data point and the corresponding label space. Given a choice for the loss function and model, a learning task gives rise to an instance of ERM. Thus, we could define a learning task also via an instance of ERM, i.e., via an objective function. Note that, for the same dataset, we obtain different learning tasks by using different choices for the features and label of a data point. These learning tasks are related, as they are based on the same dataset, and solving them jointly (via multitask learning methods) is typically preferable over solving them separately [86, 87, 207].
See also dataset, data point, feature, probability distribution, label, label space, loss function, model, ERM, objective function, multitask learning.

Least absolute shrinkage and selection operator (Lasso) The Lasso is an instance of structural risk minimization (SRM). It learns the weights $\mathbf{w}$ of a linear map $h(\mathbf{x}) = \mathbf{w}^T\mathbf{x}$ based on a training set. Lasso is obtained from linear regression by adding the scaled ℓ_1-norm $\alpha \|\mathbf{w}\|_1$ to the average squared error loss incurred on the training set.
See also SRM, weights, linear map, training set, linear regression, norm, squared error loss.

Linear classifier Consider data points characterized by numeric features $\mathbf{x} \in \mathbb{R}^d$ and a label $y \in \mathcal{Y}$ from some finite label space $\mathcal{Y}$. A linear classifier is characterized by having decision regions that are separated by hyperplanes in $\mathbb{R}^d$ [23, Ch. 2].
See also data point, feature, label, label space, classifier, decision region.

Linear map A linear map $f : \mathbb{R}^n \rightarrow \mathbb{R}^m$ is a function that satisfies additivity $f(\mathbf{x}+\mathbf{y}) = f(\mathbf{x}) + f(\mathbf{y})$ and homogeneity $f(c\mathbf{x}) = cf(\mathbf{x})$ for all vectors $\mathbf{x}, \mathbf{y} \in \mathbb{R}^n$ and scalars $c \in \mathbb{R}$. In particular, $f(\mathbf{0}) = \mathbf{0}$. Any linear map can be represented as a matrix multiplication $f(\mathbf{x}) = \mathbf{A}\mathbf{x}$ for some matrix $\mathbf{A} \in \mathbb{R}^{m\times n}$. The collection of real-valued linear maps for a given dimension n constitute a linear model which is used in many ML methods.
See also linear model, linear regression, PCA, feature vector.

Linear model Consider an ML application involving data points, each represented by a numeric feature vector $\mathbf{x} \in \mathbb{R}^d$. A linear model defines a hypothesis space consisting of all real-valued linear maps from $\mathbb{R}^d$ to $\mathbb{R}$,

$$\mathcal{H}^{(d)} := \left\{ h : \mathbb{R}^d \rightarrow \mathbb{R} \mid h(\mathbf{x}) = \mathbf{w}^\top \mathbf{x} \text{ for some } \mathbf{w} \in \mathbb{R}^d \right\}. \tag{7}$$

Each value of d defines a different hypothesis space, corresponding to the number of features used to compute the prediction $h(\mathbf{x})$. The choice of d is often guided by computational aspects (e.g., fewer features reduce computation), statistical aspects (e.g., more features typically reduce bias and thus lower risk), and interpretability. A linear model using a small number of well-chosen features is generally considered more interpretable [195, 208]. The linear model is attractive, because it can typically be trained using scalable convex optimization methods [38, 209]. Moreover, linear models often permit rigorous statistical analysis, including fundamental limits on the minimum achievable risk [33]. They are also useful for analyzing more complex, nonlinear models such as ANNs. For instance, a deep net can be viewed as the composition of a feature map—implemented by the input and hidden layers—and a linear model in the output layer. Similarly, a decision tree can be interpreted as applying a one-hot encoded feature map based on decision regions, followed by a linear model that assigns a prediction to each region. More generally, any trained model $\hat{h} \in \mathcal{H}$ that is differentiable at some $\mathbf{x}'$ can be locally approximated by a linear map $g(\mathbf{x})$. Figure 18 illustrates such a local linear approximation, defined by the gradient $\nabla\hat{h}(\mathbf{x}')$. Note that the gradient is only defined where $\hat{h}$ is differentiable. To ensure robustness in the context of trustworthy AI, one may prefer models whose associated map $\hat{h}$ is Lipschitz continuous. A classic result in mathematical analysis—Rademacher's Theorem—states that if $\hat{h}$ is Lipschitz continuous with some constant L over an open set $\Omega \subseteq \mathbb{R}^d$, then $\hat{h}$ is differentiable almost everywhere in Ω [210, Thm. 3.1].
See also model, hypothesis space, gradient, linear map, LIME, interpretability.

Linear regression Linear regression aims to learn a linear hypothesis map to predict a numeric label based on the numeric features of a data point. The quality of a linear hypothesis map is measured using the average squared error loss incurred on a set of labeled datapoints, which we refer to as the training set.
See also regression, hypothesis, map, label, feature, data point, squared error loss, labeled datapoint, training set.

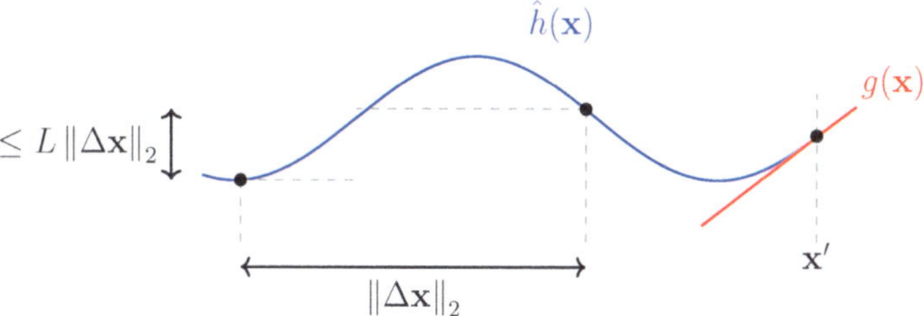

Fig. 18 A trained model $\hat{h}(\mathbf{x})$ that is differentiable at a point $\mathbf{x}'$ can be locally approximated by a linear map $g \in \mathcal{H}^{(d)}$. This local approximation is determined by the gradient $\nabla \hat{h}(\mathbf{x}')$

Local dataset The concept of a local dataset is in between the concept of a data point and a dataset. A local dataset consists of several individual data points, which are characterized by features and labels. In contrast to a single dataset used in basic ML methods, a local dataset is also related to other local datasets via different notions of similarity. These similarities might arise from probabilistic models or communication infrastructure and are encoded in the edges of an FL network.

See also dataset, data point, feature, label, ML, probabilistic model, FL network.

Local interpretable model-agnostic explanations (LIME) Consider a trained model (or learned hypothesis) $\widehat{h} \in \mathcal{H}$, which maps the feature vector of a data point to the prediction $\widehat{y} = \widehat{h}$. LIME is a technique for explaining the behavior of $\widehat{h}$, locally around a data point with feature vector $\mathbf{x}^{(0)}$ [195]. The explanation is given in the form of a local approximation $g \in \mathcal{H}'$ of $\widehat{h}$ (see Fig. 19). This approximation can be obtained by an instance of ERM with a carefully designed training set. In particular, the training set consists of data points with feature vector $\mathbf{x}$ close to $\mathbf{x}^{(0)}$ and the (pseudo-)label $\widehat{h}(\mathbf{x})$. Note that we can use a different model $\mathcal{H}'$ for the approximation from the original model $\mathcal{H}$. For example, we can use a decision tree to locally approximate a deep net. Another widely used choice for $\mathcal{H}'$ is the linear model.

Fig. 19 To explain a trained model $\widehat{h} \in \mathcal{H}$, around a given feature vector $\mathbf{x}^{(0)}$, we can use a local approximation $g \in \mathcal{H}'$

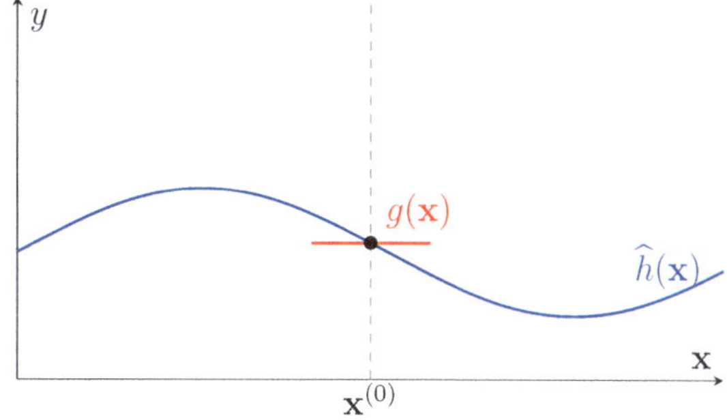

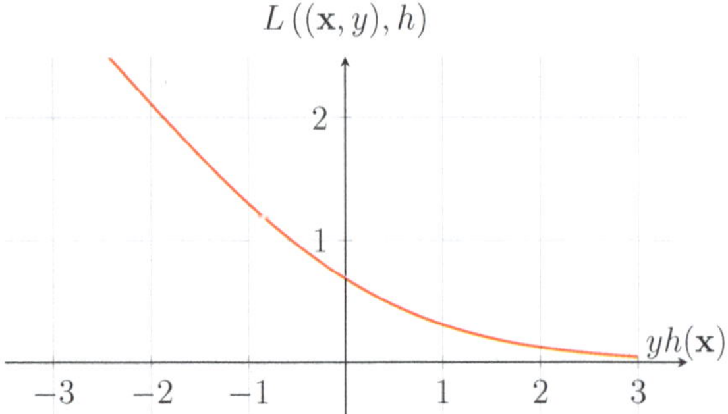

Fig. 20 The logistic loss incurred by the prediction $h(\mathbf{x}) \in \mathbb{R}$ for a data point with label $y \in \{-1, 1\}$

See also model, explanation, ERM, training set, label, decision tree, deep net, linear model.

Local model Consider a collection of devices that are represented as nodes $\mathcal{V}$ of an FL network. A local model $\mathcal{H}^{(i)}$ is a hypothesis space assigned to a node $i \in \mathcal{V}$. Different nodes might be assigned different hypothesis spaces, i.e., in general $\mathcal{H}^{(i)} \neq \mathcal{H}^{(i')}$ for different nodes $i, i' \in \mathcal{V}$.
See also device, FL network, model, hypothesis space.

Logistic loss Consider a data point characterized by the features $\mathbf{x}$ and a binary label $y \in \{-1, 1\}$. We use a real-valued hypothesis h to predict the label y from the features $\mathbf{x}$. The logistic loss incurred by this prediction is defined as (Fig. 20)

$$L((\mathbf{x}, y), h) := \log(1 + \exp(-yh(\mathbf{x}))). \tag{8}$$

Carefully note that the expression (8) for the logistic loss applies only for the label space $\mathcal{Y} = \{-1, 1\}$ and when using the thresholding rule (1).
See also data point, feature, label, hypothesis, loss, prediction, label space.

Logistic regression Logistic regression learns a linear hypothesis map (or classifier) $h(\mathbf{x}) = \mathbf{w}^T\mathbf{x}$ to predict a binary label y based on the numeric feature vector $\mathbf{x}$ of a data point. The quality of a linear hypothesis map is measured by the average logistic loss on some labeled datapoints (i.e., the training set).
See also regression, hypothesis, map, classifier, label, feature vector, data point, logistic loss, labeled datapoint, training set.

Loss ML methods use a loss function $L(\mathbf{z}, h)$ to measure the error incurred by applying a specific hypothesis to a specific data point. With a slight abuse of notation, we use the term loss for both the loss function L itself and the specific value $L(\mathbf{z}, h)$, for a data point $\mathbf{z}$ and hypothesis h.
See also ML, loss function, hypothesis, data point.

Loss function A loss function is a map

$$L : \mathcal{X} \times \mathcal{Y} \times \mathcal{H} \rightarrow \mathbb{R}_+ : \big((\mathbf{x}, y), h\big) \mapsto L((\mathbf{x}, y), h).$$

It assigns a nonnegative real number (i.e., the loss) $L((\mathbf{x}, y), h)$ to a pair that consists of a data point, with features $\mathbf{x}$ and label y, and a hypothesis $h \in \mathcal{H}$. The value $L((\mathbf{x}, y), h)$ quantifies the discrepancy between the true label y and the prediction $h(\mathbf{x})$. Lower (closer to zero) values $L((\mathbf{x}, y), h)$ indicate a smaller discrepancy between prediction $h(\mathbf{x})$ and label y. Figure 21 depicts a loss function for a given data point, with features $\mathbf{x}$ and label y, as a function of the hypothesis $h \in \mathcal{H}$.

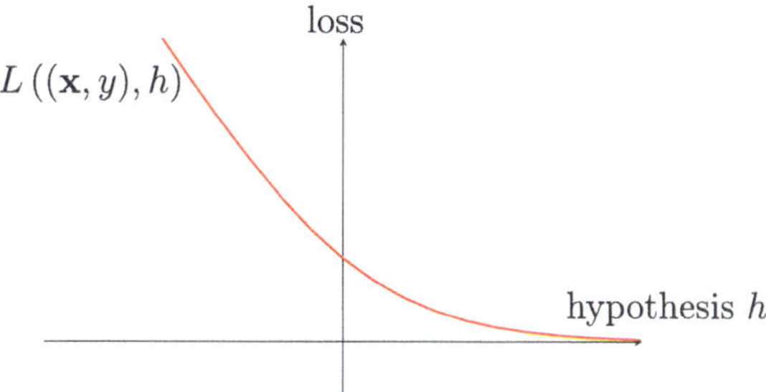

Fig. 21 Some loss function $L((\mathbf{x}, y), h)$ for a fixed data point, with feature vector $\mathbf{x}$ and label y, and a varying hypothesis h. ML methods try to find (or learn) a hypothesis that incurs minimal loss

See also loss, function, map, data point, feature, label, hypothesis, prediction, feature vector, ML.

Machine learning (ML) ML aims to predict a label from the features of a data point. ML methods achieve this by learning a hypothesis from a hypothesis space (or model) through the minimization of a loss function [23, 211]. One precise formulation of this principle is ERM. Different ML methods are obtained from different design choices for data points (i.e., their features and label), the model, and the loss function [23, Ch. 3].
See also label, feature, data point, hypothesis, hypothesis space, model, loss function, ERM.

Map We use the term map as a synonym for function.
See also function.

Maximum The maximum of a set $\mathcal{A} \subseteq \mathbb{R}$ of real numbers is the greatest element in that set, if such an element exists. A set $\mathcal{A}$ has a maximum if it is bounded above and attains its supremum (or least upper bound) [2, Sec. 1.4].
See also supremum.

Maximum likelihood Consider data points $\mathcal{D} = \{\mathbf{z}^{(1)}, \ldots, \mathbf{z}^{(m)}\}$ that are interpreted as the realizations of i.i.d. RVs with a common probability distribution $p(\mathbf{z}; \mathbf{w})$ which depends on the model parameters $\mathbf{w} \in \mathcal{W} \subseteq \mathbb{R}^n$. Maximum likelihood methods learn model parameters $\mathbf{w}$ by maximizing the probability (density) $p(\mathcal{D}; \mathbf{w}) = \prod_{r=1}^{m} p(\mathbf{z}^{(r)}; \mathbf{w})$ of the observed dataset. Thus, the maximum likelihood estimator is a solution to the optimization problem $\max_{\mathbf{w} \in \mathcal{W}} p(\mathcal{D}; \mathbf{w})$.
See also data point, realization, i.i.d., RV, probability distribution, model parameters, maximum, dataset, optimization problem.

Mean The mean of an RV $\mathbf{x}$, taking values in an Euclidean space $\mathbb{R}^d$, is its expectation $\mathbb{E}\{\mathbf{x}\}$. It is defined as the Lebesgue integral of $\mathbf{x}$ with respect to the underlying probability distribution P (e.g., see [2] or [28]), i.e.,

$$\mathbb{E}\{\mathbf{x}\} = \int_{\mathbb{R}^d} \mathbf{x}\,\mathrm{d}P(\mathbf{x}).$$

We also use the term to refer to the average of a finite sequence $\mathbf{x}^{(1)}, \ldots, \mathbf{x}^{(m)} \in \mathbb{R}^d$. However, these two definitions are essentially the same. Indeed, we can use the sequence $\mathbf{x}^{(1)}, \ldots, \mathbf{x}^{(m)} \in \mathbb{R}^d$ to construct a discrete RV $\widetilde{\mathbf{x}} = \mathbf{x}^{(I)}$, with the index I being chosen uniformly at random from the set $\{1, \ldots, m\}$. The mean of $\widetilde{\mathbf{x}}$ is precisely the average $\frac{1}{m}\sum_{r=1}^{m} \mathbf{x}^{(r)}$.
See also RV, Euclidean space, expectation, probability distribution.

Minimum Given a set of real numbers, the minimum is the smallest of those numbers. Note that for some sets, such as the set of negative real numbers, the minimum does not exist.

Model In the context of ML, the term model typically refers to the hypothesis space underlying an ML method [23, 201]. However, the term is also used in other fields but with a different meaning. For example, a probabilistic model refers to a parametrized set of probability distributions.
See also ML, hypothesis space, probabilistic model, probability distribution.

Model inversion A model inversion is a form of privacy attack on an ML system. An adversary seeks to infer sensitive attributes of individual data points by exploiting partial access to a trained model $\hat{h} \in \mathcal{H}$. This access typically consists of querying the model for predictions $\hat{h}(\mathbf{x})$ on carefully chosen inputs. Basic model inversion techniques have been demonstrated in the context of facial image classification, where images are reconstructed using the (gradient of) model outputs combined with auxiliary information such as a person's name [160].

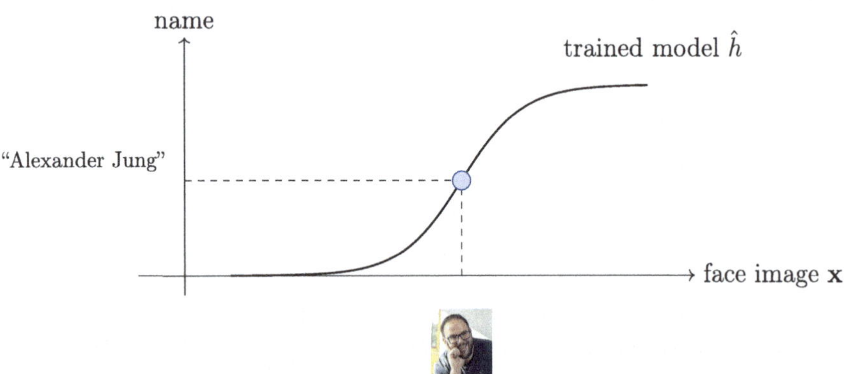

See also model, privacy attack, ML, sensitive attribute, data point, prediction, classification, gradient, trustworthy AI, privacy protection.

Model parameters Model parameters are quantities that are used to select a specific hypothesis map from a model. We can think of a list of model parameters as a unique identifier for a hypothesis map, similar to how a social security number identifies a person in Finland.
See also model, parameter, hypothesis, map.

Multitask learning Multitask learning aims at leveraging relations between different learning tasks. Consider two learning tasks obtained from the same dataset of webcam snapshots. The first task is to predict the presence of a human, while the second task is to predict the presence of a car. It might be useful to use the same deep net structure for both tasks and only allow the weights of the final output layer to be different.
See also learning task, dataset, deep net, weights.

Multivariate normal distribution The multivariate normal distribution, which is denoted $\mathcal{N}(\boldsymbol{\mu}, \boldsymbol{\Sigma})$, is a fundamental probabilistic model for numerical feature vectors of fixed dimension d. It defines a family of probability distributions over vector-valued RVs $\mathbf{x} \in \mathbb{R}^d$ [174, 196, 198]. Each distribution in this family is fully specified by its mean vector $\boldsymbol{\mu} \in \mathbb{R}^d$ and covariance matrix $\boldsymbol{\Sigma} \in \mathbb{R}^{d \times d}$. When the covariance matrix $\boldsymbol{\Sigma}$ is invertible, its probability distribution is fully characterized by the following pdf:

$$p(\mathbf{x}) = \frac{1}{\sqrt{(2\pi)^d \det(\boldsymbol{\Sigma})}} \exp\left(-\frac{1}{2}(\mathbf{x} - \boldsymbol{\mu})^T \boldsymbol{\Sigma}^{-1} (\mathbf{x} - \boldsymbol{\mu})\right).$$

Note that the pdf is only defined when $\boldsymbol{\Sigma}$ is invertible. More generally, any RV $\mathbf{x} \sim \mathcal{N}(\boldsymbol{\mu}, \boldsymbol{\Sigma})$ admits the following innovation representation:

$$\mathbf{x} = \mathbf{A}\mathbf{z} + \boldsymbol{\mu},$$

where $\mathbf{z} \sim \mathcal{N}(\mathbf{0}, \mathbf{I})$ is a standard normal vector and $\mathbf{A} \in \mathbb{R}^{d \times d}$ satisfies $\mathbf{A}\mathbf{A}^T = \boldsymbol{\Sigma}$. This innovation representation is valid even when the covariance matrix $\boldsymbol{\Sigma}$ is singular, in which case $\mathbf{A}$ is not necessarily full-rank [198, Ch. 23] . The family of multivariate normal distributions is exceptional among probabilistic models for numerical quantities at least for the following reasons. First, the family is closed under affine transformations, i.e.,

$$\mathbf{x} \sim \mathcal{N}(\boldsymbol{\mu}, \boldsymbol{\Sigma}) \text{ implies } \mathbf{B}\mathbf{x} + \mathbf{c} \sim \mathcal{N}\big(\mathbf{B}\boldsymbol{\mu} + \mathbf{c}, \mathbf{B}\boldsymbol{\Sigma}\mathbf{B}^T\big).$$

Second, the probability distribution $\mathcal{N}(\mathbf{0}, \boldsymbol{\Sigma})$ maximizes the differential entropy among all distributions with the same covariance matrix $\boldsymbol{\Sigma}$ [178].
See also probabilistic model, feature vector, probability distribution, RV, covariance matrix, pdf, standard normal vector, Gaussian RV, mean, entropy.

Mutual information (MI) The MI $I(\mathbf{x}; y)$ between two RVs $\mathbf{x}$, y defined on the same probability space is given by [178]

$$I(\mathbf{x}; y) := \mathbb{E}\left\{\log \frac{p(\mathbf{x}, y)}{p(\mathbf{x})p(y)}\right\}.$$

It is a measure of how well we can estimate y based solely on $\mathbf{x}$. A large value of $I(\mathbf{x}; y)$ indicates that y can be well predicted solely from $\mathbf{x}$. This prediction could be obtained by a hypothesis learned by an ERM-based ML method.
See also RV, probability space, prediction, hypothesis, ERM, ML.

Neighborhood The neighborhood of a node $i \in \mathcal{V}$ is the subset of nodes constituted by the neighbors of i.
See also neighbors.

Neighbors The neighbors of a node $i \in \mathcal{V}$ within an FL network are those nodes $i' \in \mathcal{V} \setminus \{i\}$ that are connected (via an edge) to node i.
See also FL network.

Networked data Networked data consists of local datasets that are related by some notion of pairwise similarity. We can represent networked data using a graph whose nodes carry local datasets and edges encode pairwise similarities. One example of networked data arises in FL applications where local datasets are generated by spatially distributed devices.
See also data, local dataset, graph, FL, device.

Node degree The degree $d^{(i)}$ of a node $i \in \mathcal{V}$ in an undirected graph is the number of its neighbors, i.e., $d^{(i)} := \left|\mathcal{N}^{(i)}\right|$.
See also graph, neighbors.

Non-smooth We refer to a function as non-smooth if it is not smooth [57].
See also function, smooth.

Norm A norm is a function that maps each (vector) element of a vector space to a nonnegative real number. This function must be homogeneous and definite, and it must satisfy the triangle inequality [212].
See also function, vector space.

Objective function An objective function is a map that assigns a numeric objective value $f(\mathbf{w})$ to each choice $\mathbf{w}$ of some variable that we want to optimize (see Fig. 22). In the context of ML, the optimization variable could be the model parameters of a hypothesis $h^{(\mathbf{w})}$. Common objective functions include the risk (i.e., expected loss) or the empirical risk (i.e., average loss over a training set). ML methods apply optimization techniques, such as gradient-based methods, to find the choice $\mathbf{w}$ with the optimal value (e.g., the minimum or the maximum) of the objective function.

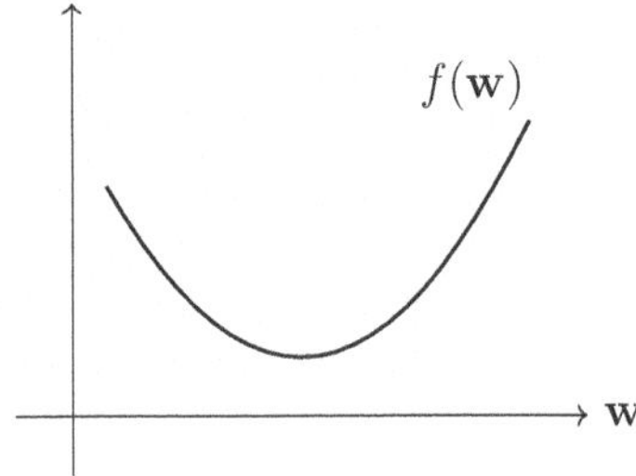

Fig. 22 An objective function maps each possible value $\mathbf{w}$ of an optimization variable, such as the model parameters of an ML model, to a value that measures the usefulness of $\mathbf{w}$

See also function, map, ML, model parameters, hypothesis, risk, loss, empirical risk, training set, gradient-based methods, minimum, maximum, model, loss function.

Online gradient descent (online GD) Consider an ML method that learns model parameters $\mathbf{w}$ from some parameter space $\mathcal{W} \subseteq \mathbb{R}^d$. The learning process uses data points $\mathbf{z}^{(t)}$ that arrive at consecutive time instants $t = 1, 2, \ldots$. Let us interpret the data points $\mathbf{z}^{(t)}$ as i.i.d. copies of an RV $\mathbf{z}$. The risk $\mathbb{E}\{L(\mathbf{z}, \mathbf{w})\}$ of a hypothesis $h^{(\mathbf{w})}$ can then (under mild conditions) be obtained as the limit $\lim_{T\to\infty}(1/T)\sum_{t=1}^{T} L\left(\mathbf{z}^{(t)}, \mathbf{w}\right)$. We might use this limit as the objective function for learning the model parameters $\mathbf{w}$. Unfortunately, this limit can only be evaluated if we wait infinitely long in order to collect all data points. Some ML applications require methods that learn online; i.e., as soon as a new data point $\mathbf{z}^{(t)}$ arrives at time t, we update the current model parameters $\mathbf{w}^{(t)}$. Note that the new data point $\mathbf{z}^{(t)}$ contributes the component $L\left(\mathbf{z}^{(t)}, \mathbf{w}\right)$ to the risk. As its name suggests, online GD updates $\mathbf{w}^{(t)}$ via a (projected) gradient step such that

$$\mathbf{w}^{(t+1)} := P_{\mathcal{W}}\Big(\mathbf{w}^{(t)} - \eta_t \nabla_{\mathbf{w}} L\left(\mathbf{z}^{(t)}, \mathbf{w}\right)\Big). \tag{9}$$

Note that (9) is a gradient step for the current component $L\left(\mathbf{z}^{(t)}, \cdot\right)$ of the risk. The update (9) ignores all the previous components $L\left(\mathbf{z}^{(t')}, \cdot\right)$, for $t' < t$. It might therefore happen that, compared to $\mathbf{w}^{(t)}$, the updated model parameters $\mathbf{w}^{(t+1)}$ increase the retrospective average loss $\sum_{t'=1}^{t-1} L\left(\mathbf{z}^{(t')}, \cdot\right)$. However, for a suitably chosen learning rate η_t, online GD can be shown to be optimal in practically relevant settings. By optimal, we mean that the model parameters $\mathbf{w}^{(T+1)}$ delivered by online GD after observing T data points $\mathbf{z}^{(1)}, \ldots, \mathbf{z}^{(T)}$ are at least as good as those delivered by any other learning method [188, 213] (Fig. 23). See also ML, model parameters, parameter space, data point, i.i.d., RV, risk, hypothesis, objective function, GD, gradient step, loss, learning rate, squared error loss.

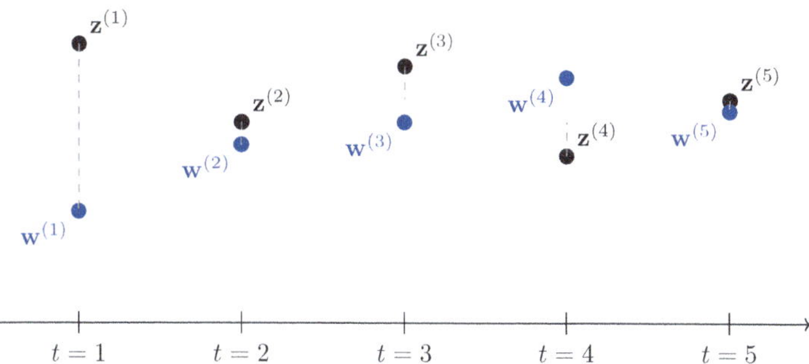

Fig. 23 An instance of online GD that updates the model parameters $\mathbf{w}^{(t)}$ using the data point $\mathbf{z}^{(t)} = x^{(t)}$ arriving at time t. This instance uses the squared error loss $L\left(\mathbf{z}^{(t)}, w\right) = (x^{(t)} - w)^2$

Optimization method An optimization method is an algorithm that reads in a representation of an optimization problem and delivers an (approximate) solution as its output [47, 57, 209].
See also algorithm, optimization problem.

Optimization problem An optimization problem is a mathematical structure consisting of an objective function $f : \mathcal{U} \rightarrow \mathcal{V}$ defined over an optimization variable $\mathbf{w} \in \mathcal{U}$, together with a feasible set $\mathcal{W} \subseteq \mathcal{U}$. The co-domain $\mathcal{V}$ is assumed to be ordered, meaning that for any two elements $\mathbf{a}, \mathbf{b} \in \mathcal{V}$, we can determine whether $\mathbf{a} < \mathbf{b}$, $\mathbf{a} = \mathbf{b}$, or $\mathbf{a} > \mathbf{b}$. The goal of optimization is to find those values $\mathbf{w} \in \mathcal{W}$ for which the objective $f(\mathbf{w})$ is extremal—i.e., minimal or maximal [47, 57, 209].
See also objective function.

Overfitting Consider an ML method that uses ERM to learn a hypothesis with the minimum empirical risk on a given training set. Such a method is overfitting the training set if it learns a hypothesis with a small empirical risk on the training set but a significantly larger loss outside the training set.
See also ML, ERM, hypothesis, minimum, empirical risk, training set, loss.

Parameter The parameter of an ML model is a tunable (i.e., learnable or adjustable) quantity that allows us to choose between different hypothesis maps. For example, the linear model $\mathcal{H} := \{h^{(\mathbf{w})} : h^{(\mathbf{w})}(x) = w_1 x + w_2\}$ consists of all hypothesis maps $h^{(\mathbf{w})}(x) = w_1 x + w_2$ with a particular choice for the parameters $\mathbf{w} = (w_1, w_2)^T \in \mathbb{R}^2$. Another example of a model parameter is the weights assigned to a connection between two neurons of an ANN.
See also ML, model, hypothesis, map, linear model, weights, ANN.

Parameter space The parameter space $\mathcal{W}$ of an ML model $\mathcal{H}$ is the set of all feasible choices for the model parameters (see Fig. 24). Many important ML methods use a model that is parametrized by vectors of the Euclidean space $\mathbb{R}^d$. Two widely used examples of parametrized models are linear models and deep nets. The parameter space is then often a subset $\mathcal{W} \subseteq \mathbb{R}^d$, e.g., all vectors $\mathbf{w} \in \mathbb{R}^d$ with a norm smaller than one. See also parameter, model, model parameters.

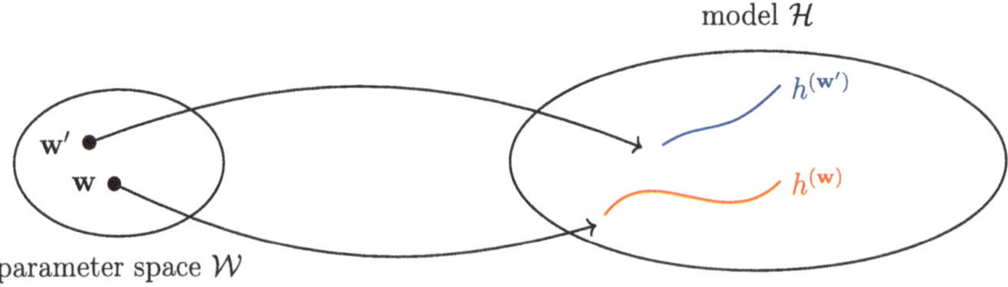

Fig. 24 The parameter space $\mathcal{W}$ of an ML model $\mathcal{H}$ consists of all feasible choices for the model parameters. Each choice $\mathbf{w}$ for the model parameters selects a hypothesis map $h^{(\mathbf{w})} \in \mathcal{H}$

Polynomial regression Polynomial regression aims at learning a polynomial hypothesis map to predict a numeric label based on the numeric features of a data point. For data points characterized by a single numeric feature, polynomial

regression uses the hypothesis space $\mathcal{H}_d^{(\text{poly})} := \{h(x) = \sum_{j=0}^{d-1} x^j w_j\}$. The quality of a polynomial hypothesis map is measured using the average squared error loss incurred on a set of labeled datapoints (which we refer to as the training set).
See also regression, hypothesis, map, label, feature, data point, hypothesis space, squared error loss, labeled datapoint, training set.

Positive semi-definite (psd) A (real-valued) symmetric matrix $\mathbf{Q} = \mathbf{Q}^T \in \mathbb{R}^{d \times d}$ is referred to as psd if $\mathbf{x}^T \mathbf{Q} \mathbf{x} \geq 0$ for every vector $\mathbf{x} \in \mathbb{R}^d$. The property of being psd can be extended from matrices to (real-valued) symmetric kernel maps $K : \mathcal{X} \times \mathcal{X} \rightarrow \mathbb{R}$ (with $K(\mathbf{x}, \mathbf{x}') = K(\mathbf{x}', \mathbf{x})$) as follows: For any finite set of feature vectors $\mathbf{x}^{(1)}, \ldots, \mathbf{x}^{(m)}$, the resulting matrix $\mathbf{Q} \in \mathbb{R}^{m \times m}$ with entries $Q_{r,r'} = K\big(\mathbf{x}^{(r)}, \mathbf{x}^{(r')}\big)$ is psd [26].
See also kernel, map, feature vector.

Prediction A prediction is an estimate or approximation for some quantity of interest. ML revolves around learning or finding a hypothesis map h that reads in the features $\mathbf{x}$ of a data point and delivers a prediction $\widehat{y} := h(\mathbf{x})$ for its label y.
See also ML, hypothesis, map, feature, data point, label.

Principal component analysis (PCA) PCA determines a linear feature map such that the new features allow us to reconstruct the original features with the minimum reconstruction error [23].
See also feature map, feature, minimum.

Privacy attack A privacy attack on an ML system aims to infer sensitive attributes of individuals by exploiting partial access to a trained ML model. One form of a privacy attack is model inversion.
See also attack, sensitive attribute, model inversion, trustworthy AI, GDPR.

Privacy funnel The privacy funnel is a method for learning privacy-friendly features of data points [154].
See also feature, data point.

Privacy leakage Consider an ML application that processes a dataset $\mathcal{D}$ and delivers some output, such as the predictions obtained for new data points. Privacy leakage arises if the output carries information about a private (or sensitive) feature of a data point (which might be a human) of $\mathcal{D}$. Based on a probabilistic model for the data generation, we can measure the privacy leakage via the MI between the output and the sensitive feature. Another quantitative measure of privacy leakage is DP. The relations between different measures of privacy leakage have been studied in the literature (see [214]).
See also ML, dataset, prediction, data point, feature, probabilistic model, data, MI, DP.

Privacy protection Consider some ML method $\mathcal{A}$ that reads in a dataset $\mathcal{D}$ and delivers some output $\mathcal{A}(\mathcal{D})$. The output could be the learned model parameters $\widehat{\mathbf{w}}$ or the prediction $\hat{h}(\mathbf{x})$ obtained for a specific data point with features $\mathbf{x}$. Many important ML applications involve data points representing humans. Each data point is characterized by features $\mathbf{x}$, potentially a label y, and a sensitive attribute s (e.g., a recent medical diagnosis). Roughly speaking, privacy protection means

that it should be impossible to infer, from the output $\mathcal{A}(\mathcal{D})$, any of the sensitive attributes of data points in $\mathcal{D}$. Mathematically, privacy protection requires non-invertibility of the map $\mathcal{A}(\mathcal{D})$. In general, just making $\mathcal{A}(\mathcal{D})$ non-invertible is typically insufficient for privacy protection. We need to make $\mathcal{A}(\mathcal{D})$ sufficiently non-invertible.
See also ML, dataset, model parameters, prediction, data point, feature, label, sensitive attribute, map.

Probabilistic model A probabilistic model interprets data points as realizations of RVs with a joint probability distribution. This joint probability distribution typically involves parameters which have to be manually chosen or learned via statistical inference methods such as maximum likelihood estimation [30].
See also model, data point, realization, RV, probability distribution, parameter, maximum likelihood.

Probability We assign a probability value, typically chosen in the interval $[0, 1]$, to each event that might occur in a random experiment [28, 137, 196, 215].

Probability density function (pdf) The pdf $p(x)$ of a real-valued RV $x \in \mathbb{R}$ is a particular representation of its probability distribution. If the pdf exists, it can be used to compute the probability that x takes on a value from a measurable set $\mathcal{B} \subseteq \mathbb{R}$ via $p(x \in \mathcal{B}) = \int_{\mathcal{B}} p(x')dx'$ [196, Ch. 3]. If the pdf of a vector-valued RV $\mathbf{x} \in \mathbb{R}^d$ exists, it allows us to compute the probability of $\mathbf{x}$ belonging to a measurable region $\mathcal{R}$ via $p(\mathbf{x} \in \mathcal{R}) = \int_{\mathcal{R}} p(\mathbf{x}')dx'_1 \ldots dx'_d$ [196, Ch. 3].
See also RV, probability distribution, probability.

Probability distribution To analyze ML methods, it can be useful to interpret data points as i.i.d. realizations of an RV. The typical properties of such data points are then governed by the probability distribution of this RV. The probability distribution of a binary RV $y \in \{0, 1\}$ is fully specified by the probabilities $p(y = 0)$ and $p(y = 1) = 1 - p(y = 0)$. The probability distribution of a real-valued RV $x \in \mathbb{R}$ might be specified by a pdf $p(x)$ such that $p(x \in [a, b]) \approx p(a)|b - a|$. In the most general case, a probability distribution is defined by a probability measure [28, 174].
See also i.i.d., realization, RV, probability, pdf.

Probability space A probability space is a mathematical model of a physical process (i.e., a random experiment) with an uncertain outcome. Formally, a probability space $\mathcal{P}$ is a triplet $(\Omega, \mathcal{F}, P)$ where

- Ω is a sample space containing all possible elementary outcomes of a random experiment;
- $\mathcal{F}$ is a sigma-algebra, i.e., a collection of subsets of Ω (called events) that satisfies certain closure properties under set operations; and
- P is a probability measure, i.e., a function that assigns a probability $P(\mathcal{A}) \in [0, 1]$ to each event $\mathcal{A} \in \mathcal{F}$. The function must satisfy $P(\Omega) = 1$ and $P\left(\bigcup_{i=1}^{\infty} \mathcal{A}_i\right) = \sum_{i=1}^{\infty} P(\mathcal{A}_i)$ for any countable sequence of pairwise disjoint events $\mathcal{A}_1, \mathcal{A}_2, \ldots$ in $\mathcal{F}$.

Probability spaces provide the foundation for defining RVs and to reason about uncertainty in ML applications [28, 174, 200].
See also probability, model, sample, function, RV, uncertainty, ML.

Projected gradient descent (projected GD) Consider an ERM-based method that uses a parametrized model with parameter space $\mathcal{W} \subseteq \mathbb{R}^d$. Even if the objective function of ERM is smooth, we cannot use basic GD, as it does not take into account contraints on the optimization variable (i.e., the model parameters). Projected GD extends basic GD to handle constraints on the optimization variable (i.e., the model parameters). A single iteration of projected GD consists of first taking a gradient step and then projecting the result back onto the parameter space (Fig. 25).

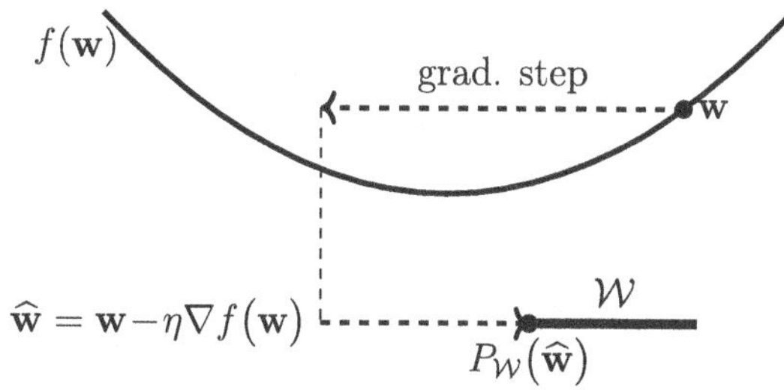

Fig. 25 Projected GD augments a basic gradient step with a projection back onto the constraint set $\mathcal{W}$

See also ERM, model, parameter space, objective function, smooth, GD, model parameters, gradient step, projection.

Projection Consider a subset $\mathcal{W} \subseteq \mathbb{R}^d$ of the d-dimensional Euclidean space. We define the projection $P_{\mathcal{W}}(\mathbf{w})$ of a vector $\mathbf{w} \in \mathbb{R}^d$ onto $\mathcal{W}$ as

$$P_{\mathcal{W}}(\mathbf{w}) = \operatorname*{argmin}_{\mathbf{w}' \in \mathcal{W}} \left\| \mathbf{w} - \mathbf{w}' \right\|_2 . \tag{10}$$

In other words, $P_{\mathcal{W}}(\mathbf{w})$ is the vector in $\mathcal{W}$ which is closest to $\mathbf{w}$. The projection is only well-defined for subsets $\mathcal{W}$ for which the above minimum exists [47].
See also Euclidean space, minimum.

Proximable A convex function for which the proximal operator can be computed efficiently is sometimes referred to as proximable or simple [41].
See also convex, function, proximal operator.

Proximal operator Given a convex function $f(\mathbf{w}')$, we define its proximal operator as [40, 58]

$$prox_{f(\cdot),\rho}(\mathbf{w}) := \operatorname*{argmin}_{\mathbf{w}' \in \mathbb{R}^d} \left[f(\mathbf{w}') + (\rho/2) \left\| \mathbf{w} - \mathbf{w}' \right\|_2^2 \right] \text{ with } \rho > 0.$$

As illustrated in Fig. 26, evaluating the proximal operator amounts to minimizing a penalized variant of $f(\mathbf{w}')$. The penalty term is the scaled squared Euclidean distance to a given vector $\mathbf{w}$ (which is the input to the proximal operator). The proximal operator can be interpreted as a generalization of the gradient step, which is defined for a smooth convex function $f(\mathbf{w}')$. Indeed, taking a gradient step with step size η at the current vector $\mathbf{w}$ is the same as applying the proximal operator of the function $\tilde{f}(\mathbf{w}') = \left(\nabla f(\mathbf{w})\right)^T(\mathbf{w}' - \mathbf{w})$ and using $\rho = 1/\eta$.

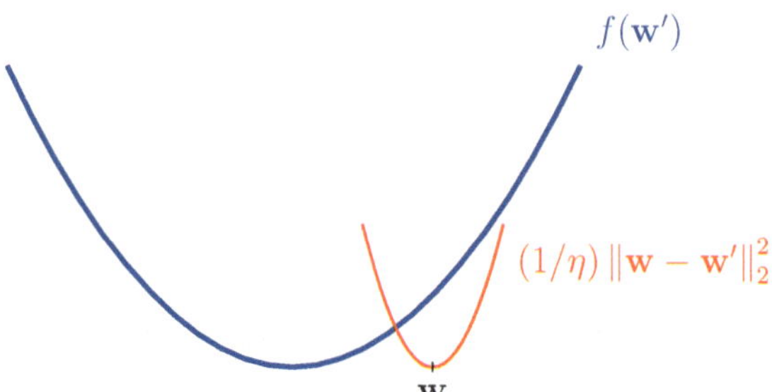

Fig. 26 A generalized gradient step updates a vector $\mathbf{w}$ by minimizing a penalized version of the function $f(\cdot)$. The penalty term is the scaled squared Euclidean distance between the optimization variable $\mathbf{w}'$ and the given vector $\mathbf{w}$

See also convex, function, generalization, gradient step, smooth, step size.

Quadratic function A function $f : \mathbb{R}^d \to \mathbb{R}$ of the form

$$f(\mathbf{w}) = \mathbf{w}^T\mathbf{Q}\mathbf{w} + \mathbf{q}^T\mathbf{w} + a,$$

with some matrix $\mathbf{Q} \in \mathbb{R}^{d\times d}$, vector $\mathbf{q} \in \mathbb{R}^d$, and scalar $a \in \mathbb{R}$.
See also function.

Random variable (RV) An RV is a function that maps from a probability space $\mathcal{P}$ to a value space [28, 174]. The probability space consists of elementary events and is equipped with a probability measure that assigns probabilities to subsets of $\mathcal{P}$. Different types of RVs include

- binary RVs, which map each elementary event to an element of a binary set (e.g., $\{-1, 1\}$ or {cat, no cat});
- real-valued RVs, which take values in the real numbers $\mathbb{R}$;
- vector-valued RVs, which map elementary events to the Euclidean space $\mathbb{R}^d$.

Probability theory uses the concept of measurable spaces to rigorously define and study the properties of (large) collections of RVs [28].
See also function, probability space, probability, Euclidean space.

Realization Consider an RV x which maps each element (i.e., outcome or elementary event) $\omega \in \mathcal{P}$ of a probability space $\mathcal{P}$ to an element a of a measurable space $\mathcal{N}$ [2, 28, 137]. A realization of x is any element $a' \in \mathcal{N}$ such that there is an element $\omega' \in \mathcal{P}$ with $x(\omega') = a'$.
See also RV, probability space.

Regression Regression problems revolve around the prediction of a numeric label solely from the features of a data point [23, Ch. 2].
See also prediction, label, feature, data point.

Regret The regret of a hypothesis h relative to another hypothesis h', which serves as a baseline, is the difference between the loss incurred by h and the loss incurred by h' [187]. The baseline hypothesis h' is also referred to as an expert. See also hypothesis, baseline, loss, expert.

Regularization A key challenge of modern ML applications is that they often use large models, which have an effective dimension in the order of billions. Training a high-dimensional model using basic ERM-based methods is prone to overfitting, i.e., the learned hypothesis performs well on the training set but poorly outside the training set. Regularization refers to modifications of a given instance of ERM in order to avoid overfitting, i.e., to ensure that the learned hypothesis performs not much worse outside the training set. There are three routes for implementing regularization: 1) Model pruning: We prune the original model $\mathcal{H}$ to obtain a smaller model $\mathcal{H}'$. For a parametric model, the pruning can be implemented via constraints on the model parameters (such as $w_1 \in [0.4, 0.6]$ for the weight of feature x_1 in linear regression). 2) Loss penalization: We modify the objective function of ERM by adding a penalty term to the training error. The penalty term estimates how much larger the expected loss (or risk) is compared to the average loss on the training set. 3) Data augmentation: We can enlarge the training set $\mathcal{D}$ by adding perturbed copies of the original data points in $\mathcal{D}$. One example for such a perturbation is to add the realization of an RV to the feature vector of a data point. Figure 27 illustrates the above three routes to regularization. These routes are closely related and sometimes fully equivalent. Data augmentation using Gaussian RVs to perturb the feature vectors in the training set of linear regression has the same effect as adding the penalty $\lambda \|\mathbf{w}\|_2^2$ to the training error (which is nothing but ridge regression). The decision on which route to use for regularization can be based on the available computational infrastructure. For example, it might be much easier to implement data augmentation than model pruning. See also ML, model, effective

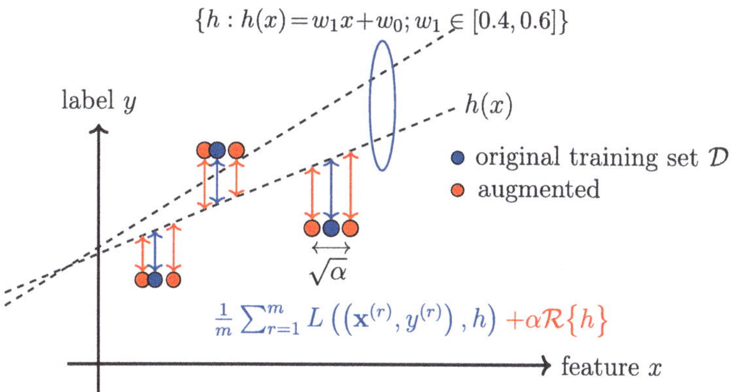

Fig. 27 Three approaches to regularization: (1) data augmentation; (2) loss penalization; and (3) model pruning (via constraints on model parameters)

dimension, ERM, overfitting, hypothesis, training set, model parameters, feature, linear regression, loss, objective function, training error, risk, data augmentation, data point, realization, RV, feature vector, Gaussian RV, ridge regression, label.

Regularized empirical risk minimization (RERM) Basic ERM learns a hypothesis (or trains a model) $h \in \mathcal{H}$ based solely on the empirical risk $\widehat{L}(h|\mathcal{D})$ incurred on a training set $\mathcal{D}$. To make ERM less prone to overfitting, we can implement regularization by including a (scaled) regularizer $\mathcal{R}\{h\}$ in the learning objective. This leads to RERM such that

$$\hat{h} \in \operatorname*{argmin}_{h \in \mathcal{H}} \widehat{L}(h|\mathcal{D}) + \alpha \mathcal{R}\{h\}. \tag{11}$$

The parameter $\alpha \geq 0$ controls the regularization strength. For $\alpha = 0$, we recover standard ERM without regularization. As α increases, the learned hypothesis is increasingly biased toward small values of $\mathcal{R}\{h\}$. The component $\alpha\mathcal{R}\{h\}$ in the objective function of (11) can be intuitively understood as a surrogate for the increased average loss that may occur when predicting labels for data points outside the training set. This intuition can be made precise in various ways. For example, consider a linear model trained using squared error loss and the regularizer $\mathcal{R}\{h\} = \|\mathbf{w}\|_2^2$. In this setting, $\alpha\mathcal{R}\{h\}$ corresponds to the expected increase in loss caused by adding Gaussian RVs to the feature vectors in the training set [23, Ch. 3]. A principled construction for the regularizer $\mathcal{R}\{h\}$ arises from approximate upper bounds on the generalization error. The resulting RERM instance is known as SRM [216, Sec. 7.2].
See also ERM, hypothesis, model, empirical risk, training set, overfitting, regularization, regularizer, parameter, objective function, loss, label, data point, linear model, squared error loss, Gaussian RV, feature vector, generalization, SRM.

Regularizer A regularizer assigns each hypothesis h from a hypothesis space $\mathcal{H}$ a quantitative measure $\mathcal{R}\{h\}$ for how much its prediction error on a training set might differ from its prediction errors on data points outside the training set. Ridge regression uses the regularizer $\mathcal{R}\{h\} := \|\mathbf{w}\|_2^2$ for linear hypothesis maps $h^{(\mathbf{w})}(\mathbf{x}) := \mathbf{w}^T\mathbf{x}$ [23, Ch. 3]. Lasso uses the regularizer $\mathcal{R}\{h\} := \|\mathbf{w}\|_1$ for linear hypothesis maps $h^{(\mathbf{w})}(\mathbf{x}) := \mathbf{w}^T\mathbf{x}$ [23, Ch. 3].
See also hypothesis, hypothesis space, prediction, training set, data point, ridge regression, map, Lasso.

Rényi divergence The Rényi divergence measures the (dis)similarity between two probability distributions [217].
See also probability distribution.

Reward A reward refers to some observed (or measured) quantity that allows us to estimate the loss incurred by the prediction (or decision) of a hypothesis $h(\mathbf{x})$. For example, in an ML application to self-driving vehicles, $h(\mathbf{x})$ could represent the current steering direction of a vehicle. We could construct a reward from the measurements of a collision sensor that indicate if the vehicle is moving toward an obstacle. We define a low reward for the steering direction $h(\mathbf{x})$ if the vehicle moves dangerously toward an obstacle.
See also loss, prediction, hypothesis, ML.

Ridge regression Ridge regression learns the weights $\mathbf{w}$ of a linear hypothesis map $h^{(\mathbf{w})}(\mathbf{x}) = \mathbf{w}^T\mathbf{x}$. The quality of a particular choice for the model parameters $\mathbf{w}$ is measured by the sum of two components. The first component is the average squared error loss incurred by $h^{(\mathbf{w})}$ on a set of labeled datapoints (i.e., the training set). The second component is the scaled squared Euclidean norm $\alpha\|\mathbf{w}\|_2^2$ with a regularization parameter $\alpha > 0$. Adding $\alpha\|\mathbf{w}\|_2^2$ to the average squared error loss is equivalent to replacing original data points by the realizations of (infinitely many) i.i.d. RVs centered around these data points (see regularization).
See also regression, weights, hypothesis, map, model parameters, squared error loss, labeled datapoint, training set, norm, regularization, parameter, data point, realization, i.i.d., RV.

Risk Consider a hypothesis h used to predict the label y of a data point based on its features $\mathbf{x}$. We measure the quality of a particular prediction using a loss function $L\left((\mathbf{x}, y), h\right)$. If we interpret data points as the realizations of i.i.d. RVs, also the $L\left((\mathbf{x}, y), h\right)$ becomes the realization of an RV. The i.i.d. assumption allows us to define the risk of a hypothesis as the expected loss $\mathbb{E}\big\{L\left((\mathbf{x}, y), h\right)\big\}$. Note that the risk of h depends on both the specific choice for the loss function and the probability distribution of the data points.
See also hypothesis, label, data point, feature, prediction, loss function, realization, i.i.d. RV, i.i.d. assumption, loss, probability distribution.

Robustness Robustness is a key requirement for trustworthy AI. It refers to the property of an ML system to maintain acceptable performance even when subjected to different forms of perturbations. These perturbations can be added to the features of a data point in order to manipulate the prediction delivered by a trained ML model. Robustness also includes the stability of ERM-based methods against perturbations of the training set. Such perturbations can occur within data poisoning attacks.
See also trustworthy AI, ML, feature, data point, prediction, model, stability, ERM, training set, data poisoning, attack.

Sample A finite sequence (or list) of data points $\mathbf{z}^{(1)}, \ldots, \mathbf{z}^{(m)}$ that is obtained or interpreted as the realization of m i.i.d. RVs with a common probability distribution $p(\mathbf{z})$. The length m of the sequence is referred to as the sample size.
See also data point, realization, i.i.d., RV, probability distribution, sample size.

Sample size The number of individual data points contained in a dataset.
See also data point, dataset.

Scatterplot A visualization technique that depicts data points by markers in a two-dimensional plane. Figure 28 depicts an example of a scatterplot. A scatterplot can enable the visual inspection of data points that are naturally represented by feature vectors in high-dimensional spaces.
See also data point, minimum, feature, maximum, label, FMI, feature vector, dimensionality reduction.

Semi-supervised learning (SSL) SSL methods use unlabeled data points to support the learning of a hypothesis from labeled datapoints [77]. This approach is particularly useful for ML applications that offer a large amount of unlabeled data points but only a limited number of labeled datapoints.
See also data point, hypothesis, labeled datapoint, ML.

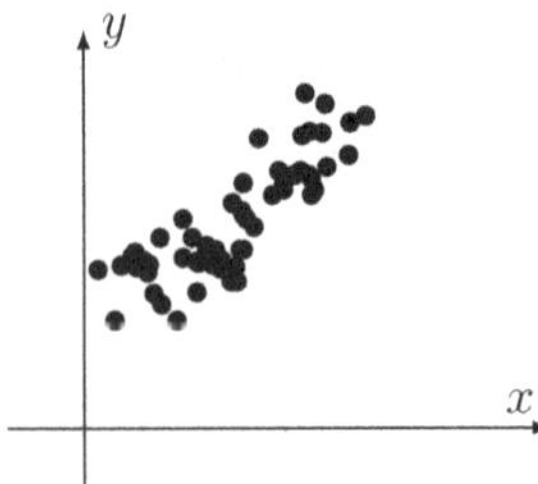

Fig. 28 A scatterplot with circle markers, where the data points represent daily weather conditions in Finland. Each data point is characterized by its minimum daytime temperature x as the feature and its maximum daytime temperature y as the label. The temperatures have been measured at the FMI weather station Helsinki Kaisaniemi during 1.9.2024–28.10.2024

Sensitive attribute ML revolves around learning a hypothesis map that allows us to predict the label of a data point from its features. In some applications, we must ensure that the output delivered by an ML system does not allow us to infer sensitive attributes of a data point. Which part of a data point is considered a sensitive attribute is a design choice that varies across different application domains.
See also ML, hypothesis, map, label, data point, feature.

Smooth A real-valued function $f : \mathbb{R}^d \to \mathbb{R}$ is smooth if it is differentiable, and its gradient $\nabla f(\mathbf{w})$ is continuous at all $\mathbf{w} \in \mathbb{R}^d$ [57, 218]. A smooth function f is referred to as β-smooth if the gradient $\nabla f(\mathbf{w})$ is Lipschitz continuous with Lipschitz constant β, i.e.,

$$\|\nabla f(\mathbf{w}) - \nabla f(\mathbf{w}')\| \leq \beta \|\mathbf{w} - \mathbf{w}'\|, \text{ for any } \mathbf{w}, \mathbf{w}' \in \mathbb{R}^d.$$

The constant β quantifies the amount of smoothness of the function f: the smaller the β, the smoother f is. Optimization problems with a smooth objective function can be solved effectively by gradient-based methods. Indeed, gradient-based methods approximate the objective function locally around a current choice $\mathbf{w}$ using its gradient. This approximation works well if the gradient does not change too rapidly. We can make this informal claim precise by studying the effect of a single gradient step with step size $\eta = 1/\beta$ (see Fig. 29). See also function, differentiable, gradient, optimization problem, objective function, gradient-based methods, gradient step, step size.

Soft clustering Soft clustering refers to the task of partitioning a given set of data points into (a few) overlapping clusters. Each data point is assigned to several different clusters with varying degrees of belonging. Soft clustering methods determine the degree of belonging (or soft cluster assignment) for each data point and each cluster. A principled approach to soft clustering is by interpreting data points as i.i.d. realizations of a GMM. We then obtain a natural choice for the degree of belonging as the conditional probability of a data point belonging to a specific mixture component.

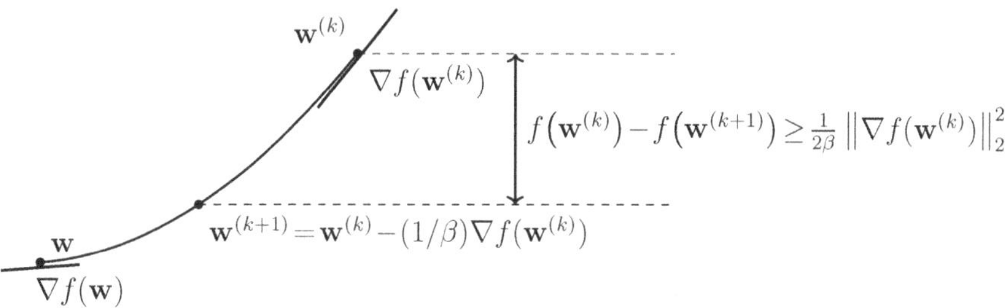

Fig. 29 Consider an objective function $f(\mathbf{w})$ that is β-smooth. Taking a gradient step, with step size $\eta = 1/\beta$, decreases the objective by at least $\frac{1}{2\beta} \left\| \nabla f(\mathbf{w}^{(k)}) \right\|_2^2$ [54, 57, 218]. Note that the step size $\eta = 1/\beta$ becomes larger for smaller β. Thus, for smoother objective functions (i.e., those with smaller β), we can take larger steps

See also clustering, data point, cluster, degree of belonging, i.i.d., realization, GMM, probability.

Squared error loss The squared error loss measures the prediction error of a hypothesis h when predicting a numeric label $y \in \mathbb{R}$ from the features $\mathbf{x}$ of a data point. It is defined as

$$L\left((\mathbf{x}, y), h\right) := \big(y - \underbrace{h(\mathbf{x})}_{=\hat{y}}\big)^2.$$

See also loss, prediction, hypothesis, label, feature, data point.

Stability Stability is a desirable property of an ML method $\mathcal{A}$ that maps a dataset $\mathcal{D}$ (e.g., a training set) to an output $\mathcal{A}(\mathcal{D})$. The output $\mathcal{A}(\mathcal{D})$ can be the learned model parameters or the prediction delivered by the trained model for a specific data point. Intuitively, $\mathcal{A}$ is stable if small changes in the input dataset $\mathcal{D}$ lead to small changes in the output $\mathcal{A}(\mathcal{D})$. Several formal notions of stability exist that enable bounds on the generalization error or risk of the method (see [201, Ch. 13]). To build intuition, consider the three datasets depicted in Fig. 30, each of which is equally likely under the same data-generating probability distribution. Since the optimal model parameters are determined by this underlying probability distribution, an accurate ML method $\mathcal{A}$ should return the same (or very similar) output $\mathcal{A}(\mathcal{D})$ for all three datasets. In other words, any useful $\mathcal{A}$ must be robust to variability in sample realizations from the same probability distribution; i.e., it must be stable. See also ML, dataset, training set, model parameters, prediction, model, data point, generalization, risk, data, probability distribution, sample, realization.

Standard normal vector A standard normal vector is a random vector $\mathbf{x} = \left(x_1, \ldots, x_d\right)^T$ whose entries are i.i.d. Gaussian RVs $x_j \sim \mathcal{N}(0, 1)$. It is a special case of a multivariate normal distribution, $\mathbf{x} \sim (\mathbf{0}, \mathbf{I})$.

See also i.i.d., Gaussian RV, multivariate normal distribution, RV.

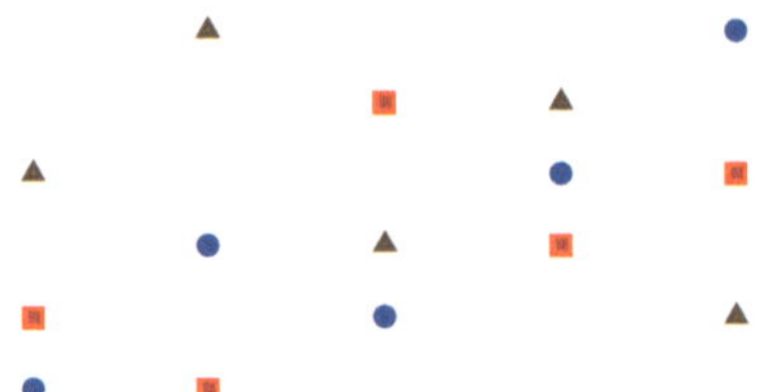

Fig. 30 Three datasets $\mathcal{D}^{(*)}$, $\mathcal{D}^{(\square)}$, and $\mathcal{D}^{(\triangle)}$, each sampled independently from the same data-generating probability distribution. A stable ML method should return similar outputs when trained on any of these datasets

Statistical aspects By statistical aspects of an ML method, we refer to (properties of) the probability distribution of its output under a probabilistic model for the data fed into the method.
See also ML, probability distribution, probabilistic model, data.

Step size See learning rate.

Stochastic A process or method is called stochastic if it involves a random component or is governed by probabilistic laws. In ML, stochastic methods often incorporate randomness for reasons such as optimization (e.g., SGD) or uncertainty modeling (e.g., probabilistic models). A stochastic process is a collection of RVs indexed by time or space, which are used to model random phenomena evolving over time (e.g., noise in sensors or financial time series).
See also ML, SGD, uncertainty, probabilistic model, RV, stochastic block model (SBM).

Stochastic block model (SBM) The SBM is a probabilistic generative model for an undirected graph $\mathcal{G} = \left(\mathcal{V}, \mathcal{E}\right)$ with a given set of nodes $\mathcal{V}$ [219]. In its most basic variant, the SBM generates a graph by first randomly assigning each node $i \in \mathcal{V}$ to a cluster index $c_i \in \{1, \ldots, k\}$. A pair of different nodes in the graph is connected by an edge with probability $p_{i,i'}$ that depends solely on the labels $c_i, c_{i'}$. The presence of edges between different pairs of nodes is statistically independent.
See also model, graph, cluster, probability, label.

Stochastic gradient descent (SGD) SGD is obtained from GD by replacing the gradient of the objective function with a stochastic approximation. A main application of SGD is to train a parametrized model via ERM on a training set $\mathcal{D}$ that is either very large or not readily available (e.g., when data points are stored in a database distributed all over the planet). To evaluate the gradient of the empirical risk (as a function of the model parameters $\mathbf{w}$), we need to compute a sum $\sum_{r=1}^{m} \nabla_{\mathbf{w}} L\left(\mathbf{z}^{(r)}, \mathbf{w}\right)$ over all data points in the training set. We obtain a stochastic approximation to the gradient by replacing the sum $\sum_{r=1}^{m} \nabla_{\mathbf{w}} L\left(\mathbf{z}^{(r)}, \mathbf{w}\right)$ with a sum $\sum_{r \in \mathcal{B}} \nabla_{\mathbf{w}} L\left(\mathbf{z}^{(r)}, \mathbf{w}\right)$ over a randomly chosen subset $\mathcal{B} \subseteq \{1, \ldots, m\}$ (see Fig. 31). We often refer to these randomly chosen data points as a batch. The batch size $|\mathcal{B}|$ is an important parameter of SGD. SGD with $|\mathcal{B}| > 1$ is referred to as mini-batch SGD [220]. See also GD, gradient,

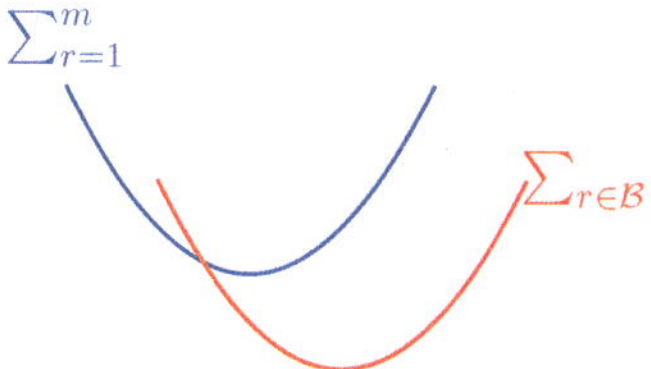

Fig. 31 SGD for ERM approximates the gradient $\sum_{r=1}^{m} \nabla_{\mathbf{w}} L\left(\mathbf{z}^{(r)}, \mathbf{w}\right)$ by replacing the sum over all data points in the training set (indexed by $r = 1, \ldots, m$) with a sum over a randomly chosen subset $\mathcal{B} \subseteq \{1, \ldots, m\}$

objective function, stochastic, model, ERM, training set, data point, empirical risk, function, model parameters, batch, parameter.

Stopping criterion Many ML methods use iterative algorithms that construct a sequence of model parameters in order to minimize the training error. For example, gradient-based methods iteratively update the parameters of a parametric model, such as a linear model or a deep net. Given a finite amount of computational resources, we need to stop updating the parameters after a finite number of iterations. A stopping criterion is any well-defined condition for deciding when to stop updating.
See also algorithm, gradient-based methods.

Strongly convex A continuously differentiable real-valued function $f(\mathbf{x})$ is strongly convex with coefficient σ if $f(\mathbf{y}) \geq f(\mathbf{x}) + \nabla f(\mathbf{x})^T(\mathbf{y} - \mathbf{x}) + (\sigma/2)\left\|\mathbf{y} - \mathbf{x}\right\|_2^2$ [57], [54, Sec. B.1.1].
See also differentiable, function, convex.

Structural risk minimization (SRM) SRM is an instance of RERM, with which the model $\mathcal{H}$ can be expressed as a countable union of submodels such that $\mathcal{H} = \bigcup_{n=1}^{\infty} \mathcal{H}^{(n)}$. Each submodel $\mathcal{H}^{(n)}$ permits the derivation of an approximate upper bound on the generalization error incurred when applying ERM to train $\mathcal{H}^{(n)}$. These individual bounds—one for each submodel—are then combined to form a regularizer used in the RERM objective. These approximate upper bounds (one for each $\mathcal{H}^{(n)}$) are then combined to construct a regularizer for RERM [201, Sec. 7.2].
See also RERM, model, generalization, ERM, regularizer, risk.

Subgradient For a real-valued function $f : \mathbb{R}^d \to \mathbb{R} : \mathbf{w} \mapsto f(\mathbf{w})$, a vector $\mathbf{a}$ such that $f(\mathbf{w}) \geq f(\mathbf{w}') + \left(\mathbf{w} - \mathbf{w}'\right)^T \mathbf{a}$ is referred to as a subgradient of f at $\mathbf{w}'$ [186, 209].
See also function.

Supremum (or least upper bound) The supremum of a set of real numbers is the smallest number that is greater than or equal to every element in the set. More formally, a real number a is the supremum of a set $\mathcal{A} \subseteq \mathbb{R}$ if (1) a is an upper bound of $\mathcal{A}$ and (2) no number smaller than a is an upper bound of $\mathcal{A}$. Every non-empty set of real numbers that is bounded above has a supremum, even if it does not contain its supremum as an element [2, Sec. 1.4].

Test set A set of data points that have been used neither to train a model (e.g., via ERM) nor in a validation set to choose between different models.
See also data point, model, ERM, validation set.

Training error The average loss of a hypothesis when predicting the labels of the data points in a training set. We sometimes refer by training error also to minimal average loss which is achieved by a solution of ERM.
See also loss, hypothesis, label, data point, training set, ERM.

Training set A training set is a dataset $\mathcal{D}$ which consists of some data points used in ERM to learn a hypothesis $\hat{h}$. The average loss of $\hat{h}$ on the training set is referred to as the training error. The comparison of the training error with the validation error of $\hat{h}$ allows us to diagnose the ML method and informs how to improve the validation error (e.g., using a different hypothesis space or collecting more data points) [23, Sec. 6.6].
See also dataset, data point, ERM, hypothesis, loss, training error, validation error, ML, hypothesis space.

Transparency Transparency is a fundamental requirement for trustworthy AI [103]. In the context of ML methods, transparency is often used interchangeably with explainability [129, 221]. However, in the broader scope of AI systems, transparency extends beyond explainability and includes providing information about the system's limitations, reliability, and intended use. In medical diagnosis systems, transparency requires disclosing the confidence level for the predictions delivered by a trained model. In credit scoring, AI-based lending decisions should be accompanied by explanations of contributing factors, such as income level or credit history. These explanations allow humans (e.g., a loan applicant) to understand and contest automated decisions. Some ML methods inherently offer transparency. For example, logistic regression provides a quantitative measure of classification reliability through the value $|h(\mathbf{x})|$. Decision trees are another example, as they allow human-readable decision rules [208]. Transparency also requires a clear indication when a user is engaging with an AI system. For example, AI-powered chatbots should notify users that they are interacting with an automated system rather than a human. Furthermore, transparency encompasses comprehensive documentation detailing the purpose and design choices underlying the AI system. For instance, model datasheets [184] and AI system cards [222] help practitioners understand the intended use cases and limitations of an AI system [223].
See also trustworthy AI, ML, explainability, AI, prediction, model, logistic regression, classification, decision tree.

Trustworthy artificial intelligence (trustworthy AI) Besides the computational aspects and statistical aspects, a third main design aspect of ML methods is their trustworthiness [224]. The EU has put forward seven key requirements (KRs) for trustworthy AI (that typically build on ML methods) [225]: 1) KR1 - Human agency and oversight; 2) KR2 - Technical robustness and safety; 3) KR3 - Privacy and data governance; 4) KR4 - Transparency; 5) KR5 - Diversity, nondiscrimination, and fairness; 6) KR6 - Societal and environmental well-being; 7) KR7 - Accountability. See also computational aspects, statistical aspects, ML, AI, robustness.

Uncertainty Uncertainty refers to the degree of confidence—or lack thereof—associated with a quantity such as a model prediction, parameter estimate, or observed data point. In ML, uncertainty arises from various sources, including noisy data, limited training samples, or ambiguity in model assumptions. Probability theory offers a principled framework for representing and quantifying such uncertainty.
See also model, prediction, parameter, data point, ML, data, sample, probability.

Validation Consider a hypothesis $\hat{h}$ that has been learned via some ML method, e.g., by solving ERM on a training set $\mathcal{D}$. Validation refers to the practice of evaluating the loss incurred by the hypothesis $\hat{h}$ on a set of data points that are not contained in the training set $\mathcal{D}$.
See also hypothesis, ML, ERM, training set, loss, data point.

Validation error Consider a hypothesis $\hat{h}$ which is obtained by some ML method, e.g., using ERM on a training set. The average loss of $\hat{h}$ on a validation set, which is different from the training set, is referred to as the validation error.
See also hypothesis, ML, ERM, training set, loss, validation set, validation.

Validation set A set of data points used to estimate the risk of a hypothesis $\hat{h}$ that has been learned by some ML method (e.g., solving ERM). The average loss of $\hat{h}$ on the validation set is referred to as the validation error and can be used to diagnose an ML method (see [23, Sec. 6.6]). The comparison between training error and validation error can inform directions for the improvement of the ML method (such as using a different hypothesis space).
See also data point, risk, hypothesis, ML, ERM, loss, validation, validation error, training error, hypothesis space.

Variance The variance of a real-valued RV x is defined as the expectation $\mathbb{E}\{(x - \mathbb{E}\{x\})^2\}$ of the squared difference between x and its expectation $\mathbb{E}\{x\}$. We extend this definition to vector-valued RVs $\mathbf{x}$ as $\mathbb{E}\{\|\mathbf{x} - \mathbb{E}\{\mathbf{x}\}\|_2^2\}$.
See also RV, expectation.

Vector space A vector space (also called linear space) is a collection of elements (called vectors) closed under vector addition and scalar multiplication, i.e.,

- If $\mathbf{x}, \mathbf{y} \in \mathcal{V}$, then $\mathbf{x} + \mathbf{y} \in \mathcal{V}$.
- If $\mathbf{x} \in \mathcal{V}$ and $c \in \mathbb{R}$, then $c\mathbf{x} \in \mathcal{V}$.
- In particular, $\mathbf{0} \in \mathcal{V}$.

The Euclidean space $\mathbb{R}^n$ is a vector space. Linear models and linear maps operate within such spaces.
See also Euclidean space, linear model, linear map.

Vertical federated learning (VFL) VFL refers to FL applications where devices have access to different features of the same set of data points [78]. Formally, the underlying global dataset is

$$\mathcal{D}^{(\text{global})} := \left\{ \left(\mathbf{x}^{(1)}, y^{(1)}\right), \ldots, \left(\mathbf{x}^{(m)}, y^{(m)}\right) \right\}.$$

We denote by $\mathbf{x}^{(r)} = \left(x_1^{(r)}, \ldots, x_{d'}^{(r)}\right)^T$, for $r = 1, \ldots, m$, the complete feature vectors for the data points. Each device $i \in \mathcal{V}$ observes only a subset $\mathcal{F}^{(i)} \subseteq \{1, \ldots, d'\}$ of features, resulting in a local dataset $\mathcal{D}^{(i)}$ with feature vectors

$$\mathbf{x}^{(i,r)} = \left(x_{j_1}^{(r)}, \ldots, x_{j_d}^{(r)}\right)^T.$$

Some of the devices might also have access to the labels $y^{(r)}$, for $r = 1, \ldots, m$, of the global dataset. One potential application of VFL is to enable collaboration between different healthcare providers. Each provider collects distinct types of measurements—such as blood values, electrocardiography, and lung X-rays—for the same patients. Another application is a national social insurance system, where health records, financial indicators, consumer behavior, and mobility data are collected by different institutions. VFL enables joint learning across these parties while allowing well-defined levels of privacy protection (Fig. 32). See also FL, device, feature, data point, dataset, feature vector, local dataset, label, data, privacy protection.

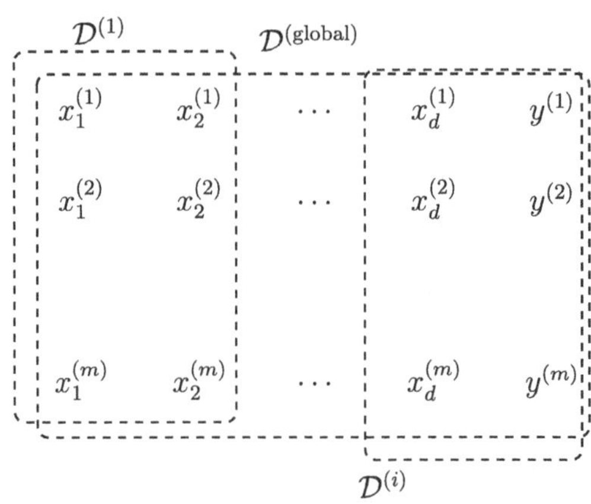

Fig. 32 VFL uses local datasets that are derived from the data points of a common global dataset. The local datasets differ in the choice of features used to characterize the data points

Weights Consider a parametrized hypothesis space $\mathcal{H}$. We use the term weights for numeric model parameters that are used to scale features or their transformations in order to compute $h^{(\mathbf{w})} \in \mathcal{H}$. A linear model uses weights $\mathbf{w} = \left(w_1, \ldots, w_d\right)^T$ to compute the linear combination $h^{(\mathbf{w})}(\mathbf{x}) = \mathbf{w}^T\mathbf{x}$. Weights are also used in ANNs to form linear combinations of features or the outputs of neurons in hidden layers.
See also hypothesis space, model parameters, feature, linear model, ANN.

Zero-gradient condition Consider the unconstrained optimization problem $\min_{\mathbf{w}\in\mathbb{R}^d} f(\mathbf{w})$ with a smooth and convex objective function $f(\mathbf{w})$. A necessary and sufficient condition for a vector $\widehat{\mathbf{w}} \in \mathbb{R}^d$ to solve this problem is that the gradient $\nabla f(\widehat{\mathbf{w}})$ is the zero vector such that

$$\nabla f(\widehat{\mathbf{w}}) = \mathbf{0} \Leftrightarrow f(\widehat{\mathbf{w}}) = \min_{\mathbf{w} \in \mathbb{R}^d} f(\mathbf{w}).$$

See also optimization problem, smooth, convex, objective function, gradient.

0/1 loss The $0/1$ loss $L^{(0/1)}((\mathbf{x}, y), h)$ measures the quality of a classifier $h(\mathbf{x})$ that delivers a prediction $\hat{y}$ (e.g., via thresholding (1)) for the label y of a data point with features $\mathbf{x}$. It is equal to 0 if the prediction is correct, i.e., $L^{(0/1)}((\mathbf{x}, y), h) = 0$ when $\hat{y} = y$. It is equal to 1 if the prediction is wrong, i.e., $L^{(0/1)}((\mathbf{x}, y), h) = 1$ when $\hat{y} \neq y$.
See also loss, classifier, prediction, label, data point, feature.

References

1. W. Rudin, *Real and Complex Analysis*, 3rd edn. (McGraw-Hill, New York, 1987)
2. W. Rudin, *Principles of Mathematical Analysis*, 3rd edn. (McGraw-Hill, New York, 1976)
3. G.H. Golub, C.F. Van Loan, *Matrix Computations*, 4th edn. (Johns Hopkins University Press, Baltimore, MD, 2013)
4. G. Golub, C. van Loan, An analysis of the total least squares problem. SIAM J. Numerical Analysis **17**(6), 883–893 (1980)
5. M. Wollschlaeger, T. Sauter, J. Jasperneite, The future of industrial communication: automation networks in the era of the internet of things and industry 4.0. IEEE Ind. Electron. Mag. **11**(1), 17–27 (2017)
6. M. Satyanarayanan, The emergence of edge computing. Computer **50**(1), 30–39 (2017). https://doi.org/10.1109/MC.2017.9
7. H. Ates, A. Yetisen, F. Güder, C. Dincer, Wearable devices for the detection of covid-19. Nat. Electron. **4**(1), 13–14 (2021). https://doi.org/10.1038/s41928-020-00533-1
8. H. Boyes, B. Hallaq, J. Cunningham, T. Watson, The industrial internet of things (IIoT): an analysis framework. Comput. Ind. **101**, 1–12 (2018). https://www.sciencedirect.com/science/article/pii/S0166361517307285
9. S. Cui, A. Hero, Z.-Q. Luo, J. Moura (eds.), *Big Data over Networks* (2016)
10. A. Barabási, N. Gulbahce, J. Loscalzo, Network medicine: a network-based approach to human disease. Nat. Rev. Genet. **12**(1), 56–68 (2011)
11. M.E.J. Newman, *Networks: An Introduction* (Oxford University Press, Oxford, 2010)
12. B. McMahan, E. Moore, D. Ramage, S. Hampson, B.A.y. Arcas, Communication-Efficient Learning of Deep Networks from Decentralized Data, in *Proceedings of the 20th International Conference on Artificial Intelligence and Statistics*, Series Proceedings of Machine Learning Research, ed. by A. Singh, J. Zhu, vol. 54 (PMLR, New York, 2017), pp. 1273–1282. https://proceedings.mlr.press/v54/mcmahan17a.html
13. T. Li, A.K. Sahu, A. Talwalkar, V. Smith, Federated learning: challenges, methods, and future directions. IEEE Signal Process. Mag. **37**(3), 50–60 (2020)
14. Y. Cheng, Y. Liu, T. Chen, Q. Yang, Federated learning for privacy-preserving AI. Commun. ACM **63**(12), 33–36 (2020)
15. N. Agarwal, A. Suresh, F. Yu, S. Kumar, H. McMahan, cpSGD: communication-efficient and differentially-private distributed SGD, in *Proceedings of the Neural Information Processing Systems (NIPS)* (2018)
16. V. Smith, C.-K. Chiang, M. Sanjabi, A. Talwalkar, Federated Multi-Task Learning, in *Advances in Neural Information Processing Systems*, vol. 30 (2017). https://proceedings.neurips.cc/paper/2017/file/6211080fa89981f66b1a0c9d55c61d0f-Paper.pdf

A. Jung, *Federated Learning*, https://doi.org/10.1007/978-981-95-1009-2

17. D.P. Bertsekas, J. Tsitsiklis, *Parallel and Distributed Computation: Numerical Methods* (Athena Scientific, New York, 2015)
18. M. van Steen, A. Tanenbaum, *Distributed Systems*, 3rd edn. (2017), self-published, open publication
19. D. Tse, P. Viswanath, *Fundamentals of Wireless Communication* (Cambridge University Press, Cambridge, 2005)
20. G. Strang, *Computational Science and Engineering* (Wellesley-Cambridge Press, Cambridge, 2007)
21. G. Strang, *Introduction to Linear Algebra*, 5th edn. (Wellesley-Cambridge Press, Cambridge, 2016)
22. H.H. Bauschke, P.L. Combettes, *Convex Analysis and Monotone Operator Theory in Hilbert Spaces* (Springer, New York, 2011)
23. A. Jung, *Machine Learning: The Basics*, 1st edn. (Springer, Singapore, 2022)
24. N. Goodall, Can you program ethics into a self-driving car? IEEE Spectr. **53**(6), 28–58 (2016)
25. J.R. Quinlan, Induction of decision trees. Mach. Learn. **1**(1), 81–106
26. B. Schölkopf, A. Smola, *Learning with Kernels: Support Vector Machines, Regularization, Optimization, and Beyond* (MIT Press, Cambridge, MA, 2002)
27. A. Juditsky, A. Nemirovski, First-order methods for nonsmooth convex large-scale optimization, I: General purpose methods, in *Optimization for Machine Learning*, ed. by S. Sra, S. Nowozin, S. Wright (MIT Press, New York, 2011), pp. 121–147
28. P. Billingsley, *Probability and Measure*, 3rd edn. (Wiley, New York, 1995)
29. A. Jung, An RKHS Approach to Estimation with Sparsity Constraints, Ph.D. dissertation, Vienna University of Technology (2011). Available online: arXiv:1311.5768
30. E.L. Lehmann, G. Casella, *Theory of Point Estimation*, 2nd edn. (Springer, New York, 1998)
31. A. Esteva, B. Kuprel, R.A. Novoa, J. Ko, S.M. Swetter, H.M. Blau, S. Thrun, Dermatologist-level classification of skin cancer with deep neural networks. Nature **542**, 115–118 (2017)
32. H. Lütkepohl, *New Introduction to Multiple Time Series Analysis* (Springer, New York, 2005)
33. M. Wainwright, *High-Dimensional Statistics: A Non-Asymptotic Viewpoint* (Cambridge University Press, Cambridge, 2019)
34. Y. SarcheshmehPour, Y. Tian, L. Zhang, A. Jung, Clustered federated learning via generalized total variation minimization. IEEE Trans. Signal Process. **71**, 4240–4256 (2023)
35. F. Chung, Spectral graph theory, in *Regional Conference Series in Mathematics*, vol. 92 (1997)
36. D. Spielman, Spectral and algebraic graph theory (2019)
37. D. Spielman, Spectral graph theory, in *Combinatorial Scientific Computing*, ed. by U. Naumann, O. Schenk (Chapman and Hall/CRC, New York, 2012)
38. T. Hastie, R. Tibshirani, J. Friedman, *The Elements of Statistical Learning*, Series. Springer Series in Statistics (Springer, New York, 2001)
39. A. Beck, *First-Order Methods in Optimization* (SIAM-Society for Industrial and Applied Mathematics, Philadelphia, 2017)
40. N. Parikh, S. Boyd, Proximal algorithms. Found. Trends Optim. **1**(3), 123–231 (2013)
41. L. Condat, A primal–dual splitting method for convex optimization involving lipschitzian, proximable and linear composite terms. J. Opt. Th. App. **158**(2), 460–479 (2013)
42. R. Peng, D.A. Spielman, An efficient parallel solver for SDD linear systems, in *Proceedings of the ACM Symposium on Theory of Computing*, New York (2014), pp. 333–342
43. N.K. Vishnoi, $Lx = b$—Laplacian solvers and their algorithmic applications. Found. Trends Theor. Comput. Sci. **8**(1–2), 1–141 (2012). http://dx.doi.org/10.1561/0400000054
44. D. Sun, K.-C. Toh, Y. Yuan, Convex clustering: model, theoretical guarantee and efficient algorithm. J. Mach. Learn. Res. **22**(9), 1–32 (2021). http://jmlr.org/papers/v22/18-694.html
45. K. Pelckmans, J.D. Brabanter, J. Suykens, B.D. Moor, Convex clustering shrinkage, in *PASCAL Workshop on Statistics and Optimization of Clustering Workshop* (2005)
46. R.T. Rockafellar, *Network Flows and Monotropic Optimization* (Athena Scientific, New York, 1998)
47. S. Boyd, L. Vandenberghe, *Convex Optimization* Cambridge, UK (2004)

48. D.P. Bertsekas, *Network Optimization: Continuous and Discrete Models* (Athena Scientific, Nashua, 1998)
49. C.G. Atkeson, A.W. Moore, S. Schaal, Locally weighted learning. Artif. Intell. Rev. **11**(1–5), 11–73 (1997). https://doi.org/10.1023/A:1006559212014
50. A. Jacot, F. Gabriel, C. Hongler, Neural tangent kernel: convergence and generalization in neural networks, in *Advances in Neural Information Processing Systems*, ed. by S. Bengio, H. Wallach, H. Larochelle, K. Grauman, N. Cesa-Bianchi, R. Garnett, vol. 31 (Curran Associates, Inc., Glasgow, 2018). https://proceedings.neurips.cc/paper_files/paper/2018/file/5a4be1fa34e62bb8a6ec6b91d2462f5a-Paper.pdf
51. S.S. Du, X. Zhai, B. Póczos, A. Singh, Gradient descent provably optimizes over-parameterized neural networks, in *7th International Conference on Learning Representations, ICLR 2019, New Orleans, LA, USA, May 6–9, 2019*. OpenReview.net (2019). https://openreview.net/forum?id=S1eK3i09YQ
52. W. E, C. Ma, L. Wu, A comparative analysis of optimization and generalization properties of two-layer neural network and random feature models under gradient descent dynamics. Sci. China Math. **63**(7), 1235–1258 (2020). https://doi.org/10.1007/s11425-019-1628-5
53. A. Paszke, S. Gross, F. Massa, A. Lerer, J. Bradbury, G. Chanan, T. Killeen, Z. Lin, N. Gimelshein, L. Antiga, A. Desmaison, A. Köpf, E. Yang, Z. DeVito, M. Raison, A. Tejani, S. Chilamkurthy, B. Steiner, L. Fang, J. Bai, S. Chintala, *PyTorch: An Imperative Style, High-Performance Deep Learning Library* (Curran Associates Inc., Red Hook, 2019)
54. D.P. Bertsekas, *Convex Optimization Algorithms* (Athena Scientific, Nashua, 2015)
55. T. Schaul, X. Zhang, Y. LeCun, No more pesky learning rates, in *Proceedings of the 30th International Conference on Machine Learning, PMLR 28(3)*, vol. 28 (Atlanta, Georgia, 2013), pp. 343–351
56. M. Andrychowicz, M. Denil, S.G. Colmenarejo, M.W. Hoffman, D. Pfau, T. Schaul, B. Shillingford, N. de Freitas, Learning to learn by gradient descent by gradient descent, in *Proceedings of the 30th International Conference on Neural Information Processing Systems*, Series NIPS'16 (Curran Associates Inc., Red Hook, 2016), pp. 3988–3996
57. Y. Nesterov, *Introductory Lectures on Convex Optimization*, Series. Applied Optimization (Kluwer Academic Publishers, Boston, MA, 2004), a basic course. http://dx.doi.org/10.1007/978-1-4419-8853-9
58. H. Bauschke, P. Combettes, *Convex Analysis and Monotone Operator Theory in Hilbert Spaces*, 2nd edn. (Springer, New York, 2017)
59. V. Istrăţescu, *Fixed point theory: An Introduction*, Ser. Mathematics and its applications, vol. 7 (Reidel, Dordrecht, 1981)
60. B. Ying, K. Yuan, Y. Chen, H. Hu, P. PAN, W. Yin, Exponential graph is provably efficient for decentralized deep training, in *Advances in Neural Information Processing Systems*, ed. by M. Ranzato, A. Beygelzimer, Y. Dauphin, P. Liang, J.W. Vaughan, vol. 34 (Curran Associates, Inc., New York, 2021), pp. 13975–13987. https://proceedings.neurips.cc/paper_files/paper/2021/file/74e1ed8b55ea44fd7dbb685c412568a4-Paper.pdf
61. S. Boyd, P. Diaconis, L. Xiao, Fastest mixing markov chain on a graph. SIAM Rev. **46**(4), 667–689 (2004)
62. D. Mills, Internet time synchronization: the network time protocol. IEEE Trans. Commun. **39**(10), 1482–1493 (1991)
63. J. Hirvonen, J. Suomela, Distributed algorithms 2020 (2023)
64. R. Diestel, *Graph Theory* (Springer, Berlin, 2005)
65. T. Li, A.K. Sahu, M. Zaheer, M. Sanjabi, A. Talwalkar, V. Smith, Federated optimization in heterogeneous networks, in *Proceedings of the Third Conference on Machine Learning and Systems, MLSys 2020, Austin, TX, USA, March 2–4, 2020*, ed. by I.S. Dhillon, D.S. Papailiopoulos, V. Sze, mlsys.org (2020). https://proceedings.mlsys.org/paper_files/paper/2020/hash/1f5fe83998a09396ebe6477d9475ba0c-Abstract.html
66. A. Tanenbaum, D. Wetherall, *Computer Networks*, 5th edn. (Prentice Hall Press, USA, 2010)

67. P. Tseng, Convergence of a block coordinate descent method for nondifferentiable minimization. J. Optim. Theory Appl. **109**(3), 475–494 (2001). https://doi.org/10.1023/A:1017501703105
68. S. Boyd, N. Parikh, E. Chu, B. Peleato, J. Eckstein, *Distributed Optimization and Statistical Learning via the Alternating Direction Method of Multipliers*, vol. 3(1) (Now Publishers, Hanover, MA, 2010)
69. J. Liu, C. Zhang, Distributed learning systems with first-order methods. Found. Trends in Databases **9**(1), 100
70. C. Wang, Y. Yang, P. Zhou, Towards efficient scheduling of federated mobile devices under computational and statistical heterogeneity. IEEE Trans Parallel Distrib Syst **32**(2), 394–410 (2021)
71. H. Feyzmahdavian, M. Johansson, Asynchronous iterations in optimization: new sequence results and sharper algorithmic guarantees. J. Mach. Learn. Res. **24**(158), 1–75 (2023)
72. E.K. Ryu, S. Boyd, A primer on monotone operator methods. Appl. Comput. Math. **15**(1), 3–43 (2016), survey
73. Q. Yang, Y. Liu, Y. Cheng, Y. Kang, T. Chen, H. Yu, *Federated Learning*, 1st edn. (Springer, Berlin, 2022)
74. S. Iyer, T. Killingback, B. Sundaram, Z. Wang, Attack robustness and centrality of complex networks.. PLoS One **8**(4), e59613 (2013)
75. D.J. Spiegelhalter, An omnibus test for normality for small samples. Biometrika **67**(2), 493–496 (2024/03/25/1980). http://www.jstor.org/stable/2335498
76. Q. Yang, Y. Liu, Y. Cheng, Y. Kang, T. Chen, H. Yu, *Horizontal Federated Learning* (Springer International Publishing, Cham, 2020), pp. 49–67. https://doi.org/10.1007/978-3-031-01585-4_4
77. O. Chapelle, B. Schölkopf, A. Zien (eds.), *Semi-Supervised Learning*. (The MIT Press, Cambridge, 2006)
78. Q. Yang, Y. Liu, Y. Cheng, Y. Kang, T. Chen, H. Yu, *Vertical Federated Learning* (Springer International Publishing, Cham, 2020), pp. 69–81. https://doi.org/10.1007/978-3-031-01585-4_5
79. H. Ludwig, N. Baracaldo (eds.), *Federated Learning: A Comprehensive Overview of Methods and Applications* (Springer, Berlin, 2022)
80. A. Shamsian, A. Navon, E. Fetaya, G. Chechik, Personalized federated learning using hypernetworks, in *Proceedings of the 38th International Conference on Machine Learning*, Series. Proceedings of Machine Learning Research, ed. by M. Meila, T. Zhang, vol. 139 (PMLR, New York, 2021), pp. 9489–9502. https://proceedings.mlr.press/v139/shamsian21a.html
81. F. Sung, Y. Yang, L. Zhang, T. Xiang, P.H. Torr, T.M. Hospedales, Learning to compare: relation network for few-shot learning, in *2018 IEEE/CVF Conference on Computer Vision and Pattern Recognition* (2018), pp. 1199–1208
82. V. Satorras, J. Bruna, Few-shot learning with graph neural networks, in *ICLR (Poster)*. OpenReview.net (2018). http://dblp.uni-trier.de/db/conf/iclr/iclr2018.html#SatorrasE18
83. A.-L. Barabási, N. Gulbahce, J. Loscalzo, Network medicine: a network-based approach to human disease. Nat. Rev. Genet. **12**(1), 56–68 (2011). https://doi.org/10.1038/nrg2918
84. A. Jung, N. Tran, Localized linear regression in networked data. IEEE Sig. Proc. Lett. **26**(7), 1090–1094 (2019)
85. D. Hallac, J. Leskovec, S. Boyd, Network lasso: clustering and optimization in large graphs, in *Proceedings of the SIGKDD* (2015), pp. 387–396
86. A. Jung, G. Hannak, N. Görtz, Graphical LASSO based model selection for time series. IEEE Sig. Proc. Letters **22**(10), 1781–1785 (2015)
87. A. Jung, Learning the conditional independence structure of stationary time series: a multitask learning approach. IEEE Trans. Signal Processing **63**(21), 5677–5690 (2015)

88. V. Kalofolias, How to learn a graph from smooth signals, in *Proceedings of the 19th International Conference on Artificial Intelligence and Statistics*, Ser. Proceedings of Machine Learning Research, ed. by A. Gretton, C.C. Robert, vol. 51. (PMLR, Cadiz,2016), pp. 920–929
89. X. Dong, D. Thanou, M. Rabbat, P. Frossard, Learning graphs from data: a signal representation perspective. IEEE Signal Process. Mag. **36**(3), 44–63 (2019)
90. J. Tropp, An introduction to matrix concentration inequalities. Found. Trends Mach. Learn. **8**(1-2), 1–230 (2015)
91. A. Jung, Clustering in partially labeled stochastic block models via total variation minimization, in *Proceedings of the 54th Asilomar Conference Signals, Systems, Computers*, Pacific Grove, CA (2020)
92. B. Bollobas, W. Fulton, A. Katok, F. Kirwan, P. Sarnak, *Random Graphs*. Cambridge Studies in Advanced Mathematics, vol. 73 (2001)
93. G. Keiser, *Optical Fiber Communication*, 4th edn. (Mc-Graw Hill, New Delhi, 2011)
94. D. Tse, P. Viswanath, *Fundamentals of Wireless Communication* (Cambridge University Press, USA, 2005)
95. M. Fiedler, Algebraic connectivity of graphs. Czechoslov. Math. J. **23**(2), 298–305 (1973)
96. S. Hoory, N. Linial, A. Wigderson, Expander graphs and their applications. Bull. Am. Math. Soc. **43**(04), 439–562 (2006)
97. Y.-T. Chow, W. Shi, T. Wu, W. Yin, Expander graph and communication-efficient decentralized optimization, in *2016 50th Asilomar Conference on Signals, Systems and Computers* (2016), pp. 1715–1720
98. S.M. Kay, *Fundamentals of Statistical Signal Processing: Estimation Theory* (Prentice Hall, Englewood Cliffs, NJ, 1993)
99. S. Chepuri, S. Liu, G. Leus, A. Hero, Learning sparse graphs under smoothness prior, in *Proceedings of the IEEE International Conference on Acoustics, Speech and Signal Processing* (2017), pp. 6508–6512
100. J. Tan, Y. Zhou, G. Liu, J.H. Wang, S. Yu, pFedSim: Similarity-Aware Model Aggregation Towards Personalized Federated Learning. arXiv e-prints, p. arXiv:2305.15706 (2023)
101. I. Goodfellow, Y. Bengio, A. Courville, *Deep Learning* (MIT Press, New York, 2016)
102. A. Ortega, P. Frossard, J. Kovačević, J.M.F. Moura, P. Vandergheynst, Graph signal processing: overview, challenges, and applications. Proc. IEEE **106**(5), 808–828 (2018)
103. H.-L. E. G. on Artificial Intelligence, Ethics guidelines for trustworthy AI. European Commission, Tech. Rep. (2019)
104. Department of Industry, Science, Energy and Resources, Australia's AI Ethics Principles, in *Government of Australia* (2024), accessed: 2024-09-30. https://www.industry.gov.au/publications/australias-artificial-intelligence-ethics-framework/australias-ai-ethics-principles
105. OECD, OECD AI principles: recommendation of the council on artificial intelligence. https://oecd.ai/en/ai-principles (2019), accessed: 2024-09-30
106. Cyberspace Administration of China, Interim measures for the management of generative artificial intelligence services (2023). https://www.chinalawtranslate.com/en/generative-ai/, accessed: 2025-05-02
107. China Academy of Information and Communications Technology (CAICT), Artificial intelligence security governance framework (2024). https://www.haynesboone.com/-/media/project/haynesboone/haynesboone/pdfs/alert-pdfs/2024/china-alert---china-publishes-the-ai-security-governance-framework.pdf, accessed: 2025-05-02
108. Ministry of Science and Technology of China, New generation artificial intelligence ethics code (2021). https://www.chinalawvision.com/2025/01/digital-economy-ai/ai-ethics-overview-china/, accessed: 2025-05-02
109. National Institute of Standards and Technology, Artificial intelligence risk management framework (AI RMF 1.0) (2023). https://nvlpubs.nist.gov/nistpubs/ai/nist.ai.100-1.pdf, accessed: 2025-05-02

110. White House Office of Science and Technology Policy, Blueprint for an AI bill of rights (2022). https://bidenwhitehouse.archives.gov/ostp/ai-bill-of-rights/, accessed: 2025-05-02
111. The White House, Executive order 14110: safe, secure, and trustworthy development and use of artificial intelligence (2023). https://www.federalregister.gov/documents/2023/11/01/2023-24283/safe-secure-and-trustworthy-development-and-use-of-artificial-intelligence, accessed: 2025-05-02
112. D. Kuss, O. Lopez-Fernandez, Internet addiction and problematic internet use: A systematic review of clinical research. World J. Psychiatry**6**(1), 143–176 (2016)
113. L. Munn, Angry by design: toxic communication and technical architectures. Humanit. Soc. Sci. Commun. **7**(1), 53 (2020). https://doi.org/10.1057/s41599-020-00550-7
114. P. Mozur, A genocide incited on facebook, with posts from myanmar's military, in *The New York Times* (2018)
115. A. Simchon, M. Edwards, S. Lewandowsky, The persuasive effects of political microtargeting in the age of generative artificial intelligence. PNAS Nexus **3**(2), pgae035 (2024)
116. J.R. Taylor, *An Introduction to Error Analysis: The study of uncertainties in physical measurements*, 2nd edn. (University Science Books, Sausalito, Calif, 1997)
117. A. Jung, A fixed-point of view on gradient methods for big data. Front. Appl. Math. Stat. **3** (2017). https://www.frontiersin.org/article/10.3389/fams.2017.00018
118. K. Pillutla, S.M. Kakade, Z. Harchaoui, Robust aggregation for federated learning. IEEE Trans. Signal Process. **70**, 1142–1154 (2022)
119. H.P. Lopuhaä, P.J. Rousseeuw, Breakdown points of affine equivariant estimators of multivariate location and covariance matrices. Ann. Stat. **19**(1), 229–248 (1991). http://www.jstor.org/stable/2241852
120. I. Akyildiz, W. Su, Y. Sankarasubramaniam, E. Cayirci, A survey on sensor networks. IEEE Commun. Magazine **40**(8), 102–114 (2002)
121. M. Parter, Small cuts and connectivity certificates: a fault tolerant approach, in *33rd International Symposium on Distributed Computing (DISC 2019)*, Ser. Leibniz International Proceedings in Informatics (LIPIcs), ed. by J. Suomela, vol. 146 (Schloss Dagstuhl–Leibniz-Zentrum für Informatik, Dagstuhl, 2019), pp. 30:1–30:16. https://drops.dagstuhl.de/entities/document/10.4230/LIPIcs.DISC.2019.30
122. S. Chechik, M. Langberg, D. Peleg, L. Roditty, Fault-tolerant spanners for general graphs, in *Proceedings of the Forty-First Annual ACM Symposium on Theory of Computing*, Ser. STOC '09. (Association for Computing Machinery, New York, 2009), pp. 435–444. https://doi.org/10.1145/1536414.1536475
123. J. Near, D. Darais, *Guidelines for Evaluating Differential Privacy Guarantees* (National Institute of Standards and Technology, Gaithersburg, MD, 2023)
124. S. Wachter, Data protection in the age of big data. Nat. Electron. **2**(1), 6–7 (2019). https://doi.org/10.1038/s41928-018-0193-y
125. P. Samarati, Protecting respondents identities in microdata release. IEEE Trans. Knowl. Data Eng. **13**(6), 1010–1027 (2001)
126. E. Comission, Regulation (eu) 2016/679 of the European parliament and of the council of 27 April 2016 on the protection of natural persons with regard to the processing of personal data and on the free movement of such data, and repealing directive 95/46/ec (general data protection regulation) (text with EEA relevance), no. 119 (2016), pp. 1–88
127. U. N. G. Assembly, *The Universal Declaration of Human Rights (UDHR)*, New York (1948)
128. J. Colin, T. Fel, R. Cadène, T. Serre, What I Cannot Predict, I Do Not Understand: A Human-Centered Evaluation Framework for Explainability Methods. Adv. Neural Inf. Proces. Syst. **35**, 2832–2845 (2022)
129. A. Jung, P. Nardelli, An information-theoretic approach to personalized explainable machine learning. IEEE Sig. Proc. Lett., **27**, 825–829 (2020)
130. L. Zhang, G. Karakasidis, A. Odnoblyudova, L. Dogruel, Y. Tian, A. Jung, Explainable empirical risk minimization. Neural Comput. Appl. **36**(8), 3983–3996 (2024). https://doi.org/10.1007/s00521-023-09269-3

131. N. Kozodoi, J. Jacob, S. Lessmann, Fairness in credit scoring: assessment, implementation and profit implications. Eur. J. Oper. Res. **297**(3), 1083–1094 (2022). https://www.sciencedirect.com/science/article/pii/S0377221721005385
132. J. Gonçalves-Sá, F. Pinheiro, *Societal Implications of Recommendation Systems: A Technical Perspective* (Springer International Publishing, Cham, 2024), pp. 47–63. https://doi.org/10.1007/978-3-031-41264-6_3
133. A. Abrol, R. Jha, Power optimization in 5g networks: a step towards green communication. IEEE Access **4**, 1355–1374 (2016)
134. M.J. Sheller, B. Edwards, G.A. Reina, J. Martin, S. Pati, A. Kotrotsou, M. Milchenko, W. Xu, D. Marcus, R.R. Colen, S. Bakas, Federated learning in medicine: facilitating multi-institutional collaborations without sharing patient data. Sci. Rep. **10**(1), 12598 (2020). https://doi.org/10.1038/s41598-020-69250-1
135. P. Amin, N.R. Anikireddypally, S. Khurana, S. Vadakkemadathil, W. Wu, Personalized health monitoring using predictive analytics, in *2019 IEEE Fifth International Conference on Big Data Computing Service and Applications (BigDataService)* (2019), pp. 271–278
136. R.B. Ash, *Probability and Measure Theory*, 2nd edn. (Academic Press, New York, 2000)
137. P.R. Halmos, *Measure Theory* (Springer, New York, 1974)
138. C. Dwork, A. Roth, The algorithmic foundations of differential privacy. Found. Trends® Theor. Comput. Sci. **9**(3–4), 211–407 (2014). http://dx.doi.org/10.1561/0400000042
139. U. Erlingsson, V. Pihur, A. Korolova, Rappor: randomized aggregatable privacy-preserving ordinal response, in *Proceedings of the 2014 ACM SIGSAC Conference on Computer and Communications Security*, ser. CCS '14 (Association for Computing Machinery, New York, 2014), pp. 1054–1067. https://doi.org/10.1145/2660267.2660348
140. Apple Machine Learning Research, Understanding aggregate trends for apple intelligence using differential privacy. https://machinelearning.apple.com/research/differential-privacy-aggregate-trends, April 2025, accessed: 2025-05-20
141. J.M. Abowd, The U.S. census bureau adopts differential privacy, in *Proceedings of the 24th ACM SIGKDD International Conference on Knowledge Discovery & Data Mining*, ser. KDD '18. (Association for Computing Machinery, New York, 2018), p. 2867. https://doi.org/10.1145/3219819.3226070
142. J.P. Near, D. Darais, N. Lefkovitz, G.S. Howarth, Guidelines for evaluating differential privacy guarantees, in *National Institute of Standards and Technology, Gaithersburg, MD, NIST Special Publication NIST SP 800-226* (2025). https://doi.org/10.6028/NIST.SP.800-226
143. S. Asoodeh, J. Liao, F.P. Calmon, O. Kosut, L. Sankar, A Better Bound Gives a Hundred Rounds: Enhanced Privacy Guarantees via f-Divergences, *arXiv e-prints*, p. arXiv:2001.05990 (2020)
144. I. Mironov, Rényi differential privacy, in *2017 IEEE 30th Computer Security Foundations Symposium (CSF)* (2017), pp. 263–275
145. S.M. Kay, *Fundamentals of statistical signal processing. Vol. 2., Detection theory*, ser. Prentice-Hall signal processing series (Prentice-Hall PTR, Upper Saddle River, NJ, 1998)
146. P. Kairouz, S. Oh, P. Viswanath, The composition theorem for differential privacy, in *Proceedings of the 32nd International Conference on Machine Learning*, ser. Proceedings of Machine Learning Research, ed. by F. Bach, D. Blei, vol. 37 (PMLR, Lille, 2015), pp. 1376–1385. https://proceedings.mlr.press/v37/kairouz15.html
147. Q. Geng, P. Viswanath, The optimal noise-adding mechanism in differential privacy. IEEE Trans. Inf. Theory **62**(2), 925–951 (2016)
148. M. Abadi, A. Chu, I. Goodfellow, H.B. McMahan, I. Mironov, K. Talwar, L. Zhang, Deep learning with differential privacy, ser. CCS '16 (Association for Computing Machinery, New York, 2016), pp. 308–318. https://doi.org/10.1145/2976749.2978318
149. L. Chua, B. Ghazi, P. Kamath, R. Kumar, P. Manurangsi, A. Sinha, C. Zhang, How private are DP-SGD implementations?" in *Proceedings of the 41st International Conference on Machine Learning*, ser. Proceedings of Machine Learning Research, ed. by R. Salakhutdinov, Z. Kolter, K. Heller, A. Weller, N. Oliver, J. Scarlett, F. Berkenkamp, vol. 235 (PMLR, Lille, 2024), pp. 8904–8918. https://proceedings.mlr.press/v235/chua24a.html

150. H. Shu, H. Zhu, Sensitivity analysis of deep neural networks, in *Proceedings of the Thirty-Third AAAI Conference on Artificial Intelligence*, ser. AAAI'19/IAAI'19/EAAI'19 (AAAI Press, New York, 2019). https://doi.org/10.1609/aaai.v33i01.33014943
151. R. Busa-Fekete, A. Munoz-Medina, U. Syed, S. Vassilvitskii, Label differential privacy and private training data release, in *Proceedings of the 40th International Conference on Machine Learning*, ser. Proceedings of Machine Learning Research, vol. 202. (PMLR, Lille, 2023), pp. 3233–3251. https://proceedings.mlr.press/v202/busa-fekete23a.html
152. B. Balle, G. Barthe, M. Gaboardi, Privacy amplification by subsampling: tight analyses via couplings and divergences, in *Proceedings of the 32nd International Conference on Neural Information Processing Systems*, ser. NIPS'18 (Curran Associates Inc., Red Hook, 2018), pp. 6280–6290
153. P. Cuff, L. Yu, Differential privacy as a mutual information constraint, in *Proceedings of the 2016 ACM SIGSAC Conference on Computer and Communications Security*, ser. CCS '16 (Association for Computing Machinery, New York, 2016), pp. 43–54. https://doi.org/10.1145/2976749.2978308
154. A. Makhdoumi, S. Salamatian, N. Fawaz, M. Médard, From the information bottleneck to the privacy funnel, in *2014 IEEE Information Theory Workshop (ITW 2014)* (2014), pp. 501–505
155. M. Mohamed, B. Shrestha, N. Saxena, Smashed: sniffing and manipulating android sensor data for offensive purposes. IEEE Trans. Inf. Forensics Secur. **12**(4), 901–913 (2017)
156. A. Turner, D. Tsipras, A. Madry, Clean-label backdoor attacks (2019). https://openreview.net/forum?id=HJg6e2CcK7
157. A. Vassilev, A. Oprea, A. Fordyce, H. Anderson, Adversarial machine learning: a taxonomy and terminology of attacks and mitigations. National Institute of Standards and Technology, Gaithersburg, MD, NIST Artificial Intelligence (AI) Report NIST AI 100-2e2023 (2024). https://doi.org/10.6028/NIST.AI.100-2e2023
158. P. Blanchard, E.M. El Mhamdi, R. Guerraoui, J. Stainer, Machine learning with adversaries: byzantine tolerant gradient descent, in *Advances in Neural Information Processing Systems*, ed. by I. Guyon, U.V. Luxburg, S. Bengio, H. Wallach, R. Fergus, S. Vishwanathan, R. Garnett, vol. 30 (Curran Associates, Inc., New York, 2017). https://proceedings.neurips.cc/paper_files/paper/2017/file/f4b9ec30ad9f68f89b29639786cb62ef-Paper.pdf
159. E. Bagdasaryan, A. Veit, Y. Hua, D. Estrin, V. Shmatikov, 'How to backdoor federated learning, in *Proceedings of the Twenty Third International Conference on Artificial Intelligence and Statistics*, ser. Proceedings of Machine Learning Research, ed. by S. Chiappa, R. Calandra, vol. 108 (PMLR, Lille, 2020), pp. 2938–2948. https://proceedings.mlr.press/v108/bagdasaryan20a.html
160. M. Fredrikson, S. Jha, T. Ristenpart, Model inversion attacks that exploit confidence information and basic countermeasures, in *Proceedings of the 22nd ACM SIGSAC Conference on Computer and Communications Security*, ser. CCS '15 (Association for Computing Machinery, New York, 2015), pp. 1322–1333. https://doi.org/10.1145/2810103.2813677
161. G. Lugosi, S. Mendelson, Robust multivariate mean estimation: The optimality of trimmed mean. Ann. Stat. **49**(1), 393–410 (2021), publisher Copyright: © Institute of Mathematical Statistics (2021)
162. X. Cao, M. Fang, J. Liu, N. Gong, Fltrust: Byzantine-robust federated learning via trust bootstrapping, in *Network and Distributed Systems Security (NDSS) Symposium 2021* (2021)
163. S.M. Stigler, The asymptotic distribution of the trimmed mean. Ann. Stat. **1**(3), 472–477 (1973). https://doi.org/10.1214/aos/1176342412
164. S. Shen, S. Tople, P. Saxena, Auror: defending against poisoning attacks in collaborative deep learning systems, in *Proceedings of the 32nd Annual Conference on Computer Security Applications*, ser. ACSAC '16 (Association for Computing Machinery, New York, 2016), pp. 508–519. https://doi.org/10.1145/2991079.2991125

165. N. Wang, Y. Xiao, Y. Chen, Y. Hu, W. Lou, Y.T. Hou, Flare: defending federated learning against model poisoning attacks via latent space representations, in *Proceedings of the 2022 ACM on Asia Conference on Computer and Communications Security*, ser. ASIA CCS '22. (Association for Computing Machinery, New York, 2022), pp. 946–958. https://doi.org/10.1145/3488932.3517395
166. M. Fang, Z. Zhang, Hairi, P. Khanduri, J. Liu, S. Lu, Y. Liu, N. Gong, Byzantine-robust decentralized federated learning, in *Proceedings of the 2024 on ACM SIGSAC Conference on Computer and Communications Security*, ser. CCS '24 (Association for Computing Machinery, New York, 2024), pp. 2874–2888. https://doi.org/10.1145/3658644.3670307
167. D. Yin, Y. Chen, R. Kannan, P. Bartlett, Byzantine-robust distributed learning: towards optimal statistical rates, in *Proceedings of the 35th International Conference on Machine Learning*, ser. Proceedings of Machine Learning Research, vol. 80 (PMLR, Lille, 2018), pp. 5650–5659. https://proceedings.mlr.press/v80/yin18a.html
168. K. Chaudhuri, C. Monteleoni, A. Sarwate, Differentially private empirical risk minimization. J. Mach. Learn. Res. **12**, 1069–1109 (2011)
169. C. Dwork, G.N. Rothblum, S. Vadhan, Boosting and differential privacy, in *2010 IEEE 51st Annual Symposium on Foundations of Computer Science* (2010), pp. 51–60
170. T.H. Cormen, C.E. Leiserson, R.L. Rivest, C. Stein, *Introduction to Algorithms* (MIT Press, Cambridge, 2022). http://ebookcentral.proquest.com/lib/aalto-ebooks/detail.action?docID=6925615
171. M. Sipser, *Introduction to the Theory of Computation*, 3rd edn. (Cengage Learning, Andover, 2013)
172. M.P. Salinas et al., A systematic review and meta-analysis of artificial intelligence versus clinicians for skin cancer diagnosis. NPJ Digit. Med. **7**(1), Art. no. 125 (2024). https://doi.org/10.1038/s41746-024-01103-x
173. G.F. Cooper, The computational complexity of probabilistic inference using Bayesian belief networks. Artif. Intell. **42**(2–3), 393–405 (1990). https://doi.org/10.1016/0004-3702(90)90060-D
174. R. Gray, *Probability, Random Processes, and Ergodic Properties*, 2nd edn. (Springer, New York, 2009)
175. A. Silberschatz, H.F. Korth, S. Sudarshan, *Database System Concepts*, 7th edn. (McGraw-Hill Education, New York, 2019). https://db-book.com/
176. E.F. Codd, A relational model of data for large shared data banks. Commun. ACM **13**(6), 377–387 (1970). https://doi.org/10.1145/362384.362685
177. European Union, Regulation (EU) 2018/1725 of the European Parliament and of the Council of 23 October 2018 on the protection of natural persons with regard to the processing of personal data by the Union institutions, bodies, offices and agencies and on the free movement of such data, and repealing Regulation (EC) No 45/2001 and Decision No 1247/2002/EC (Text with EEA relevance), L 295/39, Nov. 21, 2018. https://eur-lex.europa.eu/eli/reg/2018/1725/oj
178. T.M. Cover, J.A. Thomas, *Elements of Information Theory*, 2nd edn. (Wiley, New Jersey, 2006)
179. X. Liu, H. Li, G. Xu, Z. Chen, X. Huang, R. Lu, Privacy-enhanced federated learning against poisoning adversaries. IEEE Trans. Inf. Forensics Secur. **16**, 4574–4588 (2021)
180. J. Zhang, B. Chen, X. Cheng, H.T.T. Binh, S. Yu, Poisongan: generative poisoning attacks against federated learning in edge computing systems. IEEE Internet Things J. **8**(5), 3310–3322 (2021)
181. S. Abiteboul, R. Hull, V. Vianu, *Foundations of Databases* (Addison-Wesley Publishing Company, Reading, MA, 1995)
182. S. Hoberman, *Data Modeling Made Simple: A Practical Guide for Business and IT Professionals*, 2nd edn. (Technics Publications, Basking Ridge, 2009).
183. R. Ramakrishnan, J. Gehrke, *Database Management Systems*, 3rd edn. (McGraw-Hill, New York, 2002)

184. T. Gebru, J. Morgenstern, B. Vecchione, J. Vaughan, H. Wallach, H. Daumé, K. Crawford, Datasheets for datasets. Commun. ACM **64**(12), 86–92 (2021). https://doi.org/10.1145/3458723
185. G. Tel, *Introduction to Distributed Algorithms*, 2nd edn. (Cambridge University Press, Cambridge, 2000)
186. D. Bertsekas, A. Nedic, A. Ozdaglar, *Convex Analysis and Optimization* (Athena Scientific, Nashua, 2003)
187. N. Cesa-Bianchi, G. Lugosi, *Prediction, Learning, and Games* (Cambridge University Press, New York, 2006)
188. E. Hazan, *Introduction to Online Convex Optimization* (Now Publishers Inc., New York, 2016)
189. J. Chen, L. Song, M. Wainwright, M. Jordan, Learning to explain: an information-theoretic perspective on model interpretation, in *Proceedings of the 35th International Conference on Machine Learning*, Stockholm, Sweden (2018)
190. C. Molnar, *Interpretable Machine Learning—A Guide for Making Black Box Models Explainable* (2019). https://christophm.github.io/interpretable-ml-book/
191. R. Selvaraju, M. Cogswell, A. Das, R. Vedantam, D. Parikh, D. Batra, Grad-cam: visual explanations from deep networks via gradient-based localization, in *2017 IEEE International Conference on Computer Vision (ICCV)* (2017), pp. 618–626
192. D. Gujarati, D. Porter, *Basic Econometrics* (Mc-Graw Hill, New York, 2009)
193. Y. Dodge, *The Oxford Dictionary of Statistical Terms* (Oxford University Press, Oxford, 2003)
194. B. Everitt, *Cambridge Dictionary of Statistics* (Cambridge University Press, Cambridge, 2002)
195. M. Ribeiro, S. Singh, C. Guestrin, “Why should i trust you?”: explaining the predictions of any classifier, in *Proceedings of the 22nd ACM SIGKDD* (2016), pp. 1135–1144
196. D. Bertsekas, J. Tsitsiklis, *Introduction to Probability*, 2nd edn. (Athena Scientific, Nashua, 2008)
197. A. Papoulis, S.U. Pillai, *Probability, Random Variables, and Stochastic Processes*, 4th edn (Mc-Graw Hill, New York, 2002)
198. A. Lapidoth, *A Foundation in Digital Communication* (Cambridge University Press, New York, 2009)
199. C.E. Rasmussen, C.K.I. Williams, *Gaussian Processes for Machine Learning* (MIT Press, Cambridge, 2006)
200. S. Ross, *A First Course in Probability*, 9th edn. (Pearson Education, Boston, MA, 2014)
201. S. Shalev-Shwartz, S. Ben-David, *Understanding Machine Learning—from Theory to Algorithms* (Cambridge University Press, Cambridge, 2014)
202. J. Su, D.V. Vargas, K. Sakurai, One pixel attack for fooling deep neural networks. IEEE Trans. Evol. Comput. **23**(5), 828–841 (2019). https://doi.org/10.1109/TEVC.2019.2890858
203. R.A. Horn, C.R. Johnson, *Topics in Matrix Analysis*, Cambridge, UK (1991)
204. C. Lampert, Kernel methods in computer vision, *Foundations and Trends in Computer Graphics and Vision* (2009)
205. U. von Luxburg, A tutorial on spectral clustering. Stat. Comput. **17**(4), 395–416 (2007)
206. A.Y. Ng, M.I. Jordan, Y. Weiss, On spectral clustering: analysis and an algorithm. Adv. Neur. Inf. Proc. Syst. **14** (2001)
207. R. Caruana, Multitask learning. Mach. Learn. **28**(1), 41–75 (1997). https://doi.org/10.1023/A:1007379606734
208. C. Rudin, Stop explaining black box machine learning models for high-stakes decisions and use interpretable models instead. Nat. Mach. Intell. **1**(5), 206–215 (2019)
209. D.P. Bertsekas, *Nonlinear Programming*, 2nd edn. (Athena Scientific, Belmont, MA, 1999)
210. J. Heinonen, Lectures on lipschitz analysis, Dept. Math. Statist., Univ. Jyv.skyl., Jyv.skyl., Finland, Rep. 100, 2005. [Online]. Available: http://www.math.jyu.fi/research/reports/rep100.pdf
211. T. Hastie, R. Tibshirani, M. Wainwright, *Statistical Learning with Sparsity. The Lasso and its Generalizations* (2015)

212. R.A. Horn, C.R. Johnson, *Matrix Analysis*, 2nd edn., Cambridge, UK (2013)
213. A. Rakhlin, O. Shamir, K. Sridharan, Making gradient descent optimal for strongly convex stochastic optimization, in *Proceedings of the 29 th International Conference on Machine Learning*, Edinburgh, Scotland, UK (2012)
214. A. Ünsal, M. Önen, Information-theoretic approaches to differential privacy. ACM Comput. Surv. **56**(3), (2023). Art. no. 76. https://doi.org/10.1145/3604904
215. O. Kallenberg, *Foundations of Modern Probability* (Springer, New York, 1997)
216. S. Shalev-Shwartz, A. Tewari, Stochastic methods for l1 regularized loss minimization, in *Proceedings of the 26th Annual International Conference on Machine Learning*, ser. ICML '09, New York, NY, USA (2009), pp. 929–936
217. I. Csiszar, Generalized cutoff rates and Renyi's information measures. IEEE Trans. Inf. Theory **41**(1), 26–34 (1995)
218. S. Bubeck, Convex optimization. algorithms and complexity, in *Foundations and Trends in Machine Learning*, vol. 8 (Now Publishers, New York, 2015)
219. E. Abbe, Community detection and stochastic block models: recent developments. J. Mach. Learn. Res. **18**(177), 1–86 (2018). http://jmlr.org/papers/v18/16-480.html
220. L. Bottou, On-line learning and stochastic approximations, in *On-Line Learning in Neural Networks*, ed. by D. Saad (Cambridge University Press, New York, 1999), ch. 2, pp. 9–42
221. C. Gallese, The AI act proposal: a new right to technical interpretability? SSRN Electron. J. (2023). https://ssrn.com/abstract=4398206
222. M. Mitchell, et.al., Model cards for model reporting, in *Proceedings of the Conference on Fairness, Accountability, and Transparency*, ser. FAT* '19 (Association for Computing Machinery, New York, 2019), pp. 220–229. https://doi.org/10.1145/3287560.3287596
223. K. Shahriari, M. Shahriari, IEEE standard review—Ethically aligned design: a vision for prioritizing human wellbeing with artificial intelligence and autonomous systems, in *2017 IEEE Canada International Humanitarian Technology Conference*, pp. 197–201. https://doi.org/10.1109/IHTC.2017.8058187
224. D. Pfau, A. Jung, Engineering trustworthy AI: A developer guide for empirical risk minimization (2024). https://arxiv.org/abs/2410.19361
225. E. Commission, C. Directorate-General for Communications Networks, and Technology, *The Assessment List for Trustworthy Artificial Intelligence (ALTAI) for self assessment* (Publications Office, New York, 2020)

Index

A. Jung, *Federated Learning*, https://doi.org/10.1007/978-981-95-1009-2

The manufacturer's authorised representative in the EU is Springer Nature Customer Service Centre GmbH, Europaplatz 3, 69115 Heidelberg, Germany. If you have any concerns regarding our products, please contact ProductSafety@springernature.com

Printed and bound by CPI Group (UK) Ltd, Croydon, CR0 4YY

07/07/2026

02160907-0002